More Heat than Life: The Tangled Roots of Ecology, Energy, and Economics

"It is interesting to contemplate a tangled bank, clothed with many plants of many kinds, with birds singing on the bushes, with various insects flitting about, and with worms crawling through the damp earth, and to reflect that these elaborately constructed forms, so different from each other, and dependent upon each other in so complex a manner, have all been produced by laws acting around us."

—Charles Darwin, *The Origin of Species*, 1859

"One recognizes that there is a first agent in matter by which everything is executed in nature, which moves everything, which is the cause of all generation and all destruction; it is a fire, a matter aetherial or subtle, extremely active, which has the property of all the movement which animates the universe; it is an immense sea which contains all the sensible bodies, which it intimately penetrates and through which it works all the changes which happen."

—François Quesnay, *Essai Physique sur l'Oeconomie Animale*, 1747

"*Machine* n. apparatus in which the action of several parts is combined for the applying of mechanical force to a purpose; person like a machine in regularity or insensibility; controlling organisation in politics."

—Shorter Oxford Dictionary

Jeremy Walker

More Heat than Life: The Tangled Roots of Ecology, Energy, and Economics

palgrave
macmillan

Jeremy Walker
Climate Justice Research Centre
Social & Political Sciences, School of Communication
Faculty of Arts & Social Sciences, University of Technology Sydney
Broadway, NSW, Australia

ISBN 978-981-15-3935-0 ISBN 978-981-15-3936-7 (eBook)
https://doi.org/10.1007/978-981-15-3936-7

Cover image: In 1976, the exterior of the Montreal Biosphere caught fire. Designed by Buckminster Fuller for the US pavillion at the 1967 World Fair, the geodesic dome's internal temperature was intended to be controlled by a complex system of transparent acrylic shades. This system had proved faulty, and was not replaced after the fire. © Bettman / Getty Images.

This Palgrave Macmillan imprint is published by the registered company Springer Nature Singapore Pte Ltd.
The registered company address is: 152 Beach Road, #21-01/04 Gateway East, Singapore 189721, Singapore

This book is dedicated to Roswitha, who taught me to see the artistry of life, and without whose unfailing humour, wisdom, and dedication this book would never have been written. This book is dedicated with love to Marlene and Charlie, and to all children, to whom a living future rightfully belongs. It is offered to all who care for Country and community, to all who strive for justice here on the good Earth.

Acknowledgements

I respectfully acknowledge the past, present, and emerging elders of Country upon whose ancestral lands this work was written. It was nourished over time by the beauty and mystery of Kuringgai, Bundjalung, Barkindji, Bidjigal, Gadigal, Eora, Dharug, Darkinjung, Gundungurra, Tharawal, and Yuin country. I am indebted to Den Barber, Victor Steffenson, Oliver Costello, Lauren Tynan, Michelle Bishop, and other firesticks mob, who have shared with me a little of the knowledge of the good fire. James Goodman supported me through the earliest version of this work with good faith and humour. Jon Marshall, Devleena Ghosh, Tom Morton, and Michael Fabinyi of the UTS Climate Justice Research Centre have all offered inspiration and encouragement at various stages of the writing, as have my colleagues in the Social & Political Sciences, along with Melinda Cooper, Roel Plant, Celine Granjou, Matt Kearnes, David Fleming, Nick Coleman, Kyla Tienhaara, Shamish Gibson, Lachlan, Rowan, Lorraine and many others. I would also like to thank my doctoral and honours students, along with the tutors and undergraduates of my Environmental Studies and Environmental Communication subjects, all of whom have helped me refine the ideas and arguments of this book. Final thanks go to my editor Vishal Daryonomel, to Anushangi Weerakoon, Venkitesan VinodhKumar and the rest of the Palgrave Macmillan team for ushering this book into print with deft professionalism.

Contents

List of Figures

PART I

Introduction

CHAPTER 1

Neoliberalism, Environmentalism, and the Crisis of the 1970s

The time in which we live is characterised by a set of seemingly intractable political arrangements—a world order designed to foster 'economic freedom' and 'economic growth'—and an ever-deepening planetary crisis of ecological erosion and global heating, a crisis increasingly difficult to normalise in the consciousness of everyday life. The roots of our current condition might be traced to the crisis of American power of the early 1970s, a time when two bodies of knowledge—ecosystems ecology and the economics of the Chicago School—were transforming the institutions of the United States. The new authority accorded these incommensurable and politically charged sciences reflected the coterminous movement of environmentalism from the counter-culture to the conferences of the United Nations, and of neoliberalism from the radical fringes of right-wing economic thought to the commanding heights of governmental power.

As sources of the knowledge claims of counter-posed political cosmologies, ecology and economics, the estranged twin sciences of the *oikos* (both share this Greek root-word meaning 'household' or 'estate') were engaged in the most millenarian of anticipations. Ecologists looked to the future and warned of a coming Apocalypse. Exponential industrialisation meant mounting destruction. Without dramatic social transformation, 'business as usual' would culminate in an Earth so hopelessly polluted, depleted and over-heated that it could no longer sustain civilisation and abundant communities of multi-species life. At stake was nothing less than the regenerative capacity of the biosphere. Meanwhile, economists rallying to the

J. Walker, *More Heat than Life: The Tangled Roots of Ecology, Energy, and Economics*,
https://doi.org/10.1007/978-981-15-3936-7_1

banner of 'freedom' foretold a new Heaven and a new Earth. The liberation of individuals and business from the dead hand of government intervention would unshackle 'the invisible hand' of the market, unleashing entrepreneurial techno-innovation and opening an infinite horizon of wealth creation, freeing humanity from the dismal scarcities and servitude of the past. The struggle between these divergent visions of post-natural futures continues into the present, exemplified most dramatically in the agonistic theatre of climate and energy policy. As an offering to the predicaments of our present—which some would gather under the sign of the 'Anthropocene'—this book attempts to unearth a genealogy of the deep contradictions within and between ecology and economics, momentarily brought into sharp relief when they collided in the transformative moment of the 1970s.

There are many critical histories of economics available, less so of ecology. Yet the mutual history of the two disciplines, a topic rarely approached in the burgeoning (though too often unintegrated) literatures on neoliberalism and the climate crisis, is of crucial importance to the grave challenge of restructuring 'the economy' before it destructures 'the ecosystem' beyond all hope of timely regeneration. This task is complicated by the fact that despite their cosmopolitical opposition, the concepts of natural order and (re)production deployed in either discipline share common genealogical roots. This has been noticed by a range of scholars, although mostly with specialist concerns in mind. What remains under-recognised is the significant fact that neoclassical economics and systems ecology, the paradigmatic core disciplines of their respective fields of knowledge, both anchored their claims to the status of science in the energy physics developed by combustion engineers as the thermoindustrial revolution gathered momentum in the mid-nineteenth century. The claims to epistemic authority of both ecology and economics (and thus ultimately of environmentalists and neoliberals) can be traced to a foundational relation to the thermodynamic laws of energy and entropy, and in turn, to older concepts of equilibrium and natural law. However, these claims were made in different ways, at different times and for very different purposes. The ever present possibility for a reconciliation of the estranged twin sciences in the Earthly phenomena of heat and life—for an ecological economics which answers the practical and ethical question of 'how are we to live?'—has been perennially deferred. It is for us, for our children, and all other life to live the consequences.

Focussing on the history of particular concepts—growth and equilibrium—as they appear in each discipline, this book demonstrates the extent of the unacknowledged mutual indebtedness between the two apparently disparate fields. It brings these ideas into focus via an excavation of their histories in political theology, natural history and physics, providing an itinerary of their migration into nineteenth-century 'social physics' and the body of modern systems theories which emerged in the twentieth century. In doing so it addresses the following questions: how do we account for the uncritical commitment to infinite economic growth pursued by almost all nation-states, given the extensive empirical evidence that this is undermining the very habitability of the Earth? How does the constitution of economic knowledge lend itself to this path? How has the ecological world-view gone from a position of critical collision with economists' denial of science and nature to subordinate collusion with neoliberal 'solutions', such as financial markets for carbon and 'ecosystem services', or techno-utopian geoengineering projects to make endangered ecosystems 'resilient' to planetary heating? I pursue such questions via an historical analysis of how economics and ecology came to be constituted as separate, stand-alone sciences, from the 1870s to the 1970s.

The Laws of Nature and the Powers of the Machine

The nineteenth-century triumph of machine technology over nature, and of scientific materialism over theology, had its parallel in political economy, which sought to overcome its history as a moral discourse on wealth and poverty and become a science of statecraft in accordance with natural laws it detected operating in the market economy. Whilst concepts of 'natural law' are deeply rooted in the Western tradition, in the nineteenth century the quest to elaborate them scientifically was profoundly realised in the development of the modern physics of energy—thermodynamics—a science which arose in tandem with the fateful development of the coal-fired steam engine, and which remains foundational to the corpus of scientific materialism.

Now ubiquitous in everyday life, heat engines convert the ancient solar energy stored chemically in hydrocarbon fuels and released as heat during combustion into mechanical force, or 'work'.[1] Histories of thermodynamics

[1] For this reason I use the term 'thermoindustrial society' to characterise modern forms of social organisation in which the heat released by hydrocarbon fuel combustion is the dominant source of energy.

begin by acknowledging the young French engineer Sadi Carnot's *Reflections on the Motive Power of Fire, and on Machines fitted to develop that Power* (1825), a brilliant analysis of the efficiency limits of steam engines in converting the heat of coal combustion into mechanical work. Carnot observed that the machines never gave as good as they got: only some of the heat was converted via the working fluid into the motive force of the driving mechanism. The majority of the heat was inevitably dissipated, flowing into the cooler parts and environment of the engine. Moreover, it was precisely this dissipative flow of heat across a thermal gradient, which in the absence of fresh shovelfuls of coal would grind to a halt as the motor approached thermal equilibrium with its environment, that was the sole source of motive power. Carnot's insights were developed and synthesized with other experimental findings between the 1840s and 1860s by scientists including Julius Mayer, James Joule, William Rankine, Hermann von Helmholtz, James Clerk Maxwell, Rudolf Clausius, and Josiah Willard Gibbs. These inquiries demonstrated that light, heat, mechanical force, chemical affinity, magnetism, electricity, and the atomic structure of matter are all manifestations of a universal phenomenon we now call 'energy'. Formalised in 1847, the law of the conservation of energy (from *energeia*, a Greek term approximating 'work') states that the quantity of energy of a closed system is constant: energy can neither be created nor destroyed in all the transformations we observe, only converted from one form to another.

Revealing a universal invariance underlying all known physical and chemical phenomena, the elaboration of the 1st law of thermodynamics led to a profound confidence in the timeless order and rationality of nature, vindicating the new thermoindustrial society and its scientific achievements. By contrast, the formalisation in 1865 of the 2nd law of thermodynamics—the entropy law (from the Greek *entropia*, meaning 'turning toward', or 'transformation')—implied an irreversible historical trajectory of disorder, depletion, waste, and chaos. Just as we always observe firewood burning to ashes, and never the reverse, the 2nd law states that the entropy of a closed system will always increase, where entropy is a measure of the disorder, dissipation, or unavailability of energy to 'do work'. As it exfoliated into wider scientific and social discourse in the late nineteenth century, the spectre of entropy complicated the optimistic confidence in Progress with *fin-de-siecle* pessimism, declinist fatigue and visions of the heat death of the universe. Accounting for all phenomena involving heat, motion and work, and imposing irreversibility and rigorous limits upon all the transformations of state and material organisation observed in machines, organisms, and inanimate matter from the atomic to the cosmological scale, the laws of thermodynamics are central to the claim of scientific materialism to have identified universal principles at the foundation of all matter and material transformation. Among the most confirmed findings of modern science, contemporary physicists regard them as nothing less than 'the constitution of the universe'.[2]

These apparent paradoxes regarding the fundamental nature of nature were not resolved when economics made its bid to become an exact science on par with physics in the 1870s. When ecology sought to become a 'hard' science in the twentieth century, it similarly strove to bring biology, evolutionary history, and geochemistry into coherence with energy physics. It was not until the 1970s, however, that ecology and economics were brought into direct ontological and political conflict, although most players were unaware—and most remain so now—of the historical origins of the unresolved contradiction between the ecologists' view of the world and that of the economist, a contradiction with increasingly dire consequences for the future of life on Earth.

The neoclassical synthesis of the 1870s consolidated the style and claims of orthodox economics, as its founding authors—including Leon Walras, William Stanley Jevons, Vilfredo Pareto, and John Bates Clark—appropriated the mathematical format of the law of the conservation of energy in their portrayal of 'market forces' operating according to law-like principles of general equilibrium.[3] Modelling the 'subjective utility' sought by hedonistic individual market participants as a universal field of value analogous to 'energy' as described in the 1st law of thermodynamics, the neoclassical economists aimed to excise the 'political' from political economy and develop a pure science of economics on par with physics. This agenda was to be pursued through the construction of mathematical models of the economy as a frictionless, ahistorical market setting in which the 'price mechanism' automatically equilibrates the forces of supply and demand (Fig. 1.1). The equilibrium concept at the core of

[2] Kümmel, R. (2011). *The second law of economics: energy, entropy, and the origins of wealth.* Springer Science & Business Media, p. 113.

[3] Mirowski, P. (1989). *More heat than light: economics as social physics, physics as nature's economics.* Cambridge: Cambridge University Press.

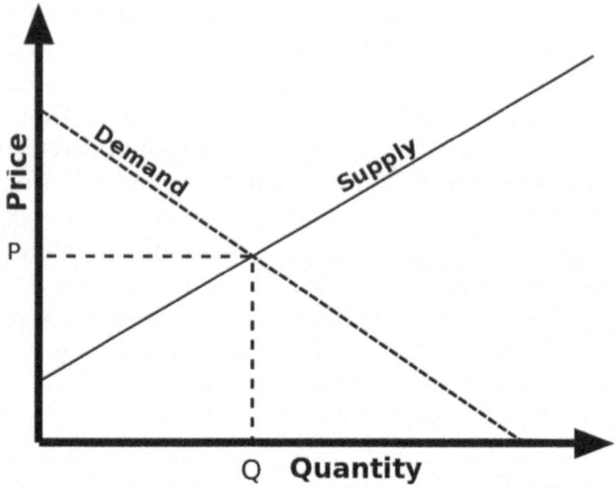

Fig. 1.1 Simple supply and demand graph. The constitutive metaphor of 'economic science': the forces of supply and demand coordinating consumption and production at the equilibrium price (PQ). (Source: Dallas Eperson, 2012. Creative Commons Attribution-Share Alike 3.0 Unported license. https://upload.wikimedia.org/wikipedia/commons/f/f2/Simple_supply_and_demand.svg)

neoclassical economics portrays the market order as universal, natural and optimal, and the economist as a scientist deserving the epistemic authority and prestige accorded to the physicist. However, this simplification of the dynamics of industrial capitalism involved some crucial omissions: the elimination of the question of the dissipation of fossil energy and natural resources in production, and even of 'land' itself from orthodox models of 'the economy'. The 2nd law of thermodynamics, despite being the most directly relevant principle of physics to our economic existence on Earth, has never been integrated into the canon of economic theory.[4]

The term 'ecology' was coined in 1866 by the Darwinian biologist and energeticist Ernst Haeckel, to distinguish as a modern science the study that had outgrown the earlier term for natural histories of multi-species inter-relations: 'oeconomy of nature'. The resemblance is more than nominal. Ecology's trajectory of scientisation mirrors the path taken by economics. Ecology has borrowed directly from the social sciences, and

[4] Georgescu-Roegen, N. (1971). *The entropy law and the economic process.* Cambridge, MA: Harvard University Press.

most heavily from political economy—a theme also treated systematically in an original way in this book. First proposed by Arthur Tansley in 1935, the 'ecosystem' has since become the central organising concept of contemporary ecology.[5] Tansley appealed to the laws of thermodynamics in applying a systems approach to the blooming, buzzing confusion of life, beginning with the proposition that 'all living organisms may be regarded as machines transforming energy from one form to another'. The 'ecosystem' is a system in the sense that communities of biological life cannot but conform to

> The great universal law of equilibrium [that] governs all the processes of which we have any knowledge, from the movements of the planets to those of molecules, atoms and electrons, from the activity of the protoplasm to the vagaries of the human mind. All things which exist are constantly tending towards a position of balance or equilibrium.[6]

In accordance with the 'systems' approach of energy physics, and inspired by Cold War cybernetics, by the mid-twentieth century the concept of the 'ecosystem' promised to unify competing schools of ecological thought around a coherent research object and bring the epistemic status and prestige of a 'hard' science to a field characterised by a past in localised natural histories. Abstracting from communities of organisms to analyse ecosystems driven by the photosynthetic capture of solar energy, ecologists showed how the 'economy of nature' was intimately bound up with the evolutionary trajectory of life's unique biochemistries, which modify geological strata and modulate the oceans and the atmosphere. As the nascent discipline grew in confidence, by the late 1960s systems ecologists offered their science as a full-spectrum framework for rational decision-making regarding human-environment relations.

If both systems ecology and neoclassical economics laid foundational claims to the achievements of energy physicists in identifying laws of nature to which there have been found no exceptions, there are crucial differences. From Vernadsky's epochal account of the photosynthetic transformation of solar energy and the non-living matter of the Earth by *The Biosphere* (1926), through to the confirmatory work of Lovelock and

[5] Tansley, A. (1935). The use and abuse of vegetational concepts and terms. *Ecology*, 16(3), 284–307.

[6] Cited in: Anker, P. (2009). *Imperial ecology: environmental order in the British Empire, 1895–1945*. Cambridge, MA: Harvard University Press, p. 31.

Margulis in the 1970s, and on to contemporary studies in climatology and Earth systems science, systems ecologists have sought to ground their account of life's complex order and planetary unfolding in the phenomena of solar radiation, heat, and energy transformation by fire and photosynthesis.[7] By contrast, orthodox economic theory has systematically excluded from its account of the economic process all the phenomena accounted for by the 2nd law of thermodynamics: the historically irreversible dissipation of energy inherent in all processes of production. This disciplinary exclusion of the physics of fossil-fuel combustion, of the solar ecology of the Earth as 'one physical system',[8] and indeed, of scientific materialism *tout court*—means that economics has no viable theory of 'production', 'growth', or 'development'. The actual physics content of economics is zero. This is of vital importance: since World War II, the period designated 'the Great Acceleration' in the Anthropocene literature, unceasing economic growth has become the goal, measure, and permanent justification of government policy.

Ecology was a term unknown by most until it exploded into public consciousness in the late 1960s. Associated with the environmental movement's apocalyptic warnings of a coming planetary catastrophe, ecology developed a critical reputation as a 'subversive science'.[9] Ecologists took on the public role of the 'sane scientist', warning that the limits to growth were fast approaching and that ecological equilibrium—the 'balance of nature'—was unravelling on a global scale. As the science of life's complex organisation, applicable from the scale of a pond to the geochemical and evolutionary history of the biosphere as whole, systems ecology should thus be recognised as the general science within which the social sciences must be brought into conformity. Since 'the phenomenal domain of ecology is broader than that covered by economics', as the renegade economist Nicholas Georgescu-Roegen put it, it seemed only logical that 'economics will have to merge into ecology.'[10] After all, this merely

[7] Vernadsky, V. ([1926] 1998). *The biosphere*. New York: Copernicus; Lovelock, J., & Margulis, L. (1974). Atmospheric homeostasis by and for the biosphere: the Gaia hypothesis. *Tellus*, 26(1–2), 2–10.

[8] Richter, D. & Billings, S. (2015). 'One physical system': Tansley's ecosystem as Earth's critical zone. *New Phytologist*, (206), 900–912.

[9] Sears, P. (1964). Ecology – a subversive subject. *BioScience*, (14), 11–13.

[10] Georgescu-Roegen, N. (1975). Energy and economic myths. *Southern Economic Journal*, 41(3), 374–381. See p. 374.

recognises (as Elie Ayache has observed in another context) the pertinent fact that

> Contrary to history or life, the market possesses a single metric, the up or down movements of market prices.[11]

As the international post-WWII consensus of 'New Deal' Keynesianism unravelled in the 1970s, along with the control of US-based oil companies over Middle Eastern oil reserves, an international network of economists that had in earlier decades described their project as 'neoliberalism' rose to prominence in the corridors of power, inaugurating what might be described in historical terms as the neoliberal era.[12] The historical and disciplinary scope of the present inquiry is ample enough that an adequate review of the literature on neoliberalism cannot be provided here.[13] What I will rather attempt in the pages that follow is to complement this scholarship in ways which seem to me of crucial importance, given the dawning realisation that the house is quite literally on fire.

Existing traditions of critical theory—institutionalist, Foucauldian, post-Marxist—have engaged with the ideas of neoliberal thinkers and the itinerary of their migration through social institutions, but have rarely (with important exceptions) systematically engaged with the problem of how neoliberalism confronts the biophysical dimensions of the world economy.[14] In much of the literature, the emergence on the political

[11] Ayache, E. (2010). *The blank swan: the end of probability.* Chichester: John Wiley & Sons, p. xvii.

[12] Friedman, M. (1951). Neoliberalism and its prospects. *Formand*, (17), 89–93. Neoliberalism as an historical period might be dated to the radical experiment in violent social engineering conducted by the MPS in collaboration with Pinochet's military dictatorship in Chile following the 1973 coup d'état.

[13] My own reference points here include: Brown, W. (2015). *Undoing the demos: neoliberalism's stealth revolution.* MIT Press; Cooper, M. (2017). *Family values: between neoliberalism and the new social conservatism.* New York: Zone Books; Slobodian, Q. (2018). *Globalists: the end of empire and the birth of neoliberalism.* Cambridge MA: Harvard University Press; Whyte, J. (2019). *The morals of the market: human rights and the rise of neoliberalism.* London: Verso.

[14] See e.g.: Cooper, M. (2008). *Life as surplus: biotechnology and capitalism in the neoliberal era.* Seattle: University of Washington Press; Heynen, N., McCarthy, J., Prudham, S., & Robbins, P. (Eds.). (2007). *Neoliberal environments: false promises and unnatural consequences.* Routledge. Castree, N. (2008). Neoliberalising nature: the logics of deregulation and reregulation. *Environment & Planning A*, 40(1), 131–152; Nelson S. (2015). Beyond the limits to growth: ecology and the neoliberal counterrevolution. *Antipode* 47(2), 461–480.

horizon of planetary ecological crisis remains external to accounts of the neoliberal counter-revolution. At stake in the struggles over knowledge and power of the 1970s, I will argue, was the rising cultural authority of ecology and the Earth system sciences, forms of knowledge now claiming to be indispensable to the urgent reform of political and economic life.

As an evolving and at times contradictory suite of doctrines extending its influence from high finance and international law to the intimate spaces of family life, neoliberalism evades characterisation as a coherent philosophy, and is only loosely captured in calls for privatisation, deregulation, fiscal austerity, and 'limited' government. Rather than attempt to pin down a defining set of propositions from an evolving series of positions, alliances, and strategies, I follow Mirowski and Plehwe's method in *The Road from Mont Pèlerin* (2009) in approaching the 'neoliberal thought collective' as a social network. This they identify with the international membership of the Mont Pèlerin Society (MPS), and with the personnel of the expansive constellation of 'think-tanks' co-ordinated by the Atlas Network, the senior executives of which are usually MPS members.[15]

Founded by Friedrich Hayek in 1947, the MPS sought to clarify the principles and strategies of a 'new liberalism' robust enough to overcome the inexorable slide into socialist unfreedom they feared in the 'unlimited democracy' of the redistributive post-war welfare state and the post-imperial order of nation-state sovereignties constituting the United Nations. Whilst not without internal tensions between the differing schools of thought represented within the MPS, which brought together Austrian economists, German ordoliberals, and American neoclassicals of the Chicago school, what united them was the need for a consistent doctrine. For economic liberalism to revive its prospects at a time when it appeared in terminal decline, it could not afford to adopt the merely negative slogan of nineteenth-century liberalism—'laissez-faire'. What was needed, according to Hayek, was a new, future-oriented liberalism, one which would inspire activism by its courage to be utopian, one robust enough to capture state power and cordon off the social-democratic 'road to serfdom'. Since the 1970s, this new liberalism has succeeded

[15] Mirowski, P., & Plehwe, D. (Eds.) (2009). *The road from Mont Pèlerin: the making of the neoliberal thought collective*. Cambridge MA: Harvard University Press, p. 4. In identifying MPS members I draw on secondary works reporting on the MPS archives, and a more current member list leaked to: DeSmogBlog (n.d.) Mont Pèlerin Society (MPS). https://www.desmogblog.com/montpelerin-society. Accessed 26 June 2019.

spectacularly in reframing economic discourse and policy, in transforming the ethos and practice of law, government, and international relations, and most tragically, in ensuring that hydrocarbon-based industry remains (for the most part) free from mandatory constraints imposed in the name of ecological survival and the maintenance of the Earth's heat balance within the range to which life as we have known it can live.

On the face of it, it's not clear why a political movement claiming to ensure that we are 'free to choose' now functions as an increasingly authoritarian project to defeat democratic attempts to respond to the climate emergency; one adept at deploying the dark arts of 'business propaganda'—to borrow a term from Hayek's mentor Ludwig von Mises[16]—and intent on capturing state power in the service of fossil-fuel corporations. If neoliberals and fossil capitalists first mobilised in the early twentieth century against the labour movement's victories in constitutionalising electoral and then social democracy, the world-historic consolidation of neoliberal technologies of rule from the 1970s cannot be fully grasped, I contend, without recognising that the spectre of a rising environmental movement was a catalyst for the mass-enrolment of transnational corporations in the neoliberal project. Without the long-term consolidated support of big business, and in particular of US-based corporations concentrated in fossil-fuels, petrochemicals, mining and other pollution-intensive industries, the intellectual output of MPS scholars would never have achieved the influence and pre-eminence it now enjoys.

In the remainder of this introduction, I sketch this argument in outline, as a prelude to the history of economic and ecological thought pursued in the book. In the chapters which follow, I offer a pre-history of the present neoliberal era, in terms of the long mutual history of interactions between economic and ecological modes of systems-thinking. This is in turn grounded in a deeper genealogy of Western accounts of social order in the metaphorical mirror of the 'economy of nature'. Offering a critique of biopolitical economy grounded in the vital facts of photosynthesis and fire, the book sets the stage for the contemporary confrontation of neoliberalism with movements for ecological survival and climate justice.

[16] von Mises, L. ([1949] 1996). *Human action*. Irvington, NY: Foundation for Economic Education, p. 272.

Crisis and Contradiction: Life Beyond the Limits of the Earth

All histories involve elective designations of the events to be counted as the critical turning points. In the aftermath of the high modernism of the twentieth century, it became common to express a profound scepticism toward the possibility of understanding 'history as a whole'. Yet any argument must be couched in a narrative, and a history of philosophy can scarcely by disentangled from the philosophy of history. Georg Hegel taught that history was driven by a dialectic of ideological struggle, with periods of complacency punctuated by the irruption of struggle between proponents of antithetical ideologies, with each party driven to distraction by alternative ideologies whose very existence contradicted their own universal ontologies. This occurred at the micro-level of local political and religious discussion and was manifest in the wars of nations—what Hegel referred to as the 'slaughter bench of history'. Through this struggle, in which ideas engaged with contradicting ideas, there was a process of exchange in which a concept or its realisation passed over into its opposite and was preserved and fulfilled by it. Thus the World Soul became conscious and rational, a rationality reflected in the order of the cosmos, and social evolution occurred.[17] Hegel's idealist philosophy of history gave contradiction between opposites in moments of profound crisis a central role in the realisation of Progress. Maintaining that history could always have been otherwise, this book is polemical in relation to Hegelian accounts of state and right which elevate the ideas of the bourgeois state beyond the material conditions of everyday life, promising future progress and bringing actual devastation in the present. Nevertheless, it turns out that a history of interactions between ecology and economics is an exercise in *dialectics*, a term which refers to

> the tension or opposition between two interacting forces or elements [...] the logic of appearances and of illusions [...] any systematic reasoning, exposition, or argument that juxtaposes opposed or contradictory ideas.[18]

[17] Hegel, G.W. ([1837] 1980). *Lectures on the philosophy of world history*. Cambridge University Press.

[18] Merriam-Webster Online Dictionary. Definition of Dialectic. http://www.m-w.com/dictionary/dialectic. Accessed 4 July 2017.

My understanding of 'contradictions' in the relation between the natural and social sciences is closer to that of Karl Marx. Marx claimed to have 'put Hegel back on his feet', offering a materialist philosophy of history in which moments of social crisis characterised by changes in the material organisation of production drove history into new stages of social organisation. What was specific to the capitalist 'stage', Marx argued, was the subordination of social relations to abstract exchange value, which tends to incorporate all phenomena into its self-referential value system, transcending the embedded times and places of the phenomenal world:

> [A]s representative of the general form of wealth—money—capital is the endless and limitless drive to go beyond its limiting barrier. Every boundary is and has to be a boundary for it.[19]

A new 'crisis of contradictions' emerged in a sharp relief in the early 1970s, transforming abiding Western conceptions of relationships between nature and society. The modern environmental movement was novel in that it re-founded older political and social economy critiques of industrial society upon the models, metaphors and findings of ecology, a previously obscure branch of the life sciences that had recently coalesced around the 'ecosystem' concept. In its orientation to the future, the ecology movement painted a grim vision of the logic of modernisation that radically reversed the growth-oriented techno-optimism of the post-WWII period.

Under the international 'Keynesianism' of the US-sponsored Bretton Woods Treaty (1944), the standardisation of Fordist industry, social insurance and the family wage, and Green revolution agriculture had witnessed the longest period of stable, widely shared increases in material consumption in history—at least amongst the privileged citizens of the West. Whilst political conflict turned on the degree to which economic growth would be best realised by state planning, competitive enterprise, or a managed compromise between them, until the late 1960s, there was an almost universal faith that 'economic growth' or 'the advance of the productive forces' was synonymous with Progress itself. At the time, US and Soviet futurists outbid one another in visions of the triumph of modern technology—and their preferred mode of social organisation—over a nature rendered the malleable servant of technoscience. In *The Year 2000* (1967), the conservative theorists Daniel Bell, Herman Kahn and Anthony Wiener

[19] Cooper (2008, p. 8).

envisioned a shift to a flexible, high-technology, service-oriented consumer economy.[20] Amidst the revolutionary upheavals of the day, Bell discerned *The Coming of Post-industrial Society* (1974): the mass labour of Fordist industry was giving way to a new dynamic of 'science-led growth', testament to the increasing 'centrality of theoretical knowledge'. Bell saw this post-industrial future emerging from the mastery of nature manifest in the harnessing of vast atomic energies and the rapid development of networked digital computers. Presaging a shift from the mass labour of factory production to specialised professional services and the creative consumption of a widening leisure class, these changes demanded a redefinition of promises of the 'end of scarcity' in terms of the new 'economics of information'.[21]

Decades later, the Wall had fallen, and the new millennium was dawning. In a moment of Hegelian euphoria, Francis Fukuyama announced the end of history:

> [...] modern natural science establishes a uniform horizon of economic production possibilities. Technology makes possible the *limitless* accumulation of wealth, and thus the satisfaction of an *ever-expanding* set of human desires. [...] the logic of modern natural science would seem to dictate a universal evolution in the direction of capitalism. [my emphasis][22]

The globalisation of financial markets and the ubiquitous penetration of information technology in the form of the World Wide Web convinced many pundits that we had entered a post-material New Economy, in which the old economy practice of converting resources into goods and services with the labour of bodies and machines had become redundant. The physical world itself appeared to vaporise into the informational ether. Bell's vision seemed to have all but come to pass.

History, as is well known, is written by the victors. Responding in the 1970s to the post-industrial theorists, Soviet academicians outbid the 'bourgeois futurologists' in imagining the future of the 'scientific-technological revolution'. By the turn of the millennium, humanity would control unlimited flows of energy, which would 'unlock the door of

[20] Bell, D., Wiener, A., & Kahn, H. (1967). *The year 2000*. London: Macmillan.

[21] Bell, D. (1974). *The coming of post-industrial society*, Heinemann: London, pp. xciv–xcvi.

[22] Fukuyama, F. (1992). *The end of history and the last man*. Avon Books, p. xv.

nature's treasure house', yielding 'infinite supplies of natural resources'.[23] With this plenitude of energy, hunger would be abolished: the world's deserts converted into gardens, synthetic food produced on an industrial scale and harvested from the seas, and the climate subjected to push-button control. Repetitive work in the Soviet bloc would be automated by 1990, and by the year 2000 citizens would live to 100 through genomic control of the aging process. Abundant leisure time would be absorbed by trips to the moon, which would have an extensive network of railways by 2030. All this would be achieved with new technology that would not degrade the biosphere but actively restore 'ecological equilibrium'.

This last piece of promissory rhetoric recognised the profound influence of one of the most ambitious attempts to anticipate world futures, the 1972 *Limits to Growth* report to the Club of Rome.[24] Like the post-industrial futurists, the Club of Rome warned that the Fordist model of industrial growth had entered a period of irreversible decline, a message brought home vividly to a public shocked by the Arab oil embargos into recognition of their dependence on imported fuel. Yet this was a crisis that went far beyond the conventional terms of productivist thought. It was rather, as Melinda Cooper puts it, 'a wholesale crisis in the realm of reproduction [...] what was at stake was no less than the continuing reproduction of the earth's biosphere and hence the future of life on earth'.[25]

Commissioned by a club of industrialists, a team of MIT computer programmers developed a model of the exponential growth trajectory of the global economy, attempting to incorporate the positive and negative feedbacks between industrial expansion and the Earth's resources and ecosystems. The report warned of the dangers of 'thermal pollution', projecting the exponential rise of atmospheric carbon dioxide meticulously recorded by Charles Keeling since 1958.[26] Adapting scenario forecasting methods developed by the US Air Force and the Shell Oil corporation, the authors did not claim to be able to predict the future with precision.[27] Yet through multiple model runs, the finding was clear: exponential growth in

[23] Modhrizinska, Y. & Stephanyan C. (1973). *The future of society*. Moscow: Progress; Kosolopav, V. (1976). *Mankind and the year 2000*. Moscow: Progress.

[24] Meadows, D., Meadows, D., Randers, J. & Behrens, W. (1972). *The limits to growth: a report to the Club of Rome*. New York: Universe.

[25] Cooper (2008, p. 16).

[26] Meadows et al. (1972, pp. 71–73).

[27] Granjou, C., Walker, J., & Salazar, J. (2017). The politics of anticipation: On knowing and governing environmental futures. *Futures*, (92), 5–11.

population and industrial output could not continue without running up against the inherent limits of arable land, fossil energy, mineral resources, and the biosphere's capacity to absorb harvest and pollution, a conclusion resolving to 'the simple fact [...] that the earth is finite'.[28]

The Club of Rome forecast that unless there was political agreement on the necessity of shifting from exponential growth to some form of steady-state economy, the limits to growth would be catastrophically transgressed around the mid-twenty-first century, causing drastic declines in human populations. The industrial economy needed to be regulated within the fixed limits set by the geophysical logic of mineral depletion and the fragile equilibria of the global ecosystem. The report called for an urgent, though undefined 'general strategy' to achieve 'global equilibrium', a situation in which 'population and capital are essentially stable, with the forces tending to increase or decrease them kept in a tightly controlled balance'.[29] The authors warned their readers that while

> [...] the concept of a society in a steady state of economic and ecological equilibrium may appear easy to grasp [...] the reality is so distant from our experience as to require a Copernican revolution of the mind.[30]

From our present fearful orbit, it need hardly be emphasised how remote we are from such a goal. Yet this proposal was widely discussed amongst the delegates to the 1972 United Nations Conference on the Human Environment at Stockholm, the first international environment summit, and the last one at which the explicit critique of economic growth would frame deliberations. Nevertheless, the consensus of the Stockholm Declaration that the 'ecological balance of the biosphere' was threatened by 'major and undesirable disturbances' due to 'dangerous levels of pollution' and the 'destruction and depletion of irreplaceable resources' has only been abundantly confirmed by the Earth systems sciences.[31]

Few American economists were equipped by training or inclination to engage with the scientific basis of such claims. Yet they were troubled by the Club of Rome's pessimistic account of growth, and the consensus of the Stockholm Declaration that '[r]ational planning constitutes an

[28] Meadows et al. (1972, p. 86).
[29] Meadows et al. (1972, p. 189).
[30] Meadows et al. (1972, p. 196).
[31] United Nations (1972). Stockholm Declaration of the UN Conference on the Human Environment. http://www.un-documents.net/unchedec.htm. Accessed 1 Feb 2018.

essential tool for reconciling any conflict between the needs of development and the need to protect and improve the environment.'[32] This was a direct challenge to the standard neoclassical model of permanent growth in 'free market' equilibrium conditions. From Robert Solow's classic 1956 paper down to the Dynamic Stochastic General Equilibrium computer models presently deployed in central banks, neoclassical models promise permanent 'equilibrium-path' growth. Excluding any necessary role for inputs of energy or natural resources to the economic process, the source of this growth is attributed to a 'residual' factor labelled 'technology'.[33] Solow's response to the Club of Rome was to argue that it had undersold the role of the price mechanism in adapting to resource scarcities. Falling supply would increase the marginal utility (and thus the marginal price) of depleted resources, presenting opportunities for entrepreneurs to get rich by innovating beyond the limits with technological substitutes:

> If it is very easy to substitute other factors for natural resources, then there is, in principle, no 'problem'. The world can, in effect, get along without natural resources [...] Exhaustion is just an event, not a catastrophe.[34]

If Solow's enormous 'if' held true, the 'price mechanism' would restore its own conditions of equilibrium, regardless of the wasted condition of the Earth. Responses to the *Limits* thesis in prestigious US economic journals converged on similar themes. Ignoring entirely the accumulation of pollution and loss of ecological abundance, the economists uniformly approached the question of resource depletion in terms of strategic minerals, confining the problem to an 'optimal rate of depletion'.[35] This approach built on a 1931 paper by Howard Hotelling modelling the profit-optimising strategy of a mine owner within the neoclassical mathematics of equilibrium. In it, Hotelling presciently observed that: '[p]roblems of exhaustible assets are peculiarly liable to become entangled

[32] UN (1972, para 16).

[33] Solow, R. (1956). A contribution to the theory of economic growth. *Quarterly Journal of Economics*, 70(1): 65–94.

[34] Solow, R. (1974). The economics of resources or the resources of economics. *American Economic Review*, 64(2), 1–14.

[35] Stiglitz, J. (1974). Growth with exhaustible natural resources. *Review of Economic Studies*, 41(128), 139–145.

with the infinite'.[36] Accordingly, concerns about the irreversible depletion of oil were dismissed with confidence in the infinite supplies of nuclear energy promised by the soon-to-be-feasible fusion reactor. Some looked further, anticipating a techno-future centuries hence in which the exhaustion of minerals had instantiated a miraculous age of 'infinite substitutability'.[37] Others claimed the problem was less the scarcity of resources than a scarcity of markets—market failures must be met with more markets. The market prices generated by the exchange of financial futures and derivatives (not the results of natural science) were the relevant data with which to know and decide the future conditions of life on Earth, because

> *Everything* depends upon how traders form their expectations about the future in situations where definite information is lacking [...] Many of the difficulties that are involved in the making of policy recommendations about the rate of depletion of exhaustible resources stem from the fact that crucial aspects of this problem are inherently uncertain, and it is not clear that an adequate class of contingent markets exists. [my emphasis][38]

Such arguments were testament to the rising influence of Chicago School finance theories of rational expectations and efficient markets, which would soon license the radical rollback of New Deal banking and financial regulation. Influenced by Hayek's precocious account of prices as information and markets as distributed computation in the face of a future inherently beyond the capacity of scientific reason to foresee or predict, the 'efficient markets hypothesis' holds that asset prices fully reflect all available 'information.'[39] Therefore any government policy to reign in 'irrational' market activity injurious to the public interest will at best be futile, and at worst, damaging to 'the economy'.

[36] Hotelling, H. (1931). The economics of exhaustible resources. *Journal of Political Economy*, 39(2), 137–175.

[37] Goeller, H., & Weinberg, A. (1978). The age of substitutability. *American Economic Review*, 68(6), 1–11.

[38] Dasgupta, P. & Heal, G. (1974). The optimal depletion of exhaustible resources. *Review of Economic Studies*, 41(128), 3–26. See p. 3.

[39] Hayek, F. (1945). The use of knowledge in society. *American Economic Review*, 35(4), 519–530; Fama, E. (1970). Efficient capital markets: a review of theory and empirical work. *Journal of Finance*, 25(2), 383–417.

Neoliberalism as Anti-environmentalism: Corporations and the Counter-revolution

The analysis of the Club of Rome crystallised into mainstream consciousness the environmental movement's vision of a looming planetary disaster, a sense of impending crisis which had been building among scientists for some time. In 1957, the oceanographer Roger Revelle noted the rising concentrations of carbon dioxide in the oceans and atmosphere, a result of burgeoning fossil-fuel combustion.

> [H]uman beings are now carrying out a large scale geophysical experiment of a kind that could not have happened in the past nor be reproduced in the future. Within a few centuries we are returning to the atmosphere and oceans the concentrated organic carbon stored in sedimentary rocks over hundreds of millions of years.[40]

A talented science organiser, Revelle would be lead author of the first public study to warn of the potentially devastating consequences of this experiment for the future heat balance of the Earth, presented to President Lyndon Johnson in 1965.[41] Revelle was also influential in the International Biological Program (1964–1974), which was tasked with examining the biological basis of the Earth's productivity as it related to human welfare. The IBP project was dominated by American ecologists deploying the 'ecosystems approach' to analyse solar energy flows and nutrient cycles through large-scale bioregions. If ecologists had previously tended to study relatively undisturbed, local ecosystems, the IBP was explicitly global and human-oriented. Ecology would increasingly be proclaimed as a basic science with profound implications for human welfare, carried forward by a new generation of practitioners 'motivated by a sense of responsibility for social action'.[42]

[40] Revelle, R. & Suess, H. (1957). Carbon dioxide exchange between atmosphere and ocean and the question of an increase of atmospheric CO_2 during the past decades. *Tellus*, 9(1), 18–27.

[41] Revelle, R., Broeker, W., Keeling, C., Craig, H. & Smagorinsky, J. (1965). Atmospheric carbon dioxide. In President's Scientific Advisory Committee, Restoring the quality of our environment: report of the environmental pollution panel (pp. 111–133). Washington DC: The White House.

[42] Smith, F. (1968). The International Biological Program and the science of ecology. *Proceedings of the National Academy of Sciences*, 60(1), 5–11. See p. 11.

The modern environmental movement made rapid inroads into public consciousness through the 1960s, and public calls for regulation of polluting industry mounted. Linking domestic environmental disasters to ecological warfare abroad, activists targeted chemical corporations for poisoning air, water, and soil at home, and for supplying napalm and dioxin for ecological warfare against the peoples of Indochina. In April 1970, 20 million US citizens took part in the first Earth Day. In September 1970, Milton Friedman (MPS) of Chicago University warned that any firm spending more on pollution controls than the absolute minimum required by law 'in order to contribute to the social objective of improving the environment', was practising 'pure and unadulterated socialism'.[43] In December 1970, President Nixon passed the National Environmental Policy Act (1970), which authorised the Environmental Protection Agency to administer a wave of new environmental statutes—the Wilderness Act (1965), the Clean Air Act (1970), the Clean Water Act (1972), the Endangered Species Act (1973), and the Control of Toxic Substances Act (1976)—at a time when most democracies were enacting similar legislation, institutionalising the environmental science-based regulation of industry with pollution taxes, prohibitions, standards, and fines. In 1971, the usually conservative *Newsweek* magazine predicted that environmental protest would soon surpass the level of protest against the Vietnam war, announcing that an 'Age of Ecology' was at hand in which 'ecologists would teach us about the web of life'. Massive sales of Eugene Odum's textbook *Fundamentals of Ecology* introduced 'the ecosystem' as an indispensable term of public discourse, and a new source of metaphors for life's complex organisation.[44]

Meanwhile, a sense of impending crisis was brewing amongst the captains of American industry. The publicity accorded to *Limits to Growth*—with its call for strict regulation and an end to growth—was disquieting to the business establishment. Corporations felt themselves under siege from militant labour, civil rights, anti-war, and feminist movements at home, from hostile nationalisations in the developing world, and from the inflation and declining returns they attributed to the New Deal welfare state and Keynesian modes of economic management. Nixon's concessions to the environmental movement catalysed a sense of betrayal among US

[43] Friedman, M. (1970, Sept 13). The social responsibility of business is to increase profits. *New York Times Magazine*.

[44] Odum, E.P. ([1959] 1971). *Fundamentals of ecology*. W.B. Saunders.

business conservatives. At stake for them was the very future of the 'American free enterprise system'. The response of US business was to launch the neoliberal counter-revolution.

In 1971, the corporate lawyer Lewis Powell, soon to be appointed to the Supreme Court by Nixon, prepared a secret memorandum for the US Chamber of Commerce. A call to arms, the tone of the memo was set by a message from Milton Friedman:

> It its crystal clear that the foundations of our free society are under wide-ranging and powerful attack—not by Communists or any conspiracy but by misguided individuals parroting one another and unwittingly serving ends they would never intentionally promote.[45]

Powell observed that the most disquieting voices of dissent came from respectable sources: 'the college campus, the pulpit, the media, the intellectual and literary journals, the arts and sciences'. The university was identified as a crucial battleground: 'The social science faculties usually include members who are unsympathetic to the free enterprise system.' The memorandum outlined a pervasive long-term strategy to marshal the resources of business against 'those who would destroy it', calling upon US corporations 'to be far more aggressive than in the past', to use 'muscle' to shape government policy and legislation 'at all levels and at every opportunity'.[46] Business ought to generously fund scholarship within universities that supported the free enterprise system, and beyond academia, organise a para-academic network of pro-business scholars, lawyers, advertisers, and public relations professionals to shape public opinion and public policy through the mass media.

In 1972, big business responded by forming the Business Roundtable, a powerful federal lobby of the CEOs of 130 of the largest US firms, chaired by the CEO of the Exxon oil company. Older business lobbies and 'business propaganda' outfits such as of the National Association of Manufacturers (formed in 1894 to counter the influence of labour unions, and later the New Deal) represented a broader sample of industries than the Roundtable, whose corporate membership was highly concentrated in energy and pollution intensive sectors such as oil and gas, chemicals and

[45] Powell, L. (1971, Aug 23). Attack on American free enterprise system – memorandum to Eugene B. Sydnor. http://law2.wlu.edu/deptimages/Powell%20Archives/PowellMemorandumTypescript.pdf. Accessed 12 Feb 2018.

[46] Powell (1971, p. 8).

autos, nuclear and coal-fired electricity, steel and aluminium refining, mining, and logging.[47] Both the National Association of Manufacturers and the Business Roundtable directed major lobbying efforts against the new environmental legislation. In 1973, wealthy donors established the Heritage Foundation to promote bold social and environmental deregulation. This was followed by the Charles Koch Foundation in 1974 (renamed the Cato Institute in 1976), founded by MPS members Murry Rothbard, Ed Crane, and the eponymous billionaire of the Koch Industries petroleum conglomerate. In 1981, Anthony Fisher, a wealthy UK businessman and devotee of Hayek, founded the Atlas Economic Research Foundation, expanding on his success in launching the Thatcher revolution from the Institute of Economic Affairs (established in 1955). Later renamed the Atlas Network, this umbrella organisation presently coordinates strategy and funding across more than 500 affiliated public policy 'research institutes' and 'non-profit' for-profit think-tanks worldwide.[48] In addition to anti-welfare, anti-labour, and anti-regulation activism, an abiding aim of the network has always been to 'strangle the environmental movement', as Ed Crane put it in the house journal of the Heritage Foundation.[49]

Whilst a detailed demonstration will have to wait for another occasion, in what follows I propose two observations of relevance to the interpretation of the broader theses of this book. Firstly, since the earliest days of the circle gathered around Ludwig von Mises at the Vienna Chamber of Commerce in the 1920s (which included Hayek, William Rappard, Fritz Machlup, and Gottfried Haberler), the project of the neoliberals to limit the impact of democracy and national government on the market process, which was for them always conceived of in terms of world orders, has been intimately connected to the geopolitical strategies of US fossil energy corporations. And none more consistently so than ExxonMobil, the major corporate descendant of John D. Rockefeller's world-spanning Standard Oil monopoly. Secondly, anti-environmentalist corporate strategists working through the neoliberal network have not merely been concerned to neutralise the effectiveness of the environmental legislation of

[47] Business Roundtable (1972). The Business Roundtable: its purpose and program. http://www.documentcloud.org/documents/3903345-Business-Roundtable-1972-Its-Purpose-and-Program.html. Accessed 30 Jan 2019.

[48] Atlas Network (2019). Our partners: global directory. https://www.atlasnetwork.org/partners/global-directory. Accessed 13 Feb 2019.

[49] Andrews, J. et al. (1990). The vision thing: conservatives take aim at the '90s. *Policy Review*, 52: 38–43.

the 1970s mentioned above, but have long anticipated the need for co-ordinated international policies to limit fossil-fuel combustion and prevent catastrophic climate change, and acted comprehensively and successfully to obstruct this, including through the mass-propagation of science denial. This is perhaps the darkest secret, and darkest victory of the neoliberals.

Fuelling Freedom's Fires: Neoliberalism and the Organisation of Fossil Capital

Ludwig von Mises described the task of economic science in the following terms:

> In the course of social events there prevails a regularity of phenomena [the interdependence of market phenomena] to which man must adjust his actions if he wishes to succeed. [...] One must study the laws of human action and social cooperation as the physicist studies the laws of nature.[50]

The image of the naturally equilibrating free-market is integral to the neoliberal constitution of political order and its contemporary technologies of government. As one scholar once summarised Mises' position: '[t]he function of the state is to use positive law to uphold natural law'.[51] This book may be read as a genealogy of the claim that 'the free economy', operates according to natural laws (somehow) analogous to those revealed by physics, and the consequent political claim that the laws of the state must be subordinate to the laws of the market revealed by economists, practitioners of a universal science that (somehow) resembles physics. For now it will be enough to note that von Mises' claims to the rigour of the physicist run directly counter to the enterprise of modern science to understand causes and effects through rigorous, testable theories built on empirical observations of material phenomena. Perhaps it is the methodological approach of the Austrian brand of economic science that has attracted the considerable investments by Koch Industries and other US fossil-fuel interests to build academic institutions and respectability for the latter-day Austrian School on American soil. After all, for von Mises,

[50] von Mises (1996, p. 2).

[51] Gonce, R. (1973). Natural law and Ludwig von Mises' praxeology and economic science. *Southern Economic Journal*, 39(4), 490–507.

> The specific method of economics is the method of imaginary constructions.

Admitting the problematic nature of this evidence-free *a priori* method, von Mises doubted whether any but he and his own disciples could practice it:

> It is, to be sure, a method difficult to handle because it can easily result in fallacious syllogisms [...] on both sides yawns the chasm of absurdity and nonsense.[52]

As Quinn Slobodian has shown, the neoliberal project of 'globalisation' was constituted in its earliest period in what he terms the Geneva School. This was a European network of liberal intellectuals which served the interests of the International Chamber of Commerce (ICC) in advocating to the League of Nations for a business-friendly order of international law and policy in the wake of the first World War.[53] These intellectuals imagined a world economic constitution which would assume the role once taken by the British Empire in guaranteeing open trade, and which would discipline nation-states by the threat of capital flight from pursuing the redistributive policies of national economy proposed by social democratic parties. Founded in 1920 by the senior executives of multinational corporations—with leading roles played by A.C. Bedford of the Standard Oil corporation, Owen Young of General Electric, and Thomas Lamont of JP Morgan—the ICC sought to reverse the nationalization of economic planning that states had imposed to fight the war of 1914–1918, and restore guarantees of international laissez-faire.[54]

With the neoliberals of the Austrian and Geneva schools, the ICC was concerned to undermine the legitimacy that labour parties and unions had assumed in the long struggle for universal suffrage. As has been amply documented in Mitchell's *Carbon Democracy* (2011) and Malm's *Fossil Capital* (2016), the mobilisation of nineteenth-century workers revolting against miserable wages and conditions in coal mines and steam-powered factories, which in Britain took the form of the Chartist movement, eventually succeeded in limiting authoritarian liberalism by establishing

[52] von Mises (1996, p. 237).

[53] Slobodian (2018).

[54] Tomashot, S. (2015). Selling peace: the history of the International Chamber of Commerce, 1919–1925. PhD thesis: Georgia State University.

universal suffrage as a basic democratic norm.[55] Co-ordinated strike action by coal-miners and allied rail, maritime and dockers unions in the nineteenth and early twentieth centuries was able to 'shut off the energy tap' to coal-fired economies, leading to electoral and labour reforms in Germany, France, Britain, Canada, Australia, and some US states. From the subterranean gloom of coal mines emerged citizens' electoral democracy and nascent forms of the welfare state.

In the United States, a progressive reform movement outraged by Ida Tarbell's *History of the Standard Oil Company* (1904) led to the breakup in 1911 of the Standard Oil monopoly by the Supreme Court under the Sherman Anti-trust Act (1890), the result of an action brought by the Federal government in response to the 'widespread impression that corporate power had been and would be used to oppress individuals and injure the public generally.'[56] At the time, Standard Oil had corporate presence in Austria-Hungary: through a subsidiary registered in Vienna, it sought control of the oil reserves of Galicia, the only major oilfield in Central Europe. Otto Neurath of the Vienna Circle, a rival of Hayek on the Austrian left, complained in 1910 of the 'business politics of the Standard Oil Company', which engaged 'in the most intense manner its destructive politics in a battle against the domestic petroleum industry.'[57]

In 1914, a strike by miners attempting to unionise at a Colorado coal mine owned by the Rockefellers ended in the Ludlow massacre. Private security and the National Guard—both paid by the company—opened fire on the striking workers and their families with machine guns, killing dozens of men, women and children. Advised by the industrial relations consultant Mackenzie King of the danger of further radicalising coal miners towards revolutionary syndicalism and general strikes, John D. Rockefeller Jr. began to contemplate subtler modes of labour management.[58] In the wake of the Ludlow crisis, the Rockefeller Foundation acquired a much expanded budget and mission. With a brief to cultivate 'scientific'

[55] Mitchell, T. (2011). *Carbon democracy: political power in the age of oil.* London: Verso; Malm, A. (2016). *Fossil capital: the rise of steam power and the roots of global warming.* London: Verso.

[56] Tarbell, I. ([1904] 2009). *The history of the Standard Oil company.* New York: Cosimo; Rowe, F. (1984) The decline of antitrust and the delusions of models. *Georgetown Law Journal*, 72: 1511–1570.

[57] Frank, A. (2009). The petroleum war of 1910: Standard Oil, Austria, and the limits of the multinational corporation. *American Historical Review*, 114(1), 16–41.

[58] Mitchell (2011, p. 406).

approaches to social problems, the Foundation made major international investments in social research, substantially influencing the direction of the social sciences in the twentieth century. Economics was chosen for particular emphasis in grants and fellowships: the Foundation's managers believed it to be the social science closest to the natural sciences.[59] In the late 1920s the Foundation provided incomes for von Mises and his protégé Hayek at the Austrian Institute of Business Cycle Research, which was housed in Vienna at the Chamber of Commerce where von Mises served as secretary.[60]

From 1923 to 1939 the Rockefeller Foundation funded an array of British universities: its grant program was then 'almost the only available source for the social sciences in Britain [...].'[61] The Foundation presided over a major expansion of the London School of Economics (LSE), chosen due to its proximity to the British government and civil service, and its role in providing administrative expertise throughout the Empire. During the mass unemployment of the Great Depression, Lionel Robbins and Hayek at the LSE offered the most orthodox voices of 'laissez-faire' opposition to Keynesian and New Deal policies, advocating for the deflationary austerity of 'non-intervention'. In the late 1930s, Hayek supervised David Rockefeller's doctoral studies, which were completed with Frank Knight at the University of Chicago—a private institution established in 1890 by foundation grants from John D. Rockefeller. In 1938, the Rockefeller Foundation supported the Colloque Walter Lippmann in Paris, which brought together intellectuals from throughout Europe to discuss the deepening crisis of liberalism. This meeting, at which the term 'neoliberalism' was first chosen to identify their shared project, was resumed in 1947 at Hayek's inaugural Mont Pèlerin conference, which also enjoyed Rockefeller support.

Between the 1930s and the 1970s, American economists widely accepted Keynesian macroeconomics and its social objectives, recognising that the bourgeois Keynes had saved capitalism from its crisis-prone financial instabilities, and democracy from the forces of Reaction and Revolution. By contrast, the Hayekian perspective which from the late 1940s informed the neoliberal revision of American economics by Milton Friedman and

[59] Fisher, D. (1980). American philanthropy and the social sciences in Britain, 1919–1939. *Sociological Review*, 28(2), 277–315.

[60] Mirowski and Plewhe (2009, p. 11).

[61] Fisher (1980, p. 300).

other MPS figures at Chicago, and which resulted in a wholesale transformation of US legislation and jurisprudence through the Chicago law and economics movement, maintained a militant critique of all forms of what von Mises called 'interventionism'. Implying that 'the market' is 'natural' and government 'artificial', the term refers to policies intended to soften the sharp edges of market society for the betterment of the majority, such as progressive taxation, public education and health provision, social insurance, wage arbitration, counter-cyclical fiscal policy, public ownership of land and utilities, competition and anti-trust law, and policies for environmental protection. An inaugural research project of the neoliberals in the Chicago law and economics faculties, the Chicago Antitrust Project (1953–1957), took aim at the legal reasoning by which Standard Oil had been declared an illegal monopoly in 1911.[62] Corporate monopoly was reimagined as merely the beneficial outworking of 'consumer sovereignty' and the efficient operation of the free market.

From its inception down to the present day, the neoliberal project has been funded by (and for) private wealth concentrated in the hydrocarbon sector, a continuity which has yet to be integrated into the growing academic literature documenting its history through archival research. Likewise, researchers and activists tracing the funding of climate denialism routinely discover connections between ExxonMobil, Koch Industries and a long list of 'think-tanks' but without connecting this finding to the literature on neoliberalism, or noticing that these think-tanks are uniformly affiliated with the Atlas Network.[63] This is a 'regime of obstruction' which may yet turn out to precede the now well-documented organisation since 1988 of fossil-fuel interests to propagate 'the Exxon position', which seeks to 'emphasize the uncertainty in the scientific conclusions regarding the potential enhanced greenhouse effect'.[64]

[62] See e.g.: McGee, J. (1956). Price discrimination and predatory effects: the Standard Oil of Indiana case. *University of Chicago Law Review*, 23, 398–473.

[63] Compare the Atlas Network (2019) US member directory with the list composed by Greenpeace (https://exxonsecrets.org/html/index.php – accessed 13 Feb 2018) and the funding networks mapped by: Brulle, R. (2014). Institutionalising delay: foundation funding and the creation of US climate change counter-movements, *Climatic Change*, (122), 681–694.

[64] Union of Concerned Scientists (2007, Jan). Smoke, mirrors and hot air: how ExxonMobil's disinformation campaign uses Big Tobacco's tactics to manufacture uncertainty on climate science. Cambridge MA; Supran, G. & Oreskes, N. (2017). Assessing ExxonMobil's climate change communications (1977–2014). *Environmental Research Letters*, (12), 1–18.

As Ben Franta has discovered, as early as 1959 there was adequate scientific warrant for the nuclear physicist Edward Teller to predict, in a keynote address to the senior executives of the US oil industry, that if the exponential rise in hydrocarbon combustion were to continue 'the earth will continue to heat up'. A carbon-dioxide driven 'greenhouse effect' would 'be sufficient to melt the icecap and submerge New York. All the coastal cities would be covered'.[65] Teller shared the stage with Robert Dunlop, president of the Sun Oil company and director of the American Petroleum Institute (API).[66] Whilst further archival research is needed, it would seem likely that several members of the API attending the conference were already sponsors of the neoliberal project. The oil baron John Howard Pew, who preceded Dunlop as president of Sun Oil, was a long term supporter, and was inducted into the MPS in 1954.[67] Indeed, as I will suggest later in the book, it would appear that US oil executives had access to accurate research linking growth in fossil-fuel combustion to catastrophic planetary heating well before this was available to the broader community of academic researchers, including the leading figures in systems ecology and ecological economics who inaugurated the scientific critique of economic growth ideology in the 1970s.

The convergence of private and public power in our time of eroding democracy, deepening inequality and accelerating global heating is precisely the achievement of the networked political infrastructure that the Mont Pèlerin Society has built across academia, business, civil society, the courts, political parties, supra-national institutions, and the corporate media. While it is mobilised for a wide array of interests and purposes in multi-national contexts, this neoliberal political machinery has long been attuned to the geopolitical strategy of fossil capital. Witness President Trump's unilateral withdrawal of the United States from the Paris Climate Treaty, his stacking of the EPA with science-denying think-tankers, and his appointment of Rex Tillerson, CEO of ExxonMobil, as the secretary of state. An indicative highwater mark of the historical entanglement of

[65] Teller, E. (1960). Energy patterns of the future. In *Energy and man: a symposium* (pp. 54–72). New York: Appleton-Century Croft; Franta, B. (2018). Early oil industry knowledge of CO_2 and global warming. *Nature Climate Change*, 8(12), 1024–1025.

[66] Franta, (2018, Jan 1). On its 100th birthday in 1959, Edward Teller warned the oil industry about global warming. *The Guardian*.

[67] Mirowski and Plehwe (2009, p. 192).

neoliberalism with fossil capital's ambition to roll back the science-based environmental laws of the 1970s might be found in a bill introduced to Congress early in the Trump administration. The text of the bill amounted to a single sentence: 'The Environmental Protection Agency shall terminate on December 31, 2018.'[68]

[68] US Congress (2017). Bill H.R.861 – To terminate the Environmental Protection Agency. 115th Congress. https://www.congress.gov/bill/115th-congress/house-bill/861/text. Accessed 1 Jan 2019.

CHAPTER 2

Oikonomia: On Metaphor, Science, and Natural Order

How shall we live? At once deeply philosophical and imminently pragmatic, this question is perhaps the most general animating the inquiry variously termed economics, political economy, moral science, or more venerably, *oikonomia*. I would go further, and suggest that it also underlies the modern science of ecology, known earlier as natural history, or 'oeconomy of nature'. Of course, it is not merely Europeans who have asked and answered this question. For many indigenous cultures, the 'we' to whom the question is addressed is 'more than human'. Kinship extends reciprocally throughout multi-species lifeworlds imbued with consciousness and value. Ethical systems of law and conduct are embedded in a wealth of detailed, profoundly observational ecological knowledge. It might be argued that it is a unique feature of Western metaphysics to have excluded empirical knowledge of the human place in the 'web of life' from the doctrines which provide the most privileged explanations of its social order, and which sanction the institutions of law that maintain it. How are we to live? This is a question which takes on a profound urgency when we consider the imminent future of the Earth: one of continuous planetary heating and cascading extinctions comparable to the most catastrophic events of geological history. A future, in short, of more heat than life.

Despite shared roots in the ancient Greek concept of the *oikos* (the community dwelling in the 'household' upon its lands and 'estate'), at some point in the history of Western thought economics as the domain of *nomos* (the laws and customs by which we apportion and 'manage the

J. Walker, *More Heat than Life: The Tangled Roots of Ecology, Energy, and Economics*,
https://doi.org/10.1007/978-981-15-3936-7_2

estate') became separate from the domain of *logos* (knowledge of the logic of life's unfolding on Earth). How are we to 'manage the estate' without such knowledge? Economics and ecology are practised as incommensurable sciences in the modern era. In the political sphere, this is manifest in the agonistic confrontation between neoliberalism and environmentalism that since the 1970s has come to define our predicament. To the extent that an intellectual history might have anything to offer, let us begin with the proposition that both sciences share conceptual roots so entangled that they ought ultimately be considered branches of the same 'tree of knowledge'. The primary task of critical analysis, for Gadamer, is the clarification of concepts: '[a] consciousness of the history of concepts becomes a duty of critical thinking'.[1] Taking the interpretive approach of a 'sociology of concepts' as my historical method, I attempt to construct a history of the modern bifurcation of *oikonomia* into the estranged twin sciences of ecology and economics.

Two lines of method are pursued in this book in pursuit of this aim. The first operates on the level of language and discourse; the second confronts the materiality of living organisms and the artefacts of thermoindustrial society. In this chapter I offer an account of the role of metaphor in the formation of the core concepts that organize disciplinary research programmes, and in the wider ideological formations and world-views that draw their truth claims from them. Metaphors, as Lakoff and Johnson have argued, are basic and ontological. Insofar as they structure our perceptions, they are inherent in the processes of cognition and social communication that organise social action.[2] Metaphor plays a key role in constituting the research objects addressed by emerging scientific disciplines and in demarcating the boundaries of disciplinary inquiry. Metaphor works by bringing together concepts from separate conceptual domains, it thus has a crucial rule in the transformative traffic of ideas across the borders. This approach will allow us to historicize the contingent mutual trajectories of the twin sciences by attending to the different ways each have deployed such foundational concepts as growth, equilibrium, law, and system. In the chapter which follows, I move from the sociology of knowledge to a materialist ontology, arguing that the common material foundations of ecological and human economic order on Earth resolve to

[1] Cited in: Müller, J. (2014). *On conceptual history: rethinking modern European intellectual history*. Oxford: Oxford University Press, p. 75.

[2] Lakoff, G., & Johnson, M. (2008). *Metaphors we live by*. University of Chicago Press.

the interplay between photosynthesis and fire: the transformative processes of energy conversion by which both the sunlit biosphere and our thermoindustrial society live and die.

Ecology and Economics as Twin Sciences and Incommensurable Paradigms

'Nature is an infinite sphere, whose centre is everywhere and whose circumference is nowhere.' Concluding a satirical essay on this aphorism of Pascal, the Argentinian writer Jorge Louis Borges made a curious suggestion: 'It may be that universal history is the history of a handful of metaphors.'[3] Working from this premise, this book will examine the career of a handful of metaphors that have played a future-making role in the recent history of the Earth; a 'fearful sphere' perhaps, but one upon whose decidedly non-infinite surface exists the only known habitation of life in an otherwise infinitely abiotic universe.

In spite of its basic incompatibility with the biophysical sciences, and indeed with other social sciences, 'economics' in its neoclassical and Austrian guises has maintained the core features of the research programme that emerged from the 'marginalist revolution' of the 1870s. This break with political economy as a philosophy of industrial society and government constituted the formative moment of economics as an abstract science of the 'laws of supply and demand' and the mechanics of market exchange. The first task before us then, is to investigate the epistemic authority of economics—'the queen of the social sciences' for Paul Samuelson—a claim part and parcel to its political hegemony as an art of government.[4] Economists' claims to practise the social science most verifiable according to rigorous standards of theory formation, ecologists might argue, is most grievously countered by its policy presumption of infinite economic growth, contrasted against the proliferating realms of empirical evidence pointing to a drastic erosion of biodiversity and ecosystem functions.

[3] Borges, J. (1964). The fearful sphere of Pascal. In *Labyrinths: Selected stories & other writings*. New Directions. For this approach I acknowledge debts to: Mirowski, P. (1989). *More heat than light: economics as social physics, physics as nature's economics*. Cambridge: Cambridge University Press, p. 1.

[4] Samuelson, P. (1970). *Economics*. New York: McGraw-Hill, p. 9.

The second task is to investigate the history of ecology up until the early 1970s when the two sciences came into open conflict. Playing David to the Goliath of economics for space in the political arena, the claim of systems ecology to superior scientific legitimacy over economics is manifest in the Club of Rome's (1972) recommendation that the preservation of 'ecological equilibrium' must replace the pursuit of 'economic growth' as the primary task of government. From the vantage point of the far-from-equilibrium present, appeals to equilibrium concepts from both ecologists and economists seem like relics of a fading past, in which the long-range stability of natural order was naively to be expected. Yet what history recalls of the 'subversive science' of ecology should raise our eyebrows. As a general theory of life, its initial trajectory of scientisation not only parallels that of economics, but also reveals that ecology has also systematically borrowed concepts and metaphors from the social sciences, and of none more so than political economy. Ecology has historically existed in a subsidiary position to economics; and it has always been something of a 'bioeconomics', as 'oeconomy of nature', the pre-modern name of the study suggests. Given these shared genealogies, we are left with the problem of how the twin sciences of the *oikos* became so estranged. Perhaps the pollinating agents of this intellectual history are, as Borges suggests: 'the different intonations given a handful of metaphors.'

Thomas Kuhn's *Structure of Scientific Revolutions* (1962) made popular the notion that alternate scientific paradigms can be incommensurable. His claim that 'proponents of competing paradigms practice their trades in different worlds' aptly describes the ontological gulf between the world-views of ecologists and economists—but the problem with this is that there is only one world, and no escaping it.[5] For Kuhn, competing theories are incommensurable if there is no common theoretical language that can be used to compare them:

> [...] the early developmental stages of most sciences have been characterised by continual competition between a number of distinct views of nature, each partly derived from, and roughly compatible with, the dictates of scientific observation and method. What differentiated these various schools was not one or another failure of method—they were all 'scientific'—but what we

[5] Kuhn, T. (1962). *The structure of scientific revolutions.* Chicago: University of Chicago Press, p. 150.

shall come to call their incommensurable ways of seeing the world and practicing science in it.[6]

Kuhns' account of incommensurability invites the following question: if two scientific paradigms claiming truths in accord with scientific method give radically incommensurable answers to the same questions, then how is it possible to determine which is true? The admission that 'ghosts of the undecidable' trouble the determination of scientific knowledge is antithetical to a Comtean or positivist view of science, in which science progresses via the straightforward accumulation and integration of knowledge over time. Drawing on the history of physics, Kuhn argued that knowledge creation was contingent upon the intellectual climate of the era and the specific cultural, linguistic, and institutional contexts which generate unique conceptual frameworks and methods. These Kuhn famously called 'paradigms', an approach anticipated in the medical sociologist Ludwik Fleck's theory of 'thought collectives'.[7]

So distinct are economics and ecology as 'paradigms', that a cursory comparison of the canon of both disciplines gives one the impression that humans occupy simultaneously two wholly separate 'economies of nature', called 'ecosystems' and 'markets' respectively. This compartmentalisation, reproduced in the common reference to nature as the external 'environment' of 'the economy', is one source of what Ulrich Beck denounced as the 'apocalypse blindness' of the social sciences.[8] Thus we have the schizophrenic situation in which the task of government is to foster optimal conditions for infinite economic growth amidst natural scientists' increasingly dire documentation of the dismantling of the biosphere and overheating of the Earth.

In the spirit of E.O. Wilson's notion of 'consilience', it would seem logical and desirable that the basic knowledge of any science ought not be in contradiction to any other. The basic proposition of Wilson's vision of consilience 'is that all tangible phenomena, from the birth of the stars to the workings of social institutions, are based on material processes that are ultimately reducible, however long and tortuous the sequences, to the laws of physics'.[9] Wilson's call for interdisciplinary coherence revives

[6] Kuhn (1962, p. 4).

[7] Fleck, L. ([1935] 2012). *Genesis and development of a scientific fact.* Chicago: University of Chicago Press.

[8] Beck, U. (1995). *Ecological politics in an age of risk.* Cambridge: Cambridge Polity Press.

[9] Wilson, E. O. (1999). *Consilience: the unity of knowledge.* Cleveland: Vintage, p. 2.

something of the project of logical positivism advanced by Rudolf Carnap, Otto Neurath and others of the Vienna Circle in the 1930s—a project which was for Hayek of a piece with 'socialism', and against which the *a priori* subjectivism of Austrian economics would be mobilised. For Carnap, science was not merely a store of confirmed facts or 'correct information' derived from rigorous empirical contact with natural phenomena, but an entirely comprehensive epistemological and moral system of knowledge that could 'encompass the whole range of human thought on all subjects'.[10] As Carnap put it, 'When we say that science is *unlimited*, we mean that there is no question whose answer is in principle unattainable to science.'[11]

In a 1983 article Dierdre McCloskey (an MPS member since 2011) argued that despite practitioners assumptions to the contrary, the mathematical superstructure of economics lacked scientific foundation: due to its reliance upon unexamined metaphors, economics ought rather to be thought of as a branch of rhetoric, the classical art of oratory and persuasion.[12] Revealing economics as a hotbed of metaphor provoked Robert Solow, famous for his model of 'equilibrium-path' growth, to defend the approach that won him the 1987 Nobel prize in Economic Science:

> What matters is that we, as scientists, write down in a precise way what we mean. Precision is one of the standards by which we measure science. And by that standard, metaphors are unscientific. If metaphor occurs in economics, so what? Its existence is incidental to the business of doing economics.[13]

Rhetoric is after all the art of persuasion, even propaganda. Consider that the Swedish Rigsbank Prize in Economic Science in Honour of Alfred Nobel conferred upon Solow is not a Nobel proper, but a parallel prize established in 1968 by banking interests opposed to parliamentary control of central banks. Nobel's original 1895 endowment did not consider economics a science, nor a worthy contribution to 'peace' or the humanities. Contrary to popular belief, there isn't really a 'Nobel Prize in economics',

[10] Midgely, M. (2002). *Science as salvation: a modern myth*. London: Routledge, p. 14.

[11] Cited in Midgely (2002, p. 14).

[12] McCloskey, D. (1983). The rhetoric of economics. *Journal of Economic Literature*, 31, 434–461.

[13] Klamer, A. & Leonard, T. (1994). So what's an economic metaphor? In P. Mirowski (Ed.), *Markets read in tooth and claw: natural images in economic thought* (pp. 20–51). Cambridge MA: Cambridge University Press, p. 21.

yet the phrase conveys a mantle of scientific legitimacy on those selected and the profession as a whole. Beginning with Hayek (in 1974) and Friedman (in 1976), MPS members have been awarded a significant proportion of the prizes, and MPS members and their students have frequently served on the prize committee. Yet it is seldom clear in the speeches given by 'laureates' to presumably well-read audiences how their work contributes to humanity or to knowledge in the human sciences.[14]

Exemplary here is Solow's classic 1956 paper, a model which squared the neoclassical circle by reconciling the theory of 'general equilibrium' with the post-WWII consensus on the need for a continuously growing economy. Solow begins, not with careful scientific statements, but with a drastic assuming away of social reality and nature. His model economy produces but a single undefined commodity *ex nihilo*, in it 'there is no scarce non-augmentable resource like land'. This is immediately followed by another remarkable assumption: '[c]onstant returns to scale seems the *natural* assumption to make in a theory of growth'. [emphasis added][15] A 'natural' theory of 'constant' growth must surely encounter the objection that no biotic entity (e.g. the 'industrial organism' implied in the biological metaphor 'growth') can live without land, nor increase in size the more it increases in size without consuming 'resources' and encountering some limit, whether spatial, epigenetic, or environmental. It is surely a matter of the most elementary empiricism that economic growth or 'production' is a cumulative, history-making material phenomena which transforms 'the environment' into the artefacts and infrastructures of 'the economy' in the process. Here is a fundamental point of incommensurability between the natural and the social sciences of the *oikos*.

Perhaps one way to explain this disciplinary non-communication would be to see it as a result of isolated and path-independent incremental developments amounting to 'progress' within the internally specified truth criteria of legitimately separate areas of knowledge. A more interesting approach to a more or less hidden mutual history was suggested some years back in a blog post by Philip Mirowski entitled 'Ecology in the mirror of economics'. As far as I am aware, Mirowski has not developed the ideas contained in the following remarks, nor has anyone else. The

[14] Karier, T. (2010). *Intellectual capital: forty years of the Nobel Prize in economics*. Cambridge MA: Cambridge University Press.

[15] Solow, R. (1956). A contribution to the theory of economic growth. *Quarterly Journal of Economics*, 70(1), 65–94.

following sentences inspired much of the approach taken here, and ought to be quoted at length:

> Perhaps the most important themes for our present purposes in this history are the following structural regularities: the long-term relationship to energetics; the trend from a 'soft' rejection of holism and organicism towards a more concerted methodological individualism; the move from a diachronic to a synchronic analysis; the heuristic role of physics in providing the primary metaphors and the mathematical formalisms over time; the relative disengagement of empirical workers within the discipline form the strictures of the core theory and the failure to uncover any qualitative 'constants'; the hardcore insistence upon the equilibrium concept (usually constrained optimisation) by the mid 20th century and some back-pedaling from that commitment by the end of the century; progressive blurring of any distinction between Nature and Society; the withdrawal into arcane technical virtuosity in order to assert the possibility of escape from the highly-charged political character of the incompletely constituted subject matter; but simultaneously a history of individual and institutional accommodations to state funding and demands to shape the research agenda; and increased reliance upon computer simulations and information processing metaphors as we approach the present. Perhaps the most striking phenomenon to the historian is the incongruous combination of assertions of arcane scientific methodologies with the simultaneous lament that the field had 'not quite yet' achieved a consensual body of knowledge. [...] All of these regularities, it appears, might equally characterize the history of ecology just as aptly as the history of orthodox economics.[16]

Recall that Kuhn explains 'incommensurability' as the result of the deployment by competing paradigms of *different* conceptual vocabularies and terminologies. Against Kuhn, I argue that economists and ecologists share an overlapping lexicon, and yet completely misunderstand one another nonetheless. Following this lead, this book attempts a work of historical contextualisation and reconstruction in order to shed light upon the processes by which 'the economy' and 'the ecosystem' were constituted as separate research objects worthy of stand-alone disciplines, drawing out further the elective affinities hinted at above. The parallel trajectory of the formation of the two sciences is not coincidental: any explanation of their

[16] Mirowski, P. (1996, Oct 6). Ecology in the mirror of economics. History of Economics Society. http://www.eh.net/lists/archives/hes/oct-1996/0042.php. Accessed 9 Nov 2005.

incommensurability must deal not only with Mirowski's mirrored 'structural regularities' but also with cases of mutual influence, if intermittent and indirect, at decisive moments of theory change. This requires a careful examination of the foundational relation of both sciences to physics, especially to thermodynamics, the universal science of heat and motion that emerged in parallel with ecology and economics in the later nineteenth century.

On the Anthropology of Classificatory Schemes

In an article advocating 'economics imperialism'—the colonisation by economists of law, political theory and anthropology whether the 'natives' like it or not—the Chicago-school economist George Stigler (MPS) repeats Wicksteed's claim that 'the fundamental laws of economic science [are] in fact, are the laws of life'.[17] The recurring claim that economic science reveals universal, context-independent 'natural laws' can be illuminated by a tradition of anthropological critique which pursues the reverse argument: nature is not autonomously discovered but always discursively ordered through concepts derived from social relations. As one author puts it,

> Every major economic and social revolution in history has been accompanied by a new explanation of the creation of life and the workings of nature. The new concept of nature is always the most important strand of the matrix that makes up the new social order. In each instance the new cosmology serves to justify the rightness and inevitability of the new way human beings are organising their world by suggesting that nature itself is organised along similar lines. Thus, every society can feel comfortable that the way it is conducting its activities is compatible with the natural order of things, and therefore, a legitimate reflection of nature's grand design.[18]

This approach has its origins in the analysis of metaphor and classification in the anthropologies of Emile Durkheim, Marcel Mauss and Mary Douglas, which relativise crude positivisms by documenting the co-constructedness of the categories of the Natural and the Social. Durkheim and his student Mauss argued in their classic essay *Primitive Classification*

[17] Stigler, G. (1984). Economics: the imperial science? *Scandinavian Journal of Economics*, 86(3), 301–313.

[18] Rifkin, J. (1998). *The biotech century*. New York: Tarcher/Putnam, p. 197.

(1903) that the Social is the source of all categories of human thought. 'The classification of things' they argued, 'reproduces the classification of men [sic]'.[19] In *The Elementary Forms of Religious Life* (1912) Durkheim argued that the structure of existing social relations was projected into the supernatural realm.[20] Like much of the anthropology of the early twentieth century, these studies were influenced by the accounts of Australian Aboriginal 'totemism' presented in Spencer and Gillen's colonial ethnography of the Aranda, *Native Tribes of Central Australia* (1899). The first peoples of Australia, reported Spencer and Gillen, organise social relations and customary obligations through a religious ontology dubbed 'totemism', in which persons are socially positioned in reciprocal relations of kinship and identity with totemic Ancestors present in a multiplicity of animal and plant species, as well as in waterholes, mountains and other natural phenomena.[21] Devoted in part to exposing the alleged errors of 'the primitive mind' at a time when the profound temporal depth and cultural continuity of the first nations of Australia was yet to be acknowledged, classical anthropology completely ignored the integration of Aboriginal 'totemism' in highly sophisticated knowledge practices of ecological management which ensured long-term abundance, through practices including 'firestick farming'.[22]

According to Durkheim and Mauss, it was through such processes of super-naturalising social relations that religion provides the stable norms crucial to social interaction, ensuring the continuity of the community's cultural and political order over time. An example familiar to Westerners might be medieval European images of God as a crowned monarch on a throne. Mary Douglas took this idea further in her study of the prohibitions of Leviticus, arguing that understandings of the order of Nature across different societies could be explained by looking at the way Society was ordered. 'The social body', she writes, 'constrains the way the physical body is perceived. The physical experience of the body, always modified by the social categories through which it is known, sustains a particular view

[19] Durkheim, E. & Mauss, M. ([1903] 1963). *Primitive classification*. Chicago: University of Chicago Press.

[20] Durkheim, E. (2001). *The elementary forms of religious life*. London: Oxford World Classics.

[21] Spencer, W. & Gillen, F. (1899). *The native tribes of Central Australia*. London: Macmillan.

[22] Walker, J. (2013). Worlds to endure: weathering disorder from Arnhem Land to Chicago. *Global Networks*, 13(3), 391–409.

of society.'[23] Douglas linked the order of taboo and sacrifice in Leviticus, with its passages on abominations, forbidden animals and the ritual uncleanliness of human body fluids with modern understandings of pollution, noting that what counted as 'dirt' was different for her and for her husband, even though they were looking at the same kitchen.

'Pollution', for Douglas, 'is matter out of place'.[24] We should approach this analysis of pollution as a cultural construct with some caution, however aptly it might describe the anthropogenic transfer since 1870 of over 600 billion tons of carbon from the lithosphere and terrestrial ecosystems to the atmosphere and oceans. A later book coauthored with Aaron Wildavsky in the early years of the Reagan revolution criticised environmentalist's demands for science-based regulation of industry according to the precautionary and polluter-pays principles, suggesting this was less a rational response to novel poisonings than an excessive demand for security of the student counter-culture of the 1960s, a generation coddled since birth by the New Deal welfare-state.[25] Writing in the journal of the Cato Institute (founded by the MPS oil billionaire Charles Koch), Wildavsky declared that 'for the libertarian [...] *there can be no pollution*' [my emphasis]. Citing the 'theorem' of Ronald Coase (MPS) that unregulated markets naturally optimise the social allocation of environmental goods and bads, Wildavsky declared a culture war against the environmental sciences:

> [...] once one understands that saying 'the environment is polluted' is equivalent to saying that 'markets are dirty', the charge that environmental pollution is ubiquitous can be understood for what it is—a fundamental attack on libertarian culture.[26]

The upshot of all of this is the claim that concepts of natural and social order are inherently political, unstable, and crisis prone. As Mirowski has put it,

[23] Douglas, M. (1973). *Natural symbols*. London: Penguin, p. 93.

[24] Douglas, M. (1970). *Purity and danger: an analysis of concepts of pollution and taboo*. New York: Routledge.

[25] Douglas, M. & Wildavsky, A. (1983). *Risk and culture: An essay on the selection of technological and environmental dangers*. University of California Press.

[26] Wildavsky, A. (1982). Pollution as moral coercion: culture, risk perception and libertarian values. *Cato Journal*, 2(1), 304–325. See p. 307.

> [...] anthropomorphic ideas of mastery and control induce a bias in cultures to project their own social categories onto their explanations of the external world; but in an infinite regress subsequent reification of notions of order prompt others in the same culture to appropriate those preceding Natural concepts and re-project them back into models and images of society.[27]

Mirowski proposes a way out of this dilemma, suggesting that 'the Natural and the Social are merely provisional designations for where explanation will halt in a crisis'.[28] Ever-present and inherently unstable, the Natural and the Social only come up for negotiation when pre-existing categories of order are in disarray. Ontological categories remain hazy background assumptions until some unpredictable disaster confounds the institutional procedures by which we assign guilt and responsibility and restore justice.

Let us up the ante here, and extend this theory of crisis to the apocalyptic limit. I would argue that the totality of environmental crises threatens the legitimacy of abiding institutional arrangements, as the power of thermoindustrial capitalism to liberate some populations from local ecological constraints threatens to undermine the basic biological and climatic conditions which render the Earth habitable. Our chaotic destructuring of the biosphere means that Nature as a stable external referent can no longer unproblematically reaffirm visions of the social order as natural, legitimate, enduring and just. Our planetary crisis represents nothing less than a standing refutation of the 'natural justice of the market' and the rationality of economic theory. Ecology reveals the apocalypse that economists have authorised and concealed in advancing the millennium of infinite progress.

What is at stake for ecologists in this? The destabilisation of nature is manifest in the pollution of their research object: the increasing rarity of pristine biotic communities thriving toward the maximum abundance and diversity of the optimal ecological 'equilibrium'. As Bill McKibben observed in *The End of Nature* (1992), global warming means that no remote sanctuary of 'wild nature' is external to social relations anymore, which is fatal to the meaning of nature in Western culture.[29] Many ecologists, faced with the social destruction of their post-natural research object,

[27] Mirowski, P. (1994). The realms of the natural. In P. Mirowski (Ed.), *Natural images in economic thought: markets read in tooth and claw* (pp. 451–483). New York: Cambridge University Press, p. 453.

[28] Mirowski (1994, p. 454).

[29] McKibben, B. (1990). *The end of nature*. London: Viking.

have been pushed into becoming political figures and social scientists of a sort. Even those who have meticulously retreated into the professional caution of the disinterested scientist risk being accused of subversive political motivations by non-scientists when commercial interests at stake. In the so far successful attempt to defend economics from logical and empirical falsification, many opportunities have been irreversibly lost. The queen still sits on the throne. In attempting to stave off the apocalyptic collapse of the liberal economists' ontology, a very real apocalypse has been systematically denied and urgently overdue reforms of law, government, and industry fatally delayed.

The Metaphorical Machinery of Science and Culture

The question of the cultural origins and authority of natural-scientific concepts leads us to the question of language and discourse, and one of its most intriguing elements: metaphor. Derived from the Greek *metapherin*, the function of metaphor is literally to 'carry over' meanings from one domain into another. Metaphor is properly at home in poetry and literature, where it is intended to stimulate the imagination. 'The moon was a ghostly galleon' initiates a chain of novel associations and images in the mind of the reader, precisely through the boldness of its incongruity. To say that the moon was *like* a ghostly galleon (a mere simile) is boring as poetry. Asserting the identity of non-identical things arrests our attention because it attacks our ordinary classification of objects. In the process of the erosion of the semantic barriers that ordinarily separate planetoids from marine transport, unrelated objects exchange some (but not all) of their qualities. The moon drifts across the sky like a sailing ship across the sea, a galleon is bathed in ghostly luminescence. Sea and sky mirror each other. Literary metaphors are suggestive, not didactic. The reader is seduced into conscious complicity in the breakdown of the order of things, invited to suspend disbelief and savour the image as one would a fine scotch.

In a more subversive role, metaphor is an indispensable part of rhetorical persuasion, exploited particularly in politics for its emotive possibilities. Consider the continuous equation by President George W. Bush of Saddam Hussein with Adolf Hitler. Beyond the visual symmetry of mustachioed dictators, this had important consequences for those who accepted the parallel. The opposition of the United Nations to the US invasion of

Iraq was akin to the appeasement of Hitler by Chamberlain in 1938, despite the fact that Iraq's military was primarily tasked with securing national control over some of the world's largest oil reserves, and scarcely capable of global Blitzkrieg. The Baath party were Nazis; and the optimism for a booming free-market 'reconstruction of Iraq' evoked the successful incorporation of West Germany and Japan into the Bretton Woods order after World War II. President Bush invoked 911 as a latter-day Pearl Harbour, and went as far as identifying an Axis of Evil that was not even close to an actual alliance. The extended suite of analogies generated by the metaphor <Hussein is Hitler> appealed to nostalgic memorials of an age of just warfare, and seemed to have intoxicated even its authors into believing that their oil war carried the imprimatur of divine justice. Here we see the deceptive possibilities of metaphor, which in invoking the familiar to explain the unfamiliar, cultivates the sloppy and emotional thinking that is the intended effect of propaganda.

So it is not surprising that philosophers have sought to rigorously exclude metaphor from the sphere of scientific communication, due to its tendency to confound precise explanation with *a priori* intuitions. Some will allow the use of *pedagogical* metaphors for the purpose of scientific instruction: to explain theories which already have non-metaphorical interpretations—'energy field', 'electron cloud' or 'wormhole' are examples—or as heuristic devices to explain theoretical claims which as yet have no non-metaphorical explanations. Ill-defined metaphors have a part to play in the discovery process, but a fertile metaphor is no substitute for the scientific labour of testing theoretical claims against carefully documented empirical phenomena. As a science matures, hypotheses are clarified and precision increases.[30] Successful prediction from formal models renders the original metaphor coincidental to the scientific process by downgrading it to an analogy external to the now non-metaphorical theory.

Broader claims have been made for the social role of metaphor. Lakoff and Johnson posit that human cognition operates irreducibly by way of metaphor, and thus our conceptual systems are in the final analysis metaphorical. Not only is the modernist dream of a pure conceptual language purged of metaphor exactly that—a dream—metaphors are inextricably bound up in the constitution of everyday reality, since they 'structure what

[30] Gentner, D. & Jeziorski, M. (1993). The shift from metaphor to analogy in Western science. In (Ed.) A. Ortony *Metaphor and thought* (pp. 447–480). Cambridge: Cambridge University Press.

we perceive, how we get around in the world and how we relate to other people'.[31] Metaphors on this view are inescapably ontological. 'Metaphors matter.' They both reveal and conceal aspects of reality in order to simplify and familiarise, and we remake the world in their image. The metaphors we live by may turn out to be the metaphors we die by.

This ontological view of metaphor has been applied to the concepts of economics by Klamer and Leonard, to ecology by Cuddington and to complexity theory by Weingart and Maasen, who show that scientists' use of metaphor is an indispensable, ubiquitous feature of the scientific process.[32] This is true not only for the 'soft' relational sciences of ecology and economics, but also of the 'hardest' of sciences, as Duhem argued:

> The history of physics shows us that the search for analogies between two distinct categories of phenomena has been the surest and most fruitful method of all the procedures put into play in the construction of physical theories.[33]

According to Mary Hesse, metaphors redescribe the phenomena of a primary system in the contextual terms of a secondary system, thus opening a two-way interaction between both domains of knowledge.[34] Mutual interactions between different knowledge contexts are thus multiplied and may generate unanticipated multidirectional knowledge transfers, and nowhere, I would argue, is this more evident and pregnant with consequence than in the history of interactions between economics, ecology and thermodynamics.

For Hesse, scientific revolutions can be described as 'metaphorical redescriptions' of nature. The upshot is that metaphor can never be truly eliminated, although the entire project of 'becoming a hard science' implicitly assumes this. Therefore, the drive amongst biological and social scientists to emulate the precision of the quantified fundamental statements of physics and chemistry risks disabling the basic task of knowledge, which is to increase insight into our experience of the world: some

[31] Lakoff and Johnson (1987, p. 2).

[32] Cuddington, K. (2001). The 'balance of nature' metaphor and equilibrium in population biology. *Biology and Philosophy*, (16), 464. Weingart, P. & Massen, S. (1997). The order of meaning: the career of chaos as a metaphor. *Configurations*, 5(3), 463–520.

[33] Cited in Mirowski (1992, p. 278).

[34] Hesse, M. (1972). The explanatory function of metaphor. In Y. Bar-Hillel (Ed.), *Logic, methodology, and philosophy of science* (pp. 249–259). Amsterdam: Elsevier.

necessary overtones of meaning are lost when a word is precisely symbolised.[35] In the nineteenth century, scientists aspired to identify 'laws of nature' underlying the 'mechanisms' of cosmic order. Nowadays, scientists restrict themselves to the construction of 'models'. Originally carrying the implication of a 'scale model' or a map, scientific models can be described as complex analogical systems attempting to reveal pattern, structure, and causality amidst the phenomenal complexity of the empirical world, across many more vectors than scale.[36]

Creating models of unmeasurable phenomena involves reasoning by analogy. Milton Friedman affirmed this when he argued that so long as predictions hold true, it is perfectly fine for economists to reason from assumptions that would not endure empirical testing, 'as if', for example, human behaviour is equivalent to the mathematical solution of a constrained optimisation problem.[37] The perennial problem is not the ubiquity of metaphor (which is unavoidable), nor the use of analogical reasoning (which is creative), but the danger of what Marxists call 'reification': the objectification of theoretical propositions:

> analogies may become elaborate things-in-themselves, and eclipse their founding metaphors. [...] Indeed most economists probably think of their work as making truth statements about the world. [...] Alertness to metaphor reminds us not only that our models are fictions, but that 'as if' reasoning—the characteristic mode of economic reasoning—is altogether incompatible with a positivist view of the world.[38]

In addition to pedagogical and heuristic metaphors, Boyd identifies another class: *constitutional* metaphors. Of the most foundational character, these are reserved for the conceptualisation of phenomena that are simply too vast or complex to directly observe: the situation confronting ecologists and economists. The term is itself based on a metaphor: constitutional metaphors frame a discursive practice in the same way that the Constitution frames legal discourse. This kind of metaphor 'constitutes, at least for a time, an irreplaceable part of the linguistic machinery of a scientific theory: cases in which there are metaphors expressing theoretical

[35] Hesse, M. (1955). *Science and the human imagination*. New York: Philosophical Library, p. 88.

[36] Klamer and Leonard (1994, p. 35).

[37] Friedman, M. (1953). *Essays in positive economics*. Chicago: University of Chicago Press.

[38] Klamer and Leonard (1994, p. 36).

claims for which no adequate literal paraphrase is known'.[39] Think of references to gene expression in terms of 'transcription' and 'coding', clearly metaphors derived from the terminology of information technology. Even if the metaphor is pointed out, it is impossible to think the theory of genes in any other way. The grandest of constitutive metaphors become 'world hypotheses', such as mechanism, or organicism: so embedded in discourse and so essential to our background thinking that they are invisible as such, and come to determine what counts for truth.[40] As they have been inhabited, renovated, extended, and lent legitimacy in thousands of speech acts and books, constitutive metaphors—such as 'growth' and 'equilibrium'—cannot be easily discarded. This leads us to the strongest position on metaphor and science, as radical now as it was when Nietzsche first proposed it in the 1870s:

> What then is truth? A movable host of metaphors, metonymies, and anthropomorphisms: in short, a sum of human relations which have been poetically and rhetorically intensified, transferred and embellished, and which, after long usage, seem to a people to be fixed, canonical and binding. Truths are illusions which we have forgotten are illusions; they are metaphors which have become worn out and drained of sensuous force, coins which have lost their embossing and are now considered as metal and no longer as coins [...].[41]

The critique of the cultural context of science is valuable not so much for desiccating knowledge claims into postmodern dust, but rather to alert us to the ways in which epistemic authority is communicated to the demos and translated into political authority. Or in other words, how some forms of knowledge acquire the 'force of law' in policy and legislation, whilst others are discounted or screened out. As Lakoff and Johnson observe, the social generation of shared meaning through metaphor is anything but a level playing field: '[W]hether in national politics or everyday interaction, people in power get to impose their metaphors.'[42]

[39] Boyd, R. (1993). Metaphor and theory change: What is metaphor for? In (Ed.) A. Ortony, *Metaphor and thought* (pp. 356–408). Cambridge: Cambridge University Press.

[40] Klamer and Leonard (1994, pp. 39–42).

[41] Nietzsche, F. (1979). *Philosophy and truth: Selections from Nietzsche's notebooks of the early 1870s*. New Jersey: Humanities Press, pp. 84–89.

[42] Lakoff and Johnson (1987, p. 157).

Whereas once natural scientists could be called upon to lend the legitimacy of scientific neutrality and objectivity to the justification of policy, this authority has come under sustained attack as global warming has manifested. Climatologists and earth scientists at prestigious public institutions such as the CSIRO in Australia, and at NASA and the EPA in the United States been actively censored by neoliberal governments seeking to delegitimise their research. Scientists publicly highlighting the implications of their research were censured for making 'political' statements. Neoliberal politicians in Australia and the US, major players in global fossil-fuel markets and among the world's largest per capita emitters of greenhouse gases, consistently pursue the claim that the science linking the mass combustion of hydrocarbons to unprecedented global heating is seriously flawed and subject to grave doubts, even that it is some kind of retrograde 'religion', or evidence of a massive conspiracy amongst thermometer readers to take away 'our freedom'.

Despite claims to access an objective nature, economics, ecology and energy physics are inescapably studies that seek to mediate between the Natural and the Social and thus cannot avoid being linked to grander metanarratives: just as metaphor supplies social meaning to natural facts, so does science fit into a grander story about the human prospect. The historical consciousness of millennium and apocalypse that enveloped premodern European cosmology has not been tamed away by the emergence of modernity and scientific rationality. The millennium was incompletely secularised into Progress ideology, then vulgarised into the economic faith that 'growth' must always increase. Yet the 'ghosts of the undecided' have returned with a vengeance in the revelation that our way of life is bringing about an ecological apocalypse which will far exceed the catastrophes of the Revelation of St. John. The hothouse future projected by the modern Earth sciences offers no *deus ex machina* of divine intervention, no New Jerusalem in which justice is restored, and no regeneration of the stricken Earth to an Edenic state, or at least in any time frame that we can contemplate. The Creation with which our ancestors were gifted will never be re-generated: the promise that the irreversibility of mass extinctions can be miraculously overcome by the perfection of biotechnology is scarcely more than a techno-utopian fantasy. The proliferation of novel species that follows the mass extinction events of the geological record occur millions of years beyond them: the restabilisation of the carbon cycle will takes place in a future geological time inaccessible to the historical time of human life. Only by ceasing to burn fossil fuels, living within the flows of energy from sun, wind, water, and photosynthesis and actively reversing the destruction of natural ecosystems can we hope to slow down the runaway train. Thus my overall narrative regarding the epistemic struggle for cultural authority between economics and ecology is directed not at scientific knowledge as such, but at the political pressure placed upon natural scientists when the actually existing state of nature threatens to falsify core beliefs in the benevolence of infinite growth and 'market forces'.

Let us conclude with a particularly powerful species of constitutive metaphor: *machine* metaphors, the unconscious work of which can be detected in the many epistemic conundrums of modernity. For Karl Marx and Lewis Mumford, machines are not only the outcomes of work, or 'objectified labour', they are in turn inextricably embedded in social relations, which the machines come to determine.[43] Reified and fetishised into discrete artefacts independent of their social context of production, machines are the 'epistemological totems' of Western claims to a superior rationality and a direct relationship with an objective nature. Machines provide technological proof that science 'works'—even the wonkiest science-deniers fly Boeings to the conferences of the Heartland Institute. Thus machines can be taken to speak with the clear voice of an objective Nature. Conversely, economic theory devotes itself to analysing Adam Smith's 'system of natural liberty that arises of its own accord' in terms of a 'price mechanism' governing a system in dynamic general equilibrium. The machine metaphor is upfront, but there is no account of the materiality of machinery. The invisible hand operates an invisible machine.

The commitment of governments to permanent 'economic growth' discloses a faith in 'technology' as autonomously productive of wealth: this belief in 'intrinsic machine fertility' is described as 'machine fetishism' by Alf Hornborg in *The Power of the Machine* (2001).[44] On Hornborg's view, that machines are 'efficient' and 'do work' is utterly dependent on their geographical and socio-economic location. Global social relations—unequal flows of natural resources in turn generated by the unequal value of human labour mediated by prices, money wages and sovereign debts—presuppose the 'productivity' of industrial technology. Hornborg's

[43] Marx, K. ([1908] 1972). Notes on machines. *Economy & Society*, 1(3), 244–254; Mumford, L. (1967). *The myth of the machine: technics and human development*. Michigan: Harcourt Brace.

[44] Hornborg, A. (2001). *The power of the machine: global inequalities of economy, technology and environment*. California: Altamira Press.

account of machines as existing in a 'dialectical penumbra' between the material world and powerful cultural constructions nevertheless accepts the stability of the laws of thermodynamics which govern the work of heat engines.

The assumption that 'technology' is the source of 'growth' remains common in the explanations of economic historians as to why so much wealth was being accumulated in Britain during the industrial revolution. The textbook thesis credits this 'growth miracle' to British technological know-how and liberal economic policies, and not to the vast energies realised by the dramatic industrial shift from solar-based energy flows (of wood, wind, water, and beasts of burden) to the mass combustion of the 'subterranean forests' preserved as coal over eons long past. Neither does it take into account the deployment of the steam-engine-rail-shipping complex by 'fossil capital' to exploit the labours and resources of colonial Others in the vast land appropriations of the British Empire. Yet this economic myth still underwrites the Promethean task of 'making poverty history' through 'export-led growth' and 'technology transfer', at a time when a handful of billionaires control more wealth than the major part of the world's human population. That individual ownership of 'labour saving devices' such as personal leafblowers appears as 'efficient' and 'rational' in the urban cultures of the global North, is on Hornborg's view due to staggering inequalities in the value of wage labour, global uniformities in the spot price of oil, and the fact that the atmosphere remains a free dumping ground for unpriced pollution. None of this, however, accounts for the mysterious aesthetic preferences of suburbanites, who fire up their leafblowers to 'clean' concrete driveways at the enormous expense of peace, birdsong, and wholly unnecessary contributions to heating away our future.

CHAPTER 3

On Photosynthesis and Pyrotechnics: Life Between Fire and Law

Life from Light, Fire from Life

From the multi-million-degree temperatures at the element-creating cores of stars, to the delicate warmth bathing the embryo in utero, all matter is transformed by heat, at specific thresholds along thermal gradients.[1] Heat is distributed throughout the infinite universe but as far as we know, fire only occurs on planet Earth. This is worth repeating: as far as we know, *fire is unique to the Earth*.[2] Life and fire are wholly intertwined phenomena. Flames rapidly dissipate as heat and light the solar energy slowly bound up in biomass by photosynthesis. Photosynthesis is the complex process by which algae, plants, and certain bacteria harvest the energy of sunlight, separating carbon from atmospheric carbon dioxide and binding it to hydrogen and oxygen from water to form the carbohydrates which store chemical energy in the molecular bonds of organic molecules (e.g. sugars, or carbohydrates), expelling oxygen as a waste product. Combustible fuel has photosynthetic origins: whether carbohydrates formed recently (wood), or anciently, long since mineralised into hydrocarbons (coal, oil, and gas). A wood fire is a tree in reverse, as is a coal fire, although with an

[1] The periodic table specifies the melting and boiling point for each element. The origins of the elements heavier than hydrogen and helium in stars was first proposed by: Eddington, A. (1920). The internal constitution of the stars. *Nature*, 106, 14–20.

[2] Scott, A., Bowman, D., Bond, W., Pyne, S. & Alexander. M. (2013). *Fire on earth: an introduction*. Wiley & Sons.

J. Walker, *More Heat than Life: The Tangled Roots of Ecology, Energy, and Economics*,
https://doi.org/10.1007/978-981-15-3936-7_3

interlude of around 300 million years. Fire is only possible because of the history of photosynthesis on the geological time scale. The early atmosphere contained almost no free oxygen. The oxygen concentrations in the atmosphere which make fire possible (16% or higher, currently at 21%), accumulated there as countless generations of photosynthesising microbial seas exhaled. The Great Oxygenation took a while: roughly, for concision, over the period from 3.4 to 0.6 billion years ago. Around 541 million years ago, complex multicelled organisms appeared, including plants on the seabed. Fire then, is an emergent effect of biological life's transformation of the Earth. The first fire—on land, amidst air—was long preceded by photosynthetic life.

The immense diversity of living organisms comprising the sunlit biosphere are ultimately dependent upon photosynthesis.[3] Plants, algae, and other photosynthesisers are autotrophs, making their own 'food' and storing it in the biomolecule adenosine triphosphate (ATP), known by biologists as the 'universal energy currency' of cellular life. The rest of us heterotrophs eat vegetable life, or other organisms that do, for energy.[4] The elements of the periodic table common on Earth's surface are recycled through the biosphere's abundantly creative and complex evolution of novel biochemistries, between the atmosphere, hydrosphere and the lithosphere. But the most important of these is the carbon cycle, which is driven by photosynthesis. Even as the sun's radiance has slowly increased through six billion years of Earth history, since the evolution of photosynthesis, the mean surface temperature has remained within the range suitable for most carbon-based biochemistry—between the freezing and boiling point of water, which is bound to carbon in the photosynthetic process. Carbon is remarkably stable from a thermal point of view, having the highest melting temperature on the periodic table, and yet is the most promiscuous of the elements in terms of its openness to novel chemical couplings and combinations. The number of compounds that contain carbon vastly exceeds all other compounds combined.

The Great Oxygenation of the atmosphere was also its Great Decarbonisation. Since its ancient evolutionary origins in microbial seas, photosynthesis has reduced the concentrations of heat-trapping methane

[3] With the exception of the extreme ecology of deep ocean thermal vents, which in the absence of light support food chains founded in heat energy captured by chemosynthetic Archaean bacteria.

[4] There are exceptions, but this is a text aimed at the general reader.

and carbon dioxide in the atmosphere by a factor of one thousand, as carbon drawn from the atmosphere by plants and micro-organisms was mineralised in the remains of organisms committed over eons to geological strata, in the form of calcium carbonates (limestones) and hydrocarbons (coal, oil, and gas). This relative thermal stability is surely the most vital of the ways in which 'the Earth's atmosphere is regulated by life on the surface so that the probability of growth of the entire biosphere is maximized', in the words of the atmospheric chemist James Lovelock and the microbiologist Lynn Margulis.[5] According to Lovelock, had the biosphere never existed the Earth would have an oxygen-less atmosphere of 98% CO_2, and a mean surface temperature of 290 °C.[6] Obviously this would not be a pleasant environment for human beings, whose lives depend on their core body temperature not departing too far either side of 37 °C for too long, a feat we accomplish at all latitudes by the social technologies of food, clothing, shelter, heating, and cooling.

Life is critically thermo-sensitive. Plant growth requires a definite quantity of heat accumulation in specific temperature ranges to reach thresholds of development such as seed sprout, flowering, and fruit set. So too does the embryological development of animals. The eggs of sea turtles which visit the Great Barrier Reef hatch in even distributions of female and male offspring when incubated in sand at 26 °C. When the temperature reaches 29 °C, nearly all of the offspring are female, a phenomenon which exposes turtles to precarious futures in a rapidly warming world. Existing coral reefs, the most productive and diverse of marine ecosystems, are unlikely to survive this century, as ocean temperatures increasingly exceed the thermal thresholds that can be tolerated by the symbiotic mutual relationship between corals and zooxanthellae, the photosynthesising microalgae that live inside corals and provide their primary source of energy.[7] The unprecedented mass bleaching of coral reefs witnessed in recent years is testament to the fact that the majority of the heat accumulated globally through anthropogenic warming is being stored in the vast thermal mass of the seas, despite our preoccupation (as land-based creatures) with air temperatures. And as the corals die, so to do the millions of individual organisms dependent on their health. Divers surveying the devastation of

[5] Margulis, L., & Lovelock, J. E. (1974). Biological modulation of the Earth's atmosphere. *Icarus*, 21(4), 471–489.

[6] Lovelock, J. (1979). *Gaia: a new look at life on Earth*. London: Oxford University Press.

[7] Veron, J., Hoegh-Guldberg, O., Lenton, T., et al. (2009). The coral reef crisis: the critical importance of <350 ppm CO_2. *Marine Pollution Bulletin*, 58(10), 1428–1436.

the Great Barrier Reef after the catastrophic bleaching event of 2016 reported that after emerging from the water, 'we just stank—we stank of the smell of rotting animals'.[8] And if this accumulation of heat were not bad enough, the accumulation of carbon dioxide absorbed from the atmosphere in seawater (in the form of carbonic acid) is now continuously lowering the long stable pH balance of the oceans. Cocolithophores, a major class of the photosynthesising phytoplankton which play a major role in the biosphere's carbon-fixation and form the basis of the marine food web, form their cell walls through calcification. Along with the corals, shellfish, krill, and sea urchins whose skeletons are also made of calcium carbonates, cocolithophores are directly threatened by ocean acidification. The major mass extinction events in the Earth's deep evolutionary and geological history can all be attributed to disruptions of the planetary carbon cycle, and the primary cause of the current disruption is anthropogenic fire.

The expansive industrial conflagration of hydrocarbons has released hundreds of millions of years' worth of mineralised carbon from geological interment, forcing the surface of the Earth to accumulate heat at a frankly terrifying rate. Among the many unprecedented ways in which 'economic growth' has altered the Earth and exceeded long stable 'planetary thresholds', this is surely the most critical to the claim that we live after the end of Nature, at the dawn of an ominous new geological 'epoch' first named 'the Anthropocene' by the atmospheric chemist Paul Crutzen.[9] However, given the infinitesimal brevity of the 'age of machines' from the perspective of geological time, and the rapidity with which the carbon cycle has been unwound, it will probably prove more accurate to speak of the Anthropocene as the 'boundary event' of a mass-extinction rather than a 'geological epoch', as these are usually measured by geologists in multi-million-year thick mineral strata.[10]

All known human cultures have intentionally utilised fire—culture is the difference between the raw and the cooked.[11] Despite the suspicion with which claims regarding human uniqueness are sometimes met in the post-humanities, fire mastery, which has progressively transformed human

[8] Wall, J. & Slezak, M. (2016, June 7). Great Barrier Reef: diving in the stench of millions of rotting animals. *The Guardian*.

[9] Crutzen, P. (2002). Geology of mankind. *Nature*, 415(6867), 23.

[10] Brannen, P. (2019, Aug 3). The Anthropocene is a joke. *The Atlantic*.

[11] There may be speculative minor exceptions, but these are not documented in the ethnographic literature to the best of my knowledge.

social existence and now the Earth itself, is surely the prime candidate. Yet whilst the importance of fire can hardly be doubted, as the editors of a recent 'pyro-geography' of *Fire on Earth* observe:

> [fire] rarely enters the discourse of relevant disciplines or appears in the standard texts of geology, human history, physics or global chemistry. Fifty years ago, the only organised inquiries lodged in applied contexts such as combustion engineering, urban fire services, and fitfully, forest and range science. [...] The outcome was an extraordinary disconnection. While fire was ubiquitous in various forms throughout the Earth, it was absent from formal inquiries about the world.[12]

This alleged absence of interest in fire may well be a symptom of the division of labour of the modern university, and the 'extraordinary disconnection' a recent phenomenon coinciding with the lack of direct experience of seeing and working with fire typical of a modern urbanite, despite the vast hidden fires powering our pyrotechnical society. After all, reflection on the Promethean mysteries of fire is as ancient and varied as the cultural knowledges transmitted down the generations around camp fires, hearth fires, kilns, and forges through time immemorial, from Timbuktu to Tenochtitlan. Among the earliest recorded of European philosophers, Heraclitus (c. 535–475 BCE) taught that '[t]he *kosmos*, which is nothing but itself, was not created by any one of the gods or of humans, but is and ever shall be ever-living fire, kindled in measure and quenched in measure'. By analogy with the phenomena of monetary exchange, then coming increasingly to determine the social relations of his Mediterranean milieu, Heraclitus proposed an invariant cosmological principle underlying the transitory phenomena of our world, one which seems to contain an early intimation of modern thermodynamics:

> All things are an exchange for fire, and fire for all things, as goods for gold and gold for goods.[13]

If Heraclitus' fire has since been generalised into heat, and then further into the universal concept of energy, this economic metaphor, as I will argue in subsequent chapters, is one of the constitutive tropes of European

[12] Scott et al. (2013).

[13] Cited in Hoffman, D. (2006). Structural logos in Heraclitus and the Sophists. *Advances in the History of Rhetoric*, 9(1), 1–32.

thought, constantly restated, transformed, updated, forgotten, and rediscovered down to the present day. In keeping with Clark and Yusoff's call for a fire-centred account of energy use,[14] this book may be read as an attempt to recover a combustion-centric history of the (human) ecology of energy, a biopolitical economy or *oikonomia* adequate to the 'Anthropocene' present of industrial pyrotechnology and global heating.

Despite its ambitious title, the scientific contributors to *Fire on Earth* generally confine themselves to the role of fire in ecosystem processes, and case studies in the contemporary practices of managing free-burning landscape fire. But as the historian Stephen Pyne acknowledges in his chapter:

> All of these expressions of human-influenced fire, even the most mammoth megafires, pale before the magnitude of industrial fire, itself the most potent of the pathologies of the pyric transition. [...] Global warming [is] the result of a radical mutation in how humans manipulate fire. [...] It is easy to understand why, because of its malfeasance with industrial combustion, humanity's fire practices overall might become suspect and tainted with illegitimacy. However, [...] anthropogenic fire is [...] what we do that no other species can. Managing fire remains the signature of our ecological agency, and in a genuine sense our ecological duty to Earth. The issue is not whether we must manage fire, but how well or how poorly we will do it.[15]

Aboriginal *Oikonomia*

Maintaining the oldest continuous cultures on Earth, the First Nations of Aboriginal Australia have consciously fulfilled this ecological duty for millennia.[16] Involving the highly skilled use of relatively cool, low-intensity fire to modulate the volatile landscapes of the great Southern 'fire continent', to foster and direct ecological succession, and to make edible plants and animals abundant and predictably available, what contemporary

[14] Clark, N., & Yusoff, K. (2014). Combustion and society: a fire-centred history of energy use. *Theory, Culture & Society*, 31(5), 203–226.

[15] Pyne, S (2013). A new epoch of fire: the Anthropocene. In (Eds.) A. Scott, et al., *Fire on Earth* (Ch. 12). Wiley and Sons.

[16] Here I must respectfully acknowledge that I am telling this story with the very incomplete knowledge of a student, from the outsiders perspective of my own Western cosmology, with its foreign language and terminology. Aboriginal people, on their own telling, have always been here. Western science dates human occupation of the Australian continent to a confirmed minimum of c. 60,000 years, with more speculative claims approaching 100,000 years.

practitioners call 'cultural burning' was common to the economic and ethical life of Indigenous Australians. Until repressed in the onslaught of colonisation, 'firestick farming', was practised across vast and vastly different Indigenous estates, from the cold rainforests of Tasmania through the deserts of the Centre to the tropical savannahs of the North.

Aboriginal firing of country is as unique as the Australian biota within what Wertime called the 'single, complex pyrotechnic tradition'[17] of human fire knowledge, for its moral ontology of fire as much for its unparalleled longevity. Preventing the ecological destruction of catastrophic wildfires by carefully reducing fuel loads, preserving the shade of the forest canopy, retaining moisture and nutrients in the soil by adding 'biochar', regenerating new plant life and the grasses favoured by grazing marsupials, the right fire, the 'good fire', is intentional and lawful, drawing its authority in practice from general principles derived from observational knowledge of Australian plant communities and the specific place-oriented knowledges of traditional custodians of particular countries. In the words of Indigenous fire practitioner Victor Steffenson, a student of Awu-Laya elders Tommy George and George Musgrave and a prominent figure in the rekindling of this widely suppressed tradition, 'there is only one fire and that is the right fire, the fire for your Country'. Where cultural burning is no longer practised,

> We've got sick Country. There's no trees, it's full of weeds, there are no grasses, no native species. We need to teach people how to burn, to bring that country back to health.[18]

Often the 'prescribed hazard reduction burns' of official fire management are, from the Indigenous perspective, just as destructive as the uncontrolled wildfires they are intended to limit—wildfires which are the consequence of rising temperatures, longer summers and shorter winters, logging of old growth trees, and a reactive policy of fire repression which accumulates leaf litter, sticks, and logs on the forest floor. Oriented to 'asset protection' through methods centred on a maximising 'fuel

[17] Wertime, T. (1973). Pyrotechnology: man's first industrial uses of fire. *American Scientist*, 61(6), 670–682.

[18] Steffensen, V. (2016, Feb 15). Indigenous fire manager: Our country needs to burn more. SBS Insight, Special Broadcasting Service. http://www.sbs.com.au/news/article/2016/02/15/our-country-needs-burn-more-indigenousfire-manager. Accessed 17 Feb 2016. For the most authoritive written account of traditional pyro-ecological knowledge to date, see: Steffensen, V. (2020). *Fire country: how Indigenous fire management could help save Australia*. Wurundjeri (Melbourne): Hardie Grant Travel.

reduction', 'prescribed' burns are too hot for healthy country, burning down the elder parent trees that Indigenous fire protects, killing all manner of wildlife, impoverishing soil, and failing to protect and foster ecological abundance. In Victor's words,

> If you burn the canopy, then you have the wrong fire. Fire [should] behave like water, trickling through the country [so] it doesn't burn everything. [...] The canopy is a whole other world. [It is] so important to us because that's the life of the flowers, the fruits, the birds, the animals [...] that top canopy is very, very sacred and the simple rule is that it never burns.[19]

Calling for a cultural shift in mainstream Australian approaches to fire, Steffenson asserts the adaptive dynamism of Aboriginal knowledge practices, the only proven and viable technology capable of restoring of Australian ecosystems and enabling them to endure into the 'Anthropocene' future.

> We need to evolve our country with fire [...] Our hot country needs shade, and wildfires take shade away. The canopy is the coping potential of our country in extreme heat. [...] We're giving the country the least possible chance to survive climate change.[20]

This is an economy of fire in which an ethics of care is central. The abiding imperative to care for country is grounded in a binding ethics of connectedness, responsibility and intra-species kinship, though practices built on detailed observational knowledge of the country which sustains the flourishing of all its inter-related and complex life. In the words of a recent scientific review of the effects accomplished by the 'ecological engineering' of traditional fire practitioners, 'pyrodiversity begets biodiversity'.[21] This a *conscious* cultural economy of fire in stark contrast to the organised irresponsibility of industrial pyroculture toward the ecological consequences of its burnings, of enclosed fires hidden behind ignition switches and powerpoints.

Recognition of the first peoples' use of fire to compose landscapes into 'nourishing terrains' was almost completely absent from Anglo-Australian academic discourse until the term 'fire-stick farming' was coined by Rhys Jones:

[19] Steffensen (2016).

[20] Steffensen (2016).

[21] Bowman, D., et al. (2016). Pyrodiversity is the coupling of biodiversity and fire regimes in food webs. *Philosophical Transactions of the Royal Society B*, 371: 20150169.

> Why, wondered Rhys Jones (1969), did the Australian Aborigines not adopt farming, as virtually all the peoples around them did? He decided that they had; or at least that they had come up with a working analogue, a mode of production so elemental, ubiquitous, and misunderstood that observers failed to recognize it for what it was. Aborigines used fire to massage the indigenous environment with such skill that they became, in effect, cultivators of that landscape.[22]

Indigenous fire practices have since become the subject of a growing literature across the natural and social sciences. While it is beyond the scope of this chapter to summarise this literature, it is now widely accepted 'that Aboriginal people's land management practices, especially their skilled and detailed use of fire, were responsible for the long-term productivity and biodiversity of this continent'.[23] Bill Gammage's *The Biggest Estate on Earth: How Aborigines Made Australia* (2011) has done more than any other work to establish this history for mainstream audiences, compiling hundreds of colonial eyewitness accounts of Aboriginal pyrocultural practices and the fecund landscapes thereby maintained prior to wholesale European land appropriation.[24] Reinterpreting early eyewitness accounts and visual records, Gammage enables the reader to 'see' what was once invisible: a land intentionally shaped by skilful pyrocultural interventions.

Marvelling at the rich lawns of 'Australia Felix', abundant with great trees, grasses, and kangaroos, and free of undergrowth, colonial diarists, explorers, and surveyors repeatedly compared the Aboriginal-managed country at the moment of invasion to 'the park lands attached to a gentleman's residence in England'.[25] They were mystified by the abrupt proximity of dense forests to almost treeless 'lawns' and grassy open woodlands, with soil 'so soft and dark you could sink a hand into it with ease'. Colonists witnessed people working the land with fire in every locality from Tasmania to Cape York, but few understood what they saw, and in any case, were too busy violently dispossessing the locals to inquire. Aliens in a land where

[22] Jones, R. (1969). Firestick farming. *Australian Natural History*, 16, 224–231; Pyne, S. (1990). Firestick history. *Journal of American History*, 76(4), 1132–1141. See p. 1132.

[23] Rose, D. B. (1996). *Nourishing terrains: Australian Aboriginal views of landscape and wilderness*. Australian Heritage Commission, p. 10.

[24] Gammage, B. (2011). *The biggest estate on Earth: how Aborigines made Australia*. Sydney: Allen & Unwin.

[25] Gammage (2011, p. 7).

many plant species regenerate in and through fire, ignorant of Aboriginal law and lore, and blind to the intricately related but highly differentiated responses of each plant, animal and insect to the frequency, intensity, or absence of fire through the cycle of the Antipodean 'seasons', the Europeans were unable to imagine the degree to which biotic communities could be nurtured and arranged in patterned mosaics across large areas of land, through the judicious deployment of relatively frequent, controlled, cool burns.

The failure to recognise the firestick as a crucial technology of Aboriginal economic organization is of a piece with the exclusion from 'economics' of the pyrotechnologies without which recognisable human societies would not exist: much less a thermoindustrial society in which combustion and control over access to fuels has historically been at the root of political, technological, and economic power. If 'the mismanagement of combustion has become a global nightmare',[26] conventional economic thought is complicit in our uneasy slumber. Its intellectual trajectory since the fossil-fuelled heat engine was hitched to mass production has been to neglect the foundational necessity of 'land' and the pyrotechnologies which have facilitated the conversion of 'land' into 'the economy'.

Recognising the importance of cultural burning is of course not to claim that people lived by fire alone. Bruce Pascoe's *Dark Emu* (2014) reveals a range of allied sciences and techno-economic infrastructures which endured for many thousands of years in Aboriginal Australia, yet remain unacknowledged in popular narrations. These include astronomical observatories for geospatial navigation, grain cropping, storage and bread baking, permanent housing and villages, and sophisticated hydrological engineering to support salt and fresh water aquaculture industries. Pascoe demolishes the 'hunter-gatherer' label as wholly inappropriate to a people at home on intimately known and cultivated estates, whose law and lore maintained a long-enduring 'system of pan-continental government that generated peace and prosperity'.[27] Given the thousands of years of investment of labour, law, and intergenerational knowledge transmission in the maintenance of rich soils, ecological abundance, cultural diversity, and political stability across the continent, the first peoples of Australia were truly and profoundly 'settled'. Rather it is the capitalist economy

[26] Pyne (1990, p. 1132).

[27] Pascoe, B. (2014). *Dark emu. Black seeds: agriculture or accident?* Broome: Magabala, p. 129.

which is parasitic and nomadic, taking 'nature's gifts' with an extractivist disregard for the carefully stewarded abundance of the ancestral estates, and opening new frontiers of violent resource appropriation when nothing of marketable value remains.

Expanding on the firestick literature, Steffensen and Pascoe's work directly challenges the economic anthropology of ecological economists such as Gowdy & Krall, who suggest that

> Hunter-gatherers lived off the flows of energy from a wide variety of plants and animals and thus the daily rhythm and orientation of the hunter-gatherer was fundamentally different than it became when humans began to exploit the stock of fertile soil and manage photosynthetic off-take through cultivation of plants and domestication of animals. Hunter–gatherers were embedded within an ecological dynamic *that was not primarily directed by them*. In a simplistic sense, there was no particular economic preoccupation for the hunter–gatherer [...] Sahlins (1968) has called hunter-gatherers 'uneconomic man.' [my emphasis][28]

The contemporary literature on Aboriginal cultural burning tends to frame it as 'environmental management', and not as an economic practice. Conversely, I would argue that a sensitivity to the *intentional* pyrocultural achievements of firestick farming allows us to more accurately generalise about what constitutes an 'economy' and a responsible policy of 'economic management'. It also opens wide the question of whether we moderns are in any way embedded in an 'ecological dynamic' which we competently 'direct'. The imperative to submit to the obscure dictates of 'free markets' in pursuit of the promise of infinite growth in the face of the ecologically disastrous results of this policy suggests otherwise—witness the wholly unprecedented and uncontrollable inferno of Australia's remnant forests over the summer of 2019–2020.

Here it is timely to revisit the ancient Greek term *oikonomia*. In language recalling the colonists' frequent invocation of a 'gentleman's estate' to describe the cared-for country they mistook for a 'state of nature', the classicist Kurt Singer tells us that the term refers to

[28] Sahlins was of course here attacking the universal claims of neoclassical economics. Gowdy, J. & Krall, L. (2013). The ultrasocial origin of the Anthropocene, *Ecological Economics*, (95), 137–147. See p. 140.

> [...] administration, management, ordering, ruling a family and its estate (*oikos*), a pattern of co-ordinated actions conceived both pragmatically and normatively. Contrary to common notions sponsored by Rousseau and Edwin Cannan, the second half of the term does not refer to 'law' (*nómos*) but points to pastoral origins: distributing, allotting (primarily grazing ground, *nomós*), managing, feeding, caring for, a flock by the herdsman (*nomeus*). The term *oikonomos* [...] belongs to a wider group of words of similar structure signifying a Warden, Steward, Guardian [...] In the Hellenistic era [...] it came to denote ordering of well-constituted wholes in general, including the cosmos and any organism. [...] This conception [...] survives in the Christian theology of Providence and in the foundations of Physiocratic thought.[29]

Singer distinguishes between *nómos* as law, and *nomós* as the pastoral allocation of grazing grounds in an ancient past, rejecting the former interpretation for the latter. This contrasts with the position of the conservative German jurist Carl Schmitt, who held that the root of *all* law and legal order can be traced to the primordial land appropriations which gave rise to the spatial order of the 'law of the land', or in his terms, the '*nomos* of the earth'. Schmitt's genealogy of law as *nomos* is mobilised against what he saw as inadequate interpretations of this Greek term. Plato's *nomos* represented 'a mere rule' over land, while Aristotle did better in interpreting *nomos* as an 'original distribution of land'. From here, Schmitt identifies the constitutive foundations of the law of the land in ancient customary rights to grazing grounds. This is implicit in the Greek word for the herdsman (*nomeus*), who was originally a *nomad*, later becoming the guardian-manager of an estate (*oikonomos*).[30]

The setting of fires to open up grazing grounds in forests, to rejuvenate rangelands, to foster soil fertility through burning crop stubbles, and to drive away feared forest predators such as wolves and bears were common practices amongst European societies, until they were repressed with the advent of urbanisation, 'scientific forestry' and capitalist agriculture.[31]

[29] Singer, K. (1958). Oikonomia: an inquiry into beginnings of economic thought and language. *Kyklos* 11(1), 29–57. See p. 55.

[30] Schmitt, C. ([1950] 2003). *The nomos of the Earth in the international law of the Jus Publicum Europaeum*. New York: Telos, pp. 48–68 & 339–340; Walker, J. & Johnson, M. (2018). On mineral sovereignty: towards a political theory of geological power. *Energy Research & Social Science*, (45), 46–66.

[31] Pyne, S. (1997). *Vestal fire: an environmental history, told through fire, of Europe and Europe's encounter with the world*. University of Washington Press.

Thus it does not stretch the imagination too far to recognise an *oikonomos* in the figure of the *djungkay*—the Kunwinjku term for the senior lawman, custodian, and manager of a clan estate and its grasslands—whose responsibilities include the timing, organisation and conduct of major ceremonial gatherings and of the *manwurrk* events—fire-drives and kangaroo hunts—which provision them.[32]

Taking the ethical responsibility of the *oikonomos* to be the art of fostering abundance and lawful social existence, of making a living and dwelling-place (an *oikos*, or estate) through social endeavour and engagements with the complex living and non-living energies of the Earth, I suggest that the renaissance of Aboriginal pyrotechnologies embodies an ethical and ontological critique of Western economic thought and praxis, of its refusal of the Earth and its inability to respond to the general crisis of homelessness augured by the hothouse of the Anthropocene. Conversely, in something of a paradox, the belated recognition of the achievement of Indigenous fire practices discloses the possibility of the realisation and rediscovery of the very thing long-sought by European 'economic science': plausibly universal and abiding principles of *oikonomia*, linking the knowledge and arts of fire and life to the moral laws by which we 'manage the estate'.

Towards a General Economy of Fire

As Georges Bataille grasped earlier than most in his reading of Vernadsky's *Biosphere* (1929), the most general statements of economic theory ought to be specified at the level of the Earth as a whole, with its openness to solar energy flows, and its material closure.[33] The economy of life is founded in life's spontaneously organised dissipation of the un-reciprocable gift of solar radiance. Vernadsky was himself so conscious of the interplay between the sun and photosynthetic life that he referred to vegetation as 'green fire'.[34] As the continuous rise of atmospheric CO_2 measured by the Keeling curve suggests, 'human activities' have become the dominant source of ignition for combustible materials, far exceeding the global

[32] Altman, J. (2009). Manwurrk (fire drive) at Namilewohwo. In J. Russell-Smith, P. Whitehead & P. Cooke (Eds.), *Culture, ecology and economy of fire management in North Australian savannas: rekindling the Wurrk tradition*. Canberra: CSIRO.

[33] Bataille, G. ([1949] 1997). The meaning of general economy; The laws of general economy. In F. Botting & S. Wilson (Eds.), *The Bataille reader*. Oxford: Blackwell.

[34] Clark, N. (2012). Rock, life, fire: speculative geophysics and the Anthropocene. *Oxford Literary Review*, 34(2), 259–276.

capacity of photosynthetic life to recapture carbon from the atmosphere. Yet the economic thought of combustion-intensive modernity is, it seems, constitutionally unable to account for fire as a precondition of everyday material culture, from the hot meal on your kiln-fired plate to the leaf-blower in your neighbour's double-brick garage. This is an incredible omission, given that

> The material fabrics of nearly all settled civilizations have by and large consisted of things that exist only because of pyrotechnology—the generation, control, and application of heat, which at sufficient temperatures can alter the properties and compositions of all materials.[35]

As Freese notes, the entry on the industrial revolution in *The Oxford Encyclopedia of Economic History* barely mentions coal, repeating the foundation myth of standard economists who attribute it to Britain's liberal economic policies and culture of technological innovation.[36] From the extensive arts of 'fire-stick farming' practised for millennia by Aboriginal Australians, through the intensive arts of contained fire developed in the towns and cities of the ancient world that literally 'cook the earth' (e.g. brickmaking, pottery, glassmaking, metallurgy), and on to the profound shift to fossil fuel combustion in the blast furnaces, heat engines, and powerplants that fired the thermoindustrial revolution, pyrotechnologies are prior to and inextricable from all hitherto known 'modes of production'.

Wood ignites at around 180 °C, and most organic materials will be destructured at 250 °C. Smelting iron and fusing it with carbon to make steel requires a furnace heated to between 1400 and 1600 °C. To make cement, limestone—a carbon compound formed of the mineralised skeletons of calcifying marine organisms—is mixed with clay and heated to 1450 °C. This is only the initial stage of a thermal engineering process which has produced enough off the stuff to encase the entire Earth in a thin concrete sphere. Crude oil is refined in cracking plants into a myriad of petrochemical products formed at specific pressures and thermal gradients, from bitumen to aviation fuel, and into the feedstock materials for a vast array of paraffins, paints, plastics, polymers, lubricants, and synthetic 'organic' chemicals. To make glass, silicon is heated to around 1100 °C,

[35] Rehder, J. (2000). *The mastery and uses of fire in antiquity*, Montreal: McGill-Queens University Press, p. 3. Cited in: Clark, N. (2015). Fiery arts: pyrotechnology and the political aesthetics of the Anthropocene. *GeoHumanities*, 1(2), 266–284.

[36] Freese, B. (2003). *Coal: a human history*, Cambridge MA: Perseus Press.

but refining it to the purity required for computer hardware and photovoltaic solar panels involves several thermo-chemical processes, the initial stage requiring a furnace heated towards 2000 °C.[37]

Yet so obscured has the industrial fire enclosed in refineries, motors, powerplants and factories become to mainstream economists, that it is possible for the author of an 'inter-disciplinary' review of *The Causes of Economic Growth* to omit from it any discussion of energy, natural resources, or of the pyrotechnologies involved at some point in the production and delivery of nearly every commodity, service, or artefact for sale on 'the market'.[38] From the crucibles and coin dies of silver and goldsmiths to the current basis of US debt imperialism in the dollar-denominated international market for crude oil, money itself has a pyrotechnical signature. Even electronic money has a thermal/carbon footprint, to the extent that the electricity running computer networks is generated by fossil-fired power stations. Yet for conventional economists, 'growth' occurs in the absence of the intentional combustion without which the history of metallurgy and engineering and the globalisation of industrial capitalism could scarcely have occurred.

Even as erudite a historian of science and economics as Philip Mirowski, whose critique of neoclassical economists' attempts to construct a 'social physics' inspires many of these pages, has rarely ventured beyond the discussion of the metaphorical appropriation by economists of the scientific terminology of 'energy' to consider the question of the thermal basis of industrial technology, much less the encounter of the industrial growth economy with the thermal limits of the Earth.[39] His *More Heat than Light* (1989), an ironic exposé of the pseudo-scientific confusion of neoclassical theory, does not engage systematically with the economic phenomena of intentionally utilised heat—or indeed with light, which every leaf and blade of grass exists to harvest.[40] At least at this point in his oeuvre, Mirowski's criterion for a valid economic theory, used to critique the efforts of Georgescu-Roegen and others to analyse the physical basis of 'production' in energy conversion, was surprisingly conventional: '[T]he

[37] Contemporary technological innovations in renewable energy offer some hope that fossil-fuel combustion is no longer necessary for such thermoindustrial processes.

[38] Szostak, R. (2009). *The causes of economic growth*. Berlin: Springer.

[39] Although see: Mirowski, P., Walker, J. & Abboud, A. (2013). Beyond denial: neoliberalism, climate change, and the left. *Overland*, 210, 80–86.

[40] Mirowski, P. (1989). *More heat than light: economics as social physics, physics as nature's economics*. Cambridge: Cambridge University Press.

theory of economic value must explain why prices are expressed as rational numbers, as well as why those prices assume the values they do.'[41]

Why are money prices numbers? Any answer must be tautology. The general purpose money of commercial societies enables numerical calculation. Money is an index of ratios of exchange, according to cultural belief and inescapable social convention a store of 'value' transferable in exchange. According to Graeber, money arose not out of the fictitious 'barter economies' imagined without ethnographic warrant from Adam Smith's armchair, but as a means by which long term obligations such as taxes and interest-bearing debts could be settled.[42] However, the reasons why any particular set of prices are formed cannot be explained with any general or calculable theory of value due to the enormous complexity of the cultural, political, juridical, historical, and geographical entanglements of economic activity. This is why, despite neoliberal claims to the contrary, there has never been a consolidation of the One True Economics from the diversity of schools and traditions. This does not mean, however, that we can escape the fact that fire and energy conversion is integral to our economic life within the Earth's 'economy of nature', as is implicit in the constant resort to monetary metaphors for energy, from Heraclitus to the molecular biologists 'universal energy currency' (ATP). Despite its enthralling universalising logic, money is a purely cultural phenomenon with no natural analogue.

Money emerged, as best we know, around 3500 BCE in the accounting conventions developed by the temple bureaucracies of Sumerian city-states.[43] From the temporal perspective of the Indigenous peoples of Australia, the longest of *longue durées*, this is fairly recent. I would tentatively suggest that for Aboriginal culture, to which money was alien, a 'theory of value' might nevertheless be found in the ethical imperative of custodial law to care for country, including through the regenerative arts of the firestick. Pyrotechnology is a far more generalized and basic feature of human socio-economic organisation than monetary exchange, and if we are to survive the climate crisis, anthropic fire must again be brought into the ethical consciousness of a 'moral science'.

[41] Mirowski, P. (1988). Energy and energetics in economic theory: a review essay. *Journal of Economic Issues*, 22(3), 811–830. See p. 827.

[42] Graeber, D. (2012). *Debt: the first 5000 years.* London: Penguin, Ch. 1.

[43] Graeber (2012, pp. 38–39).

As we approach the present, it is clear that a certain brand of 'economics' is integral to the cultural logic by which this realisation is actively prevented and suppressed. Writing from the ultra-subjectivist Austrian wing of the neoliberal thought collective, Friedrich Hayek insisted that any temptation toward contemplation of the biophysical dimensions of human social order must be resisted as a symptom of socialist error, forever *verboten* to the true economist. Economics, Hayek intoned, is the study of subjective interactions between 'men and men, *and not between men and nature*'. Since women do not get a look in, it is perhaps not surprising that Mother Nature is to be rigorously excluded, not only in her aspect as the Earth-goddess Gaia, but even in her subordinate and domesticated role as a 'standing reserve' of resources for industry:

> *all the 'physical laws of production'* which we meet, e.g. in economics, *are not physical laws* in the sense of the physical sciences but peoples' beliefs about what they can do. [...] That the objects of economic activity *cannot* be defined in objective terms but only with reference to human purposes goes without saying. [...] Economic theory [...] has *nothing* to say about iron or steel, timber or oil, or wheat or eggs as such. [my emphasis][44]

Were we to accept Hayek's *a priori* restriction of economic science to the money-oriented inter-subjectivity of 'men', it is difficult to imagine why people would engage in economic activity at all—there would be nothing to buy or sell. Without wheat or eggs, there would be no breakfast. Without iron and timber, we would not have housing and other infrastructures. Without hydrocarbons or other sources of energy we would lack motor transport, electricity, artificial light, and the myriad conveniences and entertainments provided by the prodigious inventiveness of pyrotechnologists. So determined was Hayek to purify economic theory from 'socialism' that he banished 'the economy' itself from its purview.

Here we see the ultra-relativist nihilism of the neoliberal counter-revolution against nature and science expressed in direct form. This militantly anti-empirical philosophy has been weaponised and mass marketed by the phalanx of Atlas Network affiliated thinktanks, to which we can attribute much of the organised obstruction, delay, and defeat of climate policy (e.g. the deployment of state power to restrict and phase out fossil-fuel combustion and drive transition to renewable energy sources), to the

[44] Hayek, F. (1964). *The counter-revolution of science*. New York: Free Press, p. 31.

temporary benefit of the fossil-fuel based industrial corporations which have financed the neoliberal project from its earliest days. In denying the pyrotechnology thus far foundational to economic existence, economists refuse the Earth itself, and our profound moral responsibilities to 'manage the estate' to sustain life's capacity for abundance and future regeneration.

In sum, photosynthesis is as foundational to life on Earth as pyrotechnologies have been to human social existence. Yet both phenomena have been edited out of the canon of mainstream economics, and been aggressively denied by the neoliberals. As the starting point for a more credible *oikonomia*, I would suggest that the following (highly simplified) equations disclose the obscured link between ecology and economics, the common bioenergetic foundations of the twin sciences from microcosm to macrocosm, and our existential dilemma.

Photosynthesis:

carbon dioxide + water + light energy > carbohydrates + oxygen

Fire: (e.g. combustion of carbon biomass fuels)

carbohydrates / hydrocarbons + oxygen > heat and light energy + carbon dioxide + water

These equations conform to the 1st law of thermodynamics, insofar as the quantity of matter and energy is conserved through each thermochemical transformation, equivalent and balanced on both sides of Nature's ledger. Yet whilst fire appears as the exact reverse of photosynthesis, both conform to the 2nd law: the flow of energy goes only one way, and is resolutely historical. Solar energy radiates outward from the sun to be dissipated throughout the vast cold sink of deep space. A tiny fraction of the sun's radiance warms the Earth, a lesser fraction is captured by the Earth's biosphere, a lesser fraction again comes to be stored geologically, but a growing fraction of the Earth's long accumulated mineral hydrocarbons have been burnt in the furnaces and engines of our thermoindustrial era, in the blink of an eye in geological time.

Of course, science in itself does not tell us what to do: it can never fully answer the ethical and political question of how we are to live in a world where money has become the groundless measure of all things. It remains to be seen whether the impressive technological advances in renewable energy generation and the social movements demanding climate justice will succeed in overcoming politically entrenched hydrocarbon-based corporations and industries, and replacing the fossil-fuelled infrastructures that we can no longer afford for the sake of planetary survival. Acknowledging that the fate of the Earth and 'the economy' are inextricably linked—through fire—is a crucial step toward becoming worthy ancestors, 'custodians of the future' who in knowing when to burn and when not to burn, in cultivating the 'good fire' and avoiding the truly disastrous inferno, may act according to our ecological duty to consciously manage fire for the sake of the future generations, of all the interconnected life of all countries.

PART II

Economy's Nature

CHAPTER 4

Machines, Mechanism and the Discovery of the Economy

In the basement of the London School of Economics, an object of wonder gathers dust:

> In the 1950s, Bill Phillips, an engineer turned economist, built a machine to teach his students about the workings of the economy. Levers are pulled, buttons are pressed. Sluice gates open, and liquids of different colours rush around the system in a controlled way.[1]

What follows is an attempt to explain how this concretised metaphor, for Ormerod 'the very embodiment of the economists' view of the world', came to be there. As early modern science began to displace theology and secularise society, the machine metaphors which organised post-theological concepts of natural order, exemplified in the physics of Isaac Newton, in turn exfoliated throughout the nascent social sciences. I will argue that what Ulrich Beck once termed the 'apocalypse blindness' of the social sciences,[2] manifest in the failure of economic theory, policy and institutions to adequately confront our thermal and ecological crisis, can be traced to the role of a particular machine—the carbon-fuelled heat engine—in providing the constitutive metaphors of the formative statements of modern economic thought.

[1] Ormerod, P. (1994). *The death of economics.* New York: St. Martins Press, p. 37.
[2] Beck, U. (1995). *Ecological politics in an age of risk.* Oxford: Polity Press, p. 4.

J. Walker, *More Heat than Life: The Tangled Roots of Ecology, Energy, and Economics*,
https://doi.org/10.1007/978-981-15-3936-7_4

In the opening passage of his critique of social science methodology, *The Poverty of Historicism* (1957), Karl Popper lamented that although the science of society appeared at certain periods to have advanced beyond the science of nature—'I have Plato's political theory in mind and Aristotle's collection of constitutions'—since the resounding success of physics after Galileo and Newton they had lagged far behind even the biological sciences. 'The social sciences', he mused, 'do not as yet seem to have found their Galileo'.[3] And yet in a footnote, Popper concedes an exception to his case, buttressed by frequent citations of the economist Friedrich Hayek, his intellectual mentor and Mont Pèlerin Society comrade: '[It] must be admitted that the success of mathematical economics shows that at least one social science has gone through its Newtonian revolution.'[4]

Since the 'marginalist revolution' of the 1870s, a revolt against classical and Marxist political economy with its theory of value centred on the social experience of embodied labour, 'economics' has exuded an aura of superiority over all other social sciences in rigour, precision and technical expertise. Amongst its ur-texts are Carl Menger's *Principles of Economics* (1871), Leon Walras' *Elements of Pure Economics* (1874), and William Stanley Jevons' *The Theory of Political Economy* (1871), in which Jevons defined the new science as 'the mechanics of utility and self-interest'.[5] According to Mirowski, the reason economics has been able to assume this epistemic *gravitas* is because it 'has consistently striven to be the nearest thing to a social physics in the constellation of human knowledge'.[6] But the influence of economics goes far beyond any academic claim to scientific discovery or validity in the sphere of commerce. As has been noted by many writers, canonical economic theory and its standard approaches to policy relate to core beliefs of capitalist cosmology: the primacy of individual self-interest in explaining the 'natural order' of society, a view projected into life writ large in neo-Darwinian theories of 'selfish-genes' engaged in a life or death 'struggle for existence'.[7]

Along with this comes a refusal of bio-social relationships of mutuality and inter-dependence. The philosopher Peter Sloterdijk has memorably

[3] Popper, K. ([1957] 2002). *The poverty of historicism*. London: Routledge, p. 1.

[4] Popper (2002, p. 54n).

[5] Jevons, W. S. ([1871] 1970). *The theory of political economy*. London: Penguin, p. 90.

[6] Mirowski, P. (1988). *Against mechanism: protecting economics from science*. New Jersey: Rowman & Littlefield, p. 5.

[7] Gowdy, J., Dollimore, D., Wilson, D., & Witt, U. (2013). Economic cosmology and the evolutionary challenge. *Journal of Economic Behavior & Organization*, 90, S11–S20.

described the abiotic Cartesian fantasy of methodological individualism, in which a thinker thinks his way into relation with the world from a position of absolute disconnection and aloneness, as 'placental nihilism'.[8] As he points out, every individual begins life as a 'dividual', an unborn person entirely integrated within a mother's body, wholly dependent on her for nurture and protection. To enlarge the parallel, consider that 'the economy' emerged very recently from the multi-billion year biogeological history of Mother Earth. Yet economists maintain a pattern of ontological denial of the knowledge of ecological science, and of our dependence for life upon the sustenance of the biosphere and the protection of the atmosphere. Our interest here then, is to consider how the science of society that emerged at the dawn of our Anthropocene era of thermoindustrial expansion, and that still dominates its discourse, arrived at its present condition of *biospheric nihilism*. Two functions of the machine metaphor are in evidence here. Firstly, to conceive of economics as the study of 'the mechanics of self-interest' is to portray the economist as a physicist or engineer, cool and rational, as immune to pious moralising and political bias as to utopian eschatology. Secondly, the dominant machine metaphor of economics—the 'price mechanism' automatically coordinating 'forces of supply and demand'—excludes from economic analysis the actual forces of the heat engines which 'drive' 'the economy', and from the study of which the most confirmed of the 'laws of nature' were derived.

Every new science creates its own object: it is difficult to separate the discovery of 'the economy' from its invention. Despite the frequency with which one hears politicians and journalists declaring that a particular measure or event would hurt or help 'the economy', there is no measurable natural object called 'the economy' out there in the external world. Science begins with 'metrology', the art of measurement. The conceptualisation of accurate metrics of temperature, time, velocity, mass, length, volume, and force were integral to the development of basic science as we know it, as it is through measurement that scientific theory connects with its objects. Speaking of 'the economy' as such followed the 1934 development of a comprehensive system of national accounts by Simon Kuznets at the US National Bureau of Economic Research, a response to the inability of governments to respond to the Great Depression. The enumeration of the Gross National Product made it possible to represent in dollar values

[8] Sloterdijk, P. (2011). *Bubbles: spheres vol I: microspherology*. Los Angeles: Semiotext(e), p. 387.

the scale of economic activity and (importantly) the *growth* of this new super-object comprising the totality of commerce, labour and industry. As we approach the present, we find World Bank and IMF economists addressing themselves to the incredible totality of 'the world economy', forecasting and fostering optimal rates of 'global growth'. If macroeconomics (and thus the distinction between macro- and microeconomics) dates to the crisis of the 1930s, the definition and conceptualisation of economics as 'the study of phenomena from the standpoint of price', or in other words, of the operation of 'the price mechanism' in monetised exchange, is older.

Despite the contemporary consensus on the naturalness and necessity of permanent GDP 'growth', mainstream economic thought has tended to eschew explicit engagement with the biological sciences since the Industrial Revolution, on the level of metaphor and indeed of the empirical processes by which biological life is integrated into, reproduced and transformed by economic activity. Reflecting the long unrivalled prestige of physics as the defining intellectual achievement of Western modernity, from the early development of classical political economy in the eighteenth century through to the present characterisation of research in quantitative finance as 'econophysics', economists have imagined their scientific project as the abstraction and elaboration of immanent laws of social organisation, manifest in the dynamic balance between forces of supply and demand achieved by the operation of the 'price mechanism'. It was through metaphors derived from the mechanical philosophy of physics, itself developed in close interaction and convergence with developments in mechanical engineering, that economists sought to disclose (as Hayek put it) 'the true function of the price mechanism', and thus to discover the economy as a natural object governed ultimately not by the social norms of the law of the land or the coercive authority of government, but by natural laws at work in 'the price system'.[9] The conservative impulse to naturalise unequal distributions of wealth and political power goes back to early capitalism: in his *Leviathan* (1651), Thomas Hobbes declared that '[t]he Value, or Worth of a man is as all other things, his Price; that is to say, so much as would be given for his Power'.[10]

[9] Hayek, F. (1945). The use of knowledge in society. *American Economic Review*, 4(35), 519–530.

[10] Hobbes, T. (1651). *Leviathan, or the matter, forme and power of a common-wealth ecclesiasticall and civill.* London: A. Crooke, p. 41.

A Mechanical System of the World: The Newtonian Aspirations of Social Science

Machines belong to the world of artefacts constructed by humans, and since Aristotle the key definition of Nature is simply 'the non-artificial'. This fairly obvious distinction is deeply complicated by the fact that modern scientific materialism developed the machine metaphor as its 'world hypothesis'. Explaining complex natural phenomena in terms of underlying, law-like causal mechanisms has become almost by definition the correct approach to the study of nature: organismic, vitalist, or biological metaphors are unscientific. The evolutionary biologist Stephen Jay Gould, for example, was suspicious of Lovelock and Margulis' Gaia hypothesis, which seemed to him the elaboration of 'a metaphor, not a mechanism'—as if mechanism were not itself a metaphor.[11]

The mechanistic view of nature emerged in parallel with the most important machine of the early modern period: the mechanical clock. As Otto Mayr notes, the clock was an intellectual achievement of the highest rank. Nothing of comparable ingenuity had been devised before, and up to the advent of the steam engine it remained Europe's most intellectually demanding mechanism.[12] The influence of the clock on the European habitus was profound. Installed in the bell towers of churches and town halls, clocks disciplined the labouring body away from its embeddedness in the organic temporalities of subsistence farming. With the rise of factory labour, economic activity became inseparable from mechanically regularised experiences of time: coal-fired spinning jennys could be operated by shiftworkers 'around the clock'.

As tangible artefacts, early horological masterpieces and automatons exemplified the virtues of regularity, order, and harmony through the mechanical integration of smoothly interacting parts. Early viewers were most impressed by the orderliness of clockwork, qualities previously reserved only for things eternal and divine. From the sixteenth until the mid-nineteenth century, clockwork was the key metaphor of the extraordinary intellectual debates about the nature of the world and the place of humanity within it that we now describe as the Scientific Revolution. The new mechanical philosophy of nature articulated by Copernicus, Galileo,

[11] Gould, S. J. (1988). Kropotkin was no crackpot. *Natural History*, 7(97), 12–21.

[12] Mayr, O. (1986). *Authority, liberty and automatic machinery in early modern Europe*. Baltimore: John Hopkins University Press, p. 119.

Bacon, Kepler, Descartes, Boyle, Leibniz, and Newton profoundly transformed European civilisation.

Descartes is famous as the quintessential mechanist philosopher; his name designates the entire mechanist tradition. His revolutionary reconceptualisation of Nature as a machine dates from the 10th of November 1619, when he claimed to have received a vision from the Angel of Truth, a vision of the universe as vast machine composed of a great number of lesser machines, each entirely determined by iron laws of motion. From this vision Descartes began to articulate the thesis that the principles by which engineers constructed mechanical devices were the correct ones with which to approach all problems, from the composition of the universe to the study of 'animal oeconomy' (e.g. physiology):

> There are certainly no rules in Mechanics that do not belong to Physics, of which it is a part or special case: it is no less natural for a clock composed of wheels to tell the time than for a tree grown out of a given seed to produce the corresponding fruit.[13]

The central feature of the mechanical hypothesis was its proposal to analyse the phenomenon of nature as though they were the actions of machinery. The conflation of organisms and machines was coincident with the search for the hidden 'mechanisms' of nature becoming almost the very definition of scientific inquiry. Robert Boyle reduced this method to a simple formula, 'there can be no fewer principles than the two grand ones of mechanical philosophy, matter, and motion.'[14]

It would fall to Isaac Newton, justly famous as the archetypal modern physicist, to fulfil this brief. His *Principia Mathematica* (1687) looms large in the canon of the greatest of scientific works. In it, Newton explained a host of seemingly unrelated phenomena, such as the orbital motion of planets, the acceleration of falling objects, and the movements of pendulums and tides, by positing fundamental gravitational forces at work throughout the universe. Natural laws of universal motion, Newton grandly demonstrated, could be expressed as mathematical equations. In doing so, he made the physicist a towering authority capable of grasping all the secrets of the universe.

[13] Cited in Mayr (1986, p. 62).
[14] Cited in Mayr (1986, pp. 54–55).

For later generations, Newton's brilliant career would be idealised as the dawn of the Age of Reason. The Newtonian worldview came to be identified with the epistemic power of the 'atomic-mechanistic' conception of the world to reveal the inner workings of nature, in the form of unchanging conservation principles discovered and expressed through mathematical precision, abstraction, and generalisation from observed data. Newton was not, in this sense, strictly Newtonian, believing that the universe was evolutionary, and its path of historical development corrected by divine intervention as history moved toward 'the coming of the Kingdom for which we daily pray.' His celebrated works in optics, mathematics, and physics were by no means his only pursuits. His enrolment at Cambridge was motivated as much by the scientific problems of his day as by a fascination with 'judicial astrology'—political forecasting by the stars. This was inspired by his youthful experience of the turmoil of the English Revolution—a term once expressing the astrological implication that political events reflected the orbit of the planets. Much of Newton's intellectual effort was devoted to prophecy, theology, church history, the chronology of ancient kingdoms, and the Great Work of alchemy. Newton's contemporaries reported his 'long and constant application' to 'chymistry [...] that pyrotechnical amusement': 'his fires were eternal' as he pursued the transmuting of metals. His papers on these topics were long kept from publication. Later biographers, as they became aware of the extent of Newton's hermetic exertions, were embarrassed into to explaining them away as regrettable defects of the ideal scientific figure of the Royal Society. In contrast, Betty Dobbs insists that Newton's immersion in crypto-theological texts and his ecstatic subjection of minerals to thermal transformation were essential to his project: 'it was precisely *by* the route of alchemy that Sir Isaac expected to elucidate the ultimate component parts of matter'.[15] Newton's alchemical notes were dispersed across private libraries until J.M. Keynes established a collection and donated it to Kings College. Keynes described Newton as the 'last of the magicians, of the Sumerians and the Babylonians.'[16]

The majority of Newton's career was spent not in scientific academe but in a powerful, secretive and better remunerated sinecure which drew

[15] Dobbs, B. (1983). *The foundations of Newton's alchemy.* Cambridge University Press, pp. 6–9.

[16] Keynes, J. M. (1947). Newton, the man. In *Newton, Tercentenary Celebrations.* London: Royal Society.

on his hermetic pyrotechnical skills. In 1696, Newton was appointed Warden of the Royal Mint and chief assayer to the Crown of 'royal metals' by the directors of the newly founded Bank of England. Following the model of the Swedish Rigsbank, the prototypical central bank established in 1688, the Bank of England initiated the National Debt, as private financiers deficit financed state expenditures through fractional reserve banking (e.g. issuing bonds and banknotes in excess of the bullion reserves withheld from circulation in the vaults).[17] Newton was directly involved in efforts to stabilise the silver and gold currency through its recall and recoinage. So debased had the coin of the realm become, that contemporary observers wondered if the political turmoil of the Civil War had not been driven by a near-complete breakdown of trust in monetary transactions. Under the Coin Act of 1696, which made counterfeiting a crime of high treason, Newton exercised police powers, tasked with the detection and prosecution of those dishonest smiths, the coin-clippers and counterfeiters.[18]

Retrospectively, 'Newtonian' science came to signify the austere universality of his rational mechanics, the model of the Enlightenment's scientific quest to bring to light the hidden laws of nature. According to Greene (who is clearly reading the neoclassical programme back into Smith):

> [...] the idea of creating a social science by applying the methods of natural science to the study of man and society is nearly as old as science itself. Adam Smith took Newton's conception of nature as a law-bound system of matter in motion as his model when he represented society as a collection of individuals pursuing their self-interest in an economic order governed by laws of supply and demand.[19]

The Deist philosophers of the eighteenth century attributed natural order to the 'invisible hand' of God, who imposed physical laws on nature in much the same way as moral laws were imposed upon human beings. These two domains of law were separate: human beings were free to reject moral laws prior to the balance of accounts being settled at the Day of

[17] The Bank of England would remain a private monopoly corporation by Royal Charter until 1946, when it was nationalised by Clement Attlee's Labour government.

[18] Selgin, G. (2012). Those dishonest goldsmiths. *Financial History Review*, 19(3), 269–288.

[19] Cited in: Hetherington, N. (1983). Isaac Newton's influence on Adam Smith's natural laws in economics. *Journal of the History of Ideas*, 44(3), 497–505.

Judgement, but neither human beings nor other parts of creation could contravene physical laws. The two domains were alike, in that whatever humans might think or do, the laws were divine and invariant. Saving theology from science (and scientists from the theologian's charge of atheism) involved some torturous accommodations, as d'Alembert's dialogue with Diderot attests:

> I confess that a Being who exists somewhere and yet corresponds to no point in space, a Being who, lacking extension, yet occupies space; who is present in his entirety in every part of that space, who is essentially different from matter and yet is one with matter, who follows its motion, and moves it, without himself being in motion, who acts on matter and yet is not subject to all its vicissitudes, a Being about whom I can form no idea; a Being of so contradictory a nature, is an hypothesis difficult to accept.[20]

Pierre-Simon Laplace disposed of these convolutions in his *Exposition du Système du Monde* (1796), a work on celestial mechanics. According to legend, upon being presented with this volume, Napoleon asked Laplace where God was in his system, to which Laplace replied, 'I have no need of that hypothesis.' In his 1812 *Essai sur les Philsophique les Probabilitès*, Laplace inaugurated the field of social physics—and ultimately neoclassical economics—with the following suggestion:

> Let us apply to the moral and political sciences the method founded upon observation and upon calculus, the method which has served us so well in the natural sciences.[21]

It is not certain whether it was Adolphe Quetelet or Auguste Comte who first coined the term 'social physics'. Quetelet, a statistician and demographer, was fascinated by Laplace's observation that the number of letters arriving annually in the Paris dead-letter office conformed to simple probability distributions formally similar to accepted laws of motion. He took Laplace's ideas to their full extension, generating reams of data on all kinds of subjects, such the chest sizes of Scottish males. Using the method of

[20] Stewart, J. & Kemp, J. (Eds.) (1943). Conversation between d'Alembert and Diderot, 1769. *Diderot, interpreter of nature.* International Publishers.

[21] Cited in: Lecuyer, B-P. (1994). Probabilistic thinking. In I. Cohen (Ed.), *The natural sciences and the social sciences: some critical and historical perspectives* (pp. 135–152). Dordrecht: Kluwer Academic, p. 141.

statistical probability analysis he borrowed from Gauss' theory of errors (later expressed as the 'normal law', or the bell curve), which was developed to correct differences in large numbers of astronomical observations, Quetelet came to the view that there were social laws of nature of equivalent deterministic probability as the those revealed by statistical mechanics. The science of government would be best placed upon statistical foundations. These ideas were advanced in *On Man and the Development of Human Faculties: an Essay on Social Physics* (1835) which also introduced the quotidian figure of 'sur l'homme', the average man.

Comte, who is credited with coining the terms 'biology', 'sociology', and 'positivism', was an early champion of 'physique sociale'. Comte's vision of social physics was in accord with Quetelet's mission to ground the 'science of society' on the same bases as the physical sciences, which meant applying d'Alembert's principles of analytical mechanics to the determination of regularities in society. 'Sociology' is said to have been coined by Comte because Quetelet, his rival, had already published using his preferred term: social physics. Comte saw history as the progressive unfolding of human knowledge, a theory he elaborated in the exhaustive systematisation of the sciences presented in the *Cours de Philosphie Positive* (1832). Not only did he feel that the methods of physical science were applicable to the study of social phenomena, he argued that social science had evolved from these methods and was destined to overtake the physical sciences in explanatory power. In the same work, the pregnant notion of 'the social system' was also invoked, perhaps for the first time: 'The statical study of sociology' Comte wrote, 'consists in the investigation of the laws of action and reaction of the different parts of the social system—apart, for the occasion, from the fundamental movement which is always gradually modifying them'.[22] Comte was a disciple of the utopian socialist Henri Saint-Simon, and collaborated with him in conceiving of *A Plan of the Scientific Operations Necessary for the Reorganization of Society* (1822), in which it was claimed that politics would soon become a social physics. In his hierarchy of the sciences, Comte placed sociology, which he viewed as the most complex science and the most dependent for its emergence on the development of all the others, at the pinnacle in the hierarchy.

[22] Cited in: Coser, L. (1977). *Masters of sociological thought.* New York: Harcourt Brace, pp. 11–12.

> Social science offers the attributes of a completion of the positive method. All the others are preparatory to it. Here alone can the general sense of natural law be decisively developed, by eliminating forever arbitrary wills and chimerical entities, in the most difficult case of all.[23]

When in 1840 the English mathematician William Whewell coined the word 'scientist', he indicated the profoundly new direction being taken by modern society in its intellectual relationship to the natural world. Increasingly, it was one mediated by a cadre of academically trained professionals with a shared methodology and ethic. Scientists began to assume the exclusive capability of and responsibility for producing 'positive knowledge', a reliable store of confirmed and concrete truths by which long-term human progress might be served.[24] The search for positive knowledge among this new subculture of intellectuals differed from the undertakings of gentleman amateurs or the holists of the continental *Naturphilosophie* tradition. The biogeographer Alexander von Humboldt's epic five-volume *Kosmos* (1845–1862), a poetic exposition of the harmony, beauty, diversity, and unity of nature, took the history of cultural perceptions of nature into its narrative sweep and enrolled the naturalist in the project of Enlightened resistance to despotism.[25] By contrast, scientists distinguished themselves by the elimination of subjectivity, confining themselves to disciplined objectivity, analytical theory building, empirical experimentation, and confirmation and review by trained peers.

Ilya Prigogine, the physicist renowned for studies linking evolution and biochemical emergence to far-from-equilibrium thermodynamics, argues that despite common misconceptions to the contrary, the origins of Western science are in theology. Leibniz, he says, clearly spells out the position of early scientific method. For Leibniz, the role of science was to bring man closer to knowledge of God. He supposed that for God there would be no difference between past, present, and future, as his being was eternal. The positivist sense of confidence and power in modern science came from the identification of eternal, unchanging natural laws at the core of matter and material transformation. For generations, the discovery of order within nature stirred within the bosom of European imperialists

[23] Cited in Coser (1977, p. 12).

[24] Worster, D. (1977). *Nature's economy: the roots of ecology*. San Francisco: Sierra Club, p. 130.

[25] von Humboldt, A. (1845–1862). *Kosmos*. Stuttgart: Cotta.

such as the French anthropologist Levy-Bruhl 'a feeling of intellectual superiority [...] so deeply anchored that we even do not see how it could be shaken.'[26] A most avid account of positive science's ambition to usurp the place of God and envision the totality of the universe was given by Laplace:

> We may regard the present state of the universe as the effect of its past and the cause of its future. An intellect which at a certain moment would know all forces that set nature in motion, and all positions of all items of which nature is composed, if this intellect were also vast enough to submit these data to analysis, it would embrace in a single formula the movements of the greatest bodies of the universe and those of the tiniest atom; for such an intellect nothing would be uncertain and the future just like the past would be present before its eyes.[27]

The goal of the scientist was to eliminate time, to gain the perspective of immortal deity by transcending the circumscribed evental time of biological life. This is why, Prigogine says, classical and basic science has no time, although time is phenomenologically present to us at all times.[28] With the exclusion of temporality, classical science presents the world as an automaton, a perpetual motion machine. As Isabelle Stengers and Prigogine have elsewhere commented:

> To the extent to which dynamics has become and still is the model of science [...] it is still the prophetic announcement of a description of the world seen from a divine or demonic point of view. It is the science of Newton, the new Moses to whom the truth of the world is unveiled; it is a revealed science that seems alien to any social or historical context identifying it as the result of the activity of human society.[29]

It seems to me that all of these remarks apply directly to the intellectual underpinnings of the image of 'the free market' at the core of orthodox economic discourse, an image of natural order which has been deployed by the neoliberal project to thwart attempts to maintain a relatively

[26] Cited in: Prigogine, I. & Stengers, I. (1984). *Order out of chaos: man's new dialogue with nature*. London: Heinemann, p. 292.

[27] Cited in Mirowski (1989, p. 27).

[28] Prigogine, I. (1986). The chaotic universe. In L. Wijers (Ed.), *Art meets science and spirituality in a changing economy*. London: Academy Editions, p. 95.

[29] Prigogine and Stengers (1984, pp. 74–76).

egalitarian society through democratic legislation and regulatory institutions, and more recently, to undermine efforts to preserve the ecological and climate commons.

The origins of Progress ideology have been attributed variously to the secularisation of Christian eschatology or to the promise of incrementally increasing knowledge yielded by the success of the rational-empirical method in early astronomy. In the case of economics, the current faith in infinite growth *and* the contradictory idea that economies timelessly tend toward an equilibrium state akin to a physical law—assumptions latent in all exponents of general equilibrium theories from Adam Smith down to the present day—suggests both explanations are credible.

Constituting the Machinery of Political Economy

The new mechanical world picture provided a series of metaphors for talking about systems of government. With the rise of a centralised, absolute sovereign state in tandem with an increasingly commercial ruling class in seventeenth-century Europe, clockwork became a key metaphor in the rhetorical justification of authoritarian forms of political order. In an early description of the state as a political machine, John Trenchard in his *Short History of Standing Armies* (1698), argued that '[A] government is a mere piece of clockwork, and having such springs and wheels, must act after such a manner: and there the art is to constitute it so that it must move to the public advantage.' The secret is 'to make the interest of the governors and the governed the same [...] and then our government will act mechanically, and a rogue will as naturally be hanged as a clock strike twelve when the hour has come'.[30] The quasi-military hierarchy of Prussian bureaucracies would exemplify the political machine, with its ability to unquestioningly transmit the commands of sovereign power into concerted mass action.

Political liberalism, in its efforts to check the arbitrary exercise of royal authority and devolve sovereign decisions over war, taxation, and debt to a law-making body of propertied men, began designing constitutional systems to ensure the 'rule of law', institutionalising divisions of powers to maintain a balance of forces between the different estates and social

[30] Cited in: Wootton, D. (2006). Liberty, metaphor, and mechanism: checks and balances and the origins of modern constitutionalism (pp. 209–275). In D. Womersley (Ed.), *Liberty and American experience in the eighteenth century*, Amagi.

classes.[31] 'Newtonianism' influenced philosophers such as Locke and Montesquieu in developing the liberal theory of constitutional government. According to Raymond Aron (an MPS member), 'Montesquieu's essential idea is not the separation of powers in the juridical sense but what might be called the *equilibrium of the social forces* as a condition of political freedom.' [my emphasis][32] The framers of the 1788 Constitution of the United States, authored a century after the *Principia*, reasoned in a certain conception of political mechanics according to Woodrow Wilson:

> The admirable expositions of *The Federalist* read like thoughtful applications of Montesquieu to the political needs and circumstances of America. They are full of theory of checks and balances [...] Politics is turned to mechanics under his touch.[33]

'Equilibrium', on this view, is the constitutional metaphor of the Constitution.

The historian Donald Worster has dated the earliest contemporary uses of the word 'oeconomy' to around 1530, when it referred not to a thing ('the economy'), but to the political administration of a moral community, or of a state's resources for orderly production. Christian theologians had an older tradition of using the word *oeconomia* to describe the 'dispensations' of God toward Man and Creation through successive ages or covenants.[34] By the seventeenth century the term was used routinely by theistic naturalists to describe the order of Creation. 'Oeconomy of nature' remained the term used to refer to the totality of biological and abiotic nature until 1866, when Haeckel proposed the scientific term Öekologie to bring the oeconomy of nature into secular alignment with the geology of Hutton and Lyell, the biology of Darwin, and the energy physics of von Helmholtz and Clausius. The origin of economics in 'moral science'—concerning the divine covenant aligning a law-governed Creation with the 'natural law' of a just society as presented in Thomas Aquinas' *Summa Theologiae*—is fully consonant with this usage. The secularisation of

[31] Mayr (1986, p. 163).

[32] Cited in: Cohen, I. (1994). An analysis of interactions between the natural sciences and the social sciences. In I. Cohen (Ed.), *The natural sciences and the social sciences* (pp. 1–99). Dordrecht: Kluwer Academic, p. 61.

[33] Wilson, W. (1908). *Constitutional government in the United States.* Columbia University Press.

[34] Worster (1977).

oeconomy progressed in tandem with the rising sovereignty of the nation-state: as the divine right of kings gave way to liberalism and Enlightenment humanism, moral science gave way to political economy.

Since its inception, the study of political economy has been concerned to envision some fundamental, measurable unit, or substance of *value* with which to isolate the economic realm from the mire of complexity and present it for analysis as an independent fact of social existence, an object of government distinct from the offices of government. The problem of value theory emerges in Aristotle, who distinguished between *oikonomia* and *chresmatica*. The former term refers to the productive arts of stewardship on the freeman's estates, and the latter, disparagingly, to profiting from unproductive and disordering speculation on changing prices and interest rates in the agora. In the ancient world, debts were secured against the person: failure to pay would result in the debtor's enslavement. Containing a basic ambiguity between the moral sense of value as that which is good and just in and of itself, and the fluctuating values of money prices, controversies over value theory are as old as the phenomena of money, merchant trade, and interest-bearing debt. Insofar as the writing systems used by the temple scribes of the early Mesopotamian city states were developed to account for grain taxes, wages, tributes, and debt service, the problem of value theory may well be as old as written speculation on social life itself.

Within the modern history of economic thought, shifts in the reigning theory of value mark the rise and fall of different research programmes. Measurement controversies in value theory often point to deep problems of metaphoric interpretation, and the metaphors which later become invisible to theorists are often first envisioned with visual metaphors. One of the remarkable effects of the discovery (or invention) of a scientific object, according to Susan Buck-Morss, is the rapidity with which that object takes on independent agency once it has been 'visualised': '[b]ecause the economy is not found as an empirical object among other worldly things, in order for it to be "seen" by the human perceptual apparatus it has to undergo a process, crucial for science, of representational mapping'.[35] Once this occurred, 'the economy' could be seen to act in the world, causing events and creating effects. Importantly, 'the economy' is thus always to some extent a representation. As Klamer and Leonard put it: '[w]e may

[35] Buck-Morss, S. (1995). Envisioning capital: political economy on display. In L. Cooke & P. Wollen (Eds.) *Visual display: culture beyond appearances.* Seattle: Bay Press, p. 116.

attach a name ('the economy') to the unimaginably various and complex activities of a nation's economic life, but we have not thereby ensured that it is a thing'.[36]

Historians of the origins of economics often point to the first systematic attempt to reveal this heretofore-unknown object in a single chart: the *Tableau économique* published in 1758 by the Physiocrat François Quesnay. Reflecting the Enlightenment preoccupation with natural law, the term *physiocracy* translates as 'the rule of nature'. Inspired by William Petty's dictum that land was the mother and labour the father of wealth, the Physiocrats criticised the mercantilists' direct equation of national wealth with the quantity of metal money flowing through the state exchequers. As the sum of gold and silver in Europe was relatively fixed (excepting the inflows from the mines of the New World) international trade was a zero-sum game, and wealth could only be increased in one nation by draining the coffers of another.

Prior to the eighteenth century, the dominant choice of metaphor for the discussion of national wealth was physiological. Hobbes and Rousseau, for example, inherited the medieval idea of a commonwealth is a 'body politick', with various parts of the body corresponding to social classes or state institutions. Quesnay was a physician who, influenced by William Harvey's discovery of the circulation of the blood, believed that the 'health' of the national economy resolved to the unobstructed circulation of a living value substance (denoted as *blé*, or wheat) derived ultimately from the fertility of the soil. Like blood, the life giving fundamental of agriculture was essential to the health of all the other organs of the economy. The farmer's ability to produce food surplus to her own requirements made possible all the other occupations: as the biological fertility of the Earth is the source of value, only the careful fostering of the (re)generative potential of the soil offered the possibility of increased social wealth. Quesnay mentioned in this respect the importance of burning crop stubbles for nutrient cycling: 'smoking the earth' maintained soil fertility, as 'the salts and sulfurs of mixes which are released return to the

[36] Klamer, A. & Leonard, T. (1994). So what's an economic metaphor? In P. Mirowski (Ed.), *Natural images in economic thought: markets read in tooth and claw*. Cambridge: Cambridge University Press, p. 31.

womb of the earth where they in turn contribute anew to the reproduction of the mixes'.[37]

The 'economic picture' provided by Quesnay's *Tableau*, a graphic representation of the French national economy, was novel in that it included both monetary circulation ***and*** biophysical production in an analysis of national income. Quesnay's economic picture assumed a revelatory importance among the Physiocrats, who, influenced by Cartesianism, were inclined to view the universe as a 'gigantic machine regulated by natural laws of divine origin',[38] albeit a machine of a decidedly vitalist character. 'The natural laws of the order of society', Quesnay wrote, 'are precisely the physical laws of the ***perpetual reproduction*** of the goods necessary for man's subsistence, conservation and well-being' [my italics].[39] According to Simon Kuznets, the *Tableau économique* was the first in a series which would culminate in the system of national accounts he developed for the US National Bureau of Economic Research in the depths of the Great Depression.[40] This measure of the 'size' of the economy is expressed in the form of a single GDP number totalling the dollar values of all monetised economic activity, and an annual percentage 'growth rate' benchmarked only against previous measurements, has naturalised exponential production as something akin to a law of nature, and its fostering the prime task of modern government. That perpetual GDP growth provides an index of improving social welfare, a presumption rejected by Kuznets himself, stands in stark contrast to Quesnay's concern for the 'perpetual reproduction' of the fields, pastures and vineyards of the agrarian estate.

Quesnay's chart was addressed to an elite audience of the kings' advisers and taxation officials. Its reading of productivity was reassuring to an absolute monarchy with much to fear from the rising power of the bourgeois commercial classes; the *Tableau* notably assigned the negative value of 'sterility' to industrial manufactures and trade. In developing the labour theory of value, both Adam Smith and Karl Marx acknowledged their predecessors' importance in shifting economic analysis 'from the sphere of

[37] Christensen, P. (1994). Fire, motion, and productivity: the proto-energetics of nature and economy in François Quesnay (pp. 249–288). In P. Mirowski (Ed.), *Natural images in economic thought*. Cambridge University Press, p. 258.

[38] Buck-Morss (1995, p. 118).

[39] Cited in: Ingrao, B. & Israel, G. (1990). *The invisible hand: economic equilibrium in the history of science*, Cambridge MA: MIT Press, p. 43.

[40] Kuznets, S. (1948). National income: a new version. *Review of Economics and Statistics*, 151–179.

circulation into the sphere of direct production'.[41] Both mocked the Physiocrats for what they saw as a rustic obsession with farming, manifest in the latter's insistence that agricultures' harnessing of solar energy through plant growth—'the free gift of nature'—was the ultimate origin of value, which labour and manufacturing merely transformed. Yet given that Quesnay was writing prior to fossil-fuelled industrialisation, and well before the development of thermodynamics and the Earth systems sciences, his solar-thermal bioeconomics now appears remarkably prescient:

> One recognizes that there is a first agent in matter by which everything is executed in nature, which moves everything, which is the cause of all generation and all destruction; it is a fire, a matter aetherial or subtle, extremely active, which has the property of all the movement which animates the universe; it is an immense sea which contains all the sensible bodies, which it intimately penetrates and through which it works all the changes which happen.[42]

This is a combustion-centric, bioenergetic *oikonomia* which speaks far more directly to our 'Anthropocene' present than what passes for 'economic science' in the neoliberal dispensation.

The Machinery of Freedom: The Steam-Governor and the System of Natural Liberty

With the rise of the industrial bourgeoisie in late-eighteenth- and early-nineteenth-century Britain came a renovated notion of the natural laws governing the economy of nature. Concentrating new machine metaphors derived from the proliferating steam engines of the thermoindustrial revolution, this broke with the organic and natalist implications of the Physiocratic world picture, and rejected the clockwork authoritarianism of early political theory in the name of liberty. The liberal economists established a dynamic, radically novel concept of social order, which Mayr terms the *liberal theory of self-regulation*.

[41] Marx, K. (1861–1863). The Physiocrats. *Economic manuscripts: theories of surplus value.* https://www.marxists.org/archive/marx/works/1863/theories-surplus-value/ch02.htm. Accessed 10 March 2019.

[42] Quesnay, F. (1747). *Essai physique sur l'oeconomie animale* (vol 3). Paris: Guillaume Cavelier, p. 133. (Trans: Christensen 1994, p. 266).

The idea of an automatically balancing dynamic mechanism was implicit in the liberal critique of mercantilist trade theory, with its demands for a rigorous policy of naval policing and protective tariffs to ensure that colonial and international trade led to a net inflow of precious metal currencies (or a positive 'trade balance'). Liberals such as David Hume argued that increased flows of metal currency into national circulation did not correspond to increasing wealth, as currency inflows had destabilising effects on local industry that counteracted their benefits. The agriculture and industry of Imperial Spain, after all, had been laid waste by inflationary influxes of silver and gold from the New World. Trade accounts would better balance 'automatically'.

More than any other thinker, Adam Smith is associated with the doctrine of economic liberalism, rightly or wrongly claimed by academic economists and neoliberals as their intellectual ancestor, the 'father of capitalism' and the first truly modern economist. In the *Wealth of Nations* (1776), Smith dismissed the arbitrary sovereignty of mercantilism and the agricultural bioeconomics of Physiocracy. For Smith the proper system of government would be one which withdrew from forms of economic administration that restricted or protected any particular trade or industry, as this 'retards, instead of accelerating, the progress of society toward wealth and greatness'. Discharging the sovereign from 'the duty of superintending the industry of private people and of directing it towards the employments most suitable to the interest of the society', Smith imagined that once 'all systems of preference, and restraint [were] completely taken away, the obvious and simple system of natural liberty establishes itself of its own accord'.[43] The chief concerns of the liberal movement—the denationalisation of commerce, the fostering of industrial production, the devolution of government to a body of propertied men, the constitutional protection of liberal rights and freedoms including the fortification of exclusive property in land—were reflected in the machine metaphor that eventually came to dominate the subject of political economy, as it in turn came to be dominated by British liberals. This was the metaphor of an automatic balance, or dynamic equilibrium.

The transition from the clock to the steam engine as the epistemological totem of modernity brought with it a revolutionary re-imagining of political and economic systems, suggesting their fundamental unity with

[43] Cited in: Bigelow, G. (2003). *Fiction, famine, and the rise of economics in Victorian Britain and Ireland*. Cambridge: Cambridge University Press, p. 47.

the laws of nature revealed by scientific materialism. The idea that a society of autonomous individuals motivated by pecuniary self-interest could produce automatically the most perfectly regulated and harmonious of social orders can, according to Mayr, be traced to the dramatic technological advances that accompanied combustion engineering:

> The eighteenth century machine was a product of the Newtonian universe with its multiplicity of forces, disparate sources of motion, and reversible mechanism. Engineering problems in the design of steam engines had led to the discovery that under certain conditions, dynamic systems can be capable of regulating themselves, maintaining themselves *in equilibrium upon their own resources*—without the external intervention of higher forces. [my emphasis][44]

Adam Smith likened the system of natural liberty to 'a great, immense machine, whose regular and harmonious movements produce a thousand agreeable effects'.[45] Although Smith himself never used the phrase 'market mechanism', which was retroactively mapped onto his more famous metaphor of 'the invisible hand', the association of individual liberty and automatic machinery was not lost on his contemporaries, or the generations which followed.[46] Mayr argues that the mental image of self-governing 'equilibrium' that led to the acceptance of liberal doctrines among the literate British population (with its particular enthusiasms for popular mechanics), should be traced less to the inspiration Smith took for his 'system of natural liberty' from a youthful immersion in Newton's celestial mechanics, but rather to the exemplary analogue provided by the centrifugal steam-engine governor, devised in 1785 by Smith's friend and fellow Glaswegian James Watt (Fig. 4.1).

The steady output of early steam engines was dependent on the tacit skills of the engine-driver, who would carefully time the shovelling of coal into the furnace to maintain a relatively constant level of heat. If the furnace became too hot, expansive pressures built up in the boiler, and the engine could accelerate beyond control. Boilers often exploded with extreme violence, an occupational hazard for those working in proximity to steam-power that was not substantially eliminated until design safety

[44] Mayr (1986, p. 139).

[45] Smith, A. ([1759] 1976). *The theory of moral sentiments.* Indianapolis: Liberty Fund, p. 31.

[46] Mayr (1986, p. 139).

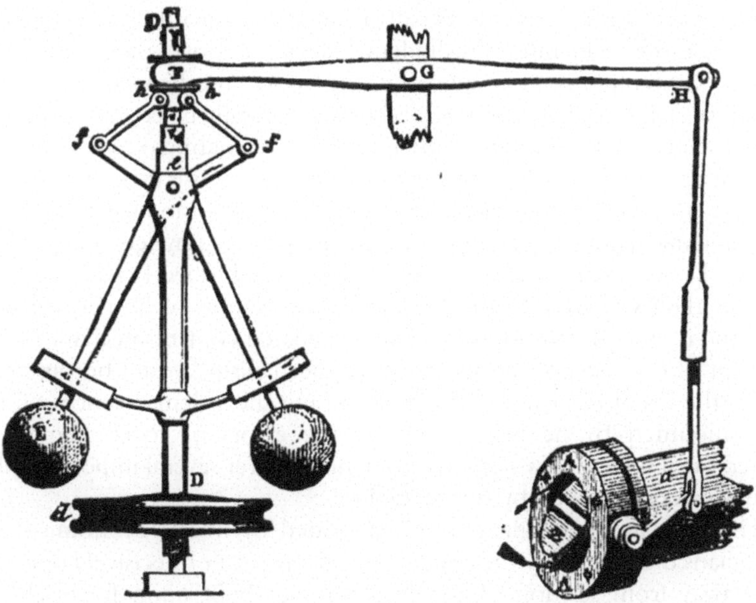

Fig. 4.1 James Watt's centrifugal steam-engine governor and throttle valve. (Image from Routledge, R. (1900). *Discoveries & inventions of the nineteenth century* (13e). Public Domain. https://commons.wikimedia.org/w/index.php?curid=231047)

regulations were codified in early twentieth-century social legislation. Amongst other decisive improvements in the efficiencies of existing engine designs for which he is credited with launching the Age of Steam, Watt resolved the problem of over-speeding engines with an ingenious device. Adapted from an invention of Huygens which modulated the transmission of rotary force from windmills to grindstones, Watt's governor automatically adjusted and controlled the heat, pressure, and speed of steam engines, maintaining a near constant output of mechanical work. As the speed of the engine increases, the central spindle of the governor rotates at a faster rate, and centrifugal forces push the two flyballs upwards and outwards. This motion is translated by connecting rods to a throttle valve, reducing its aperture. The rate at which the working fluid (steam) enters the cylinder is thus reduced, and the speed of the prime mover falls. As the speed of the prime mover falls, so the rotation of the governor slows, and the reverse effect occurs: the flyballs fall and the throttle valve re-opens, releasing more steam into the cylinder which in turn drives the engine faster.

The contemporary metaphor of the 'price mechanism' governing 'forces of supply and demand' is rarely meant literally by economists in the sense the market *is* a machine: rather it is a self-regulating or self-stabilising 'system'. In other words, a 'system of feedback' to use cybernetic terminology, itself derived mechanical engineering. A formative text in the annals of mid-twentieth-century cybernetics was the physicist James Clerk Maxwell's essay *On Governors* (1868), which established key principles of control theory.[47] Despite the paradoxical conflation of human freedom with the inhuman determinism of an automated control mechanism, the concept of the socially optimising 'price mechanism' would become central to the scientific claims of economists, and the ardent defences of 'liberty' mounted by neoliberals against the errors of social democrats, postcolonial nationalists and environmentalists. Yet several important phenomena were occluded by this metaphor.

The 'system of natural liberty' advocated by liberal economists and politicians did not imply the eradication of slavery, nor decisively open the road away from serfdom. Capitalism was not built upon freely chosen employment, but by 'free' unpaid, coerced labour. The rising wealth of Smith's Glasgow was to a large extent the result of investments in African slavery, Jamaican sugar, American tobacco, and cotton. Smith struggled to reconcile his advocacy for 'natural liberty' with a pessimistic view of social history and human nature, in which slavery and the domination of the impoverished seemed more or less intractable. Indeed, it is possible to read a 'searing indictment of commercial society' in Adam Smith's critique of the conspiracies hatched against the public by businessmen meeting in taverns, in his critique of gigantic corporations (such as the East India Company) that subjected whole nations to a despotic dominion of asset stripping and wealth removal, and in his recorded views on slavery:

> Slavery is more severe in proportion to the culture of society. Freedom and opulence contribute to the misery of the slaves. The perfection of freedom is their greatest bondage. And as they are the most numerous part of

[47] Mayr, O. (1971). Maxwell and the origins of cybernetics. *Isis*, 62(4), 425–444.

> mankind, no [...] person will wish for liberty in a country where this institution is established.[48]

The estate-holders sitting in the Scottish Parliament had instituted certain liberal rights and freedoms for their fellow subjects, but this was with telling exceptions. The 1701 Act establishing *habeus corpus*—the right to a timely trial and protection from arbitrary punishment and imprisonment—explicitly excluded coal miners. Colliers and their families were held in a state of inter-generational serfdom, bound by a 1606 statute to labour for the master of the estates to which they were confined. So damp, dirty, dangerous, and despised an occupation was coal mining, an industry that grew steadily as the great forests of Caledonia were felled for fuelwood and cleared for wool exports, that coal masters jealously guarded their captive labour supply. Colliers could not leave without an explicit authority signed by the coal-estate owner transferring them to another coal master. A collier absent without leave could be charged, imprisoned, and punished: *for theft*. Children, women, and men were listed as part of the property in legal documents leasing or selling coal mines. Vagabonds and beggars could be seized and pressed into servitude. It was not until 1799 that an Act was passed that finally emancipated the Scottish coal miners, and then only with punitive ordinances forbidding 'combination'.[49] Thus it is likely that the prototypes and early installations of Watt and Boulton's patent steam engine were fuelled with coal hewn and hauled by enslaved fellow Britons.

This leads us to consider the definitive material processes of the heat-engine, long screened out of economic theory by the forgotten analogy of the 'free operation of the market mechanism' to the dynamic 'equilibrium' maintained by the steam-governor: the continuous and irreversible combustion of fuel. Heat engines maintain a form of thermal self-regulation analogous to 'homeostasis', which in living organisms relates to their (far more complex) physiological capacity to regulate their internal environment and to maintain a stable, constant condition—life. As open thermodynamic systems, however, both engines and organisms cannot be closed to their environments and continue to 'work', and contra Mayr, certainly

[48] Cited in: Pack, S. (1996). Slavery, Adam Smith's economic vision, and the invisible hand. *History of Economic Ideas*, 4(1/2): 253–269. See p. 258.

[49] Barrowman, J. (1897). Slavery in the coal-mines of Scotland. Paper presented to the Annual General Meeting of the Federated Institution of Mining Engineers.

not 'upon their own resources'. Mammals maintain their body heat at higher average temperatures than the surrounding environment by metabolising energy from large amounts of food (compared with cold-blooded reptiles), and excreting wastes to their environment. Cut its food supply and a wombat will eventually reach thermal equilibrium with its environment, but it will also be dead. Heat engines similarly require constant access to a stream of fresh energy inputs in the form of carbon-based fuels, which are dissipated into waste flows in the form of ambient heat, exhaust particles, and carbon dioxide. Fuel supply exhausted, the engine ceases to work and cools to ambient temperatures.

Engines and organisms are best described, following Prigogine, as 'dissipative structures', in that they operate in *far from equilibrium* states of dynamic order by entropically dissipating flows of available energy from their environment, and returning waste products to the environment.[50] The availability of fuels for thermoindustrial machinery is ultimately a question of socially constructed regimes of access. If in the first instance this comes down to the capacity to pay the market price for refined hydrocarbons, in the final analysis access to depletable fuels requires land appropriation. As old oil and gas fields are pumped out, new ones must be opened. This rarely occurs with the freely chosen consent of local citizens and the legal oversight of open, participatory, deliberative democracy. So great is the capacity of transnational fossil energy corporations to shape the political, legal, and economic conditions of the many national jurisdictions in which they operate that the investigative journalist Steve Coll describes ExxonMobil—an integral funder of the equally transnational Atlas Network—as a vast 'private empire' through which American power is exercised globally.[51] Frequently, hydrocarbon resources are appropriated through the 'piratical economy' or 'primitive accumulation' of neocolonial plunder—witness the violent environments visited by oil companies and their client-states upon the people of the Middle East or the Niger Delta. Far from the neoclassical image of smoothly expanding 'production' driven by voluntary exchange in conditions of general equilibrium, as Hornborg argues, the thermoindustrial economy is more

[50] Prigogine, I., Nicolas, G., & Bablyoantz, A. (1972). Thermodynamics of evolution. *Physics Today*, 23(11 & 12), 22–28 & 38–44.

[51] Coll, S. (2012). *Private empire: ExxonMobil and American power*. Penguin.

accurately characterised as a 'dissipative structure'[52] that maintains and expands its socio-technical infrastructures 'far from equilibrium' by irreversibly depleting imported resources, exporting disorder and pollution, and progressively eroding the regenerative capacity and thermal stability of the biosphere.

[52] Hornborg, A. (2001). *The power of the machine: global inequalities of economy, technology and environment*. California: Altamira Press, pp. 42–48.

CHAPTER 5

The Market as Moral Law: Providence, Starvation, and Liberal Empire

Natural Law Without Land: The World-Historic *Nomos* of British Liberalism

One of the avowed intentions of the founders of neoclassical economics was to develop an account of economic phenomena that neutralised the class polemics associated with classical political economy and its labour theory of value, as popular movements for mass democracy, social justice, and national liberation gathered momentum through the nineteenth century. Leaving aside clear differences in the political context and character of their doctrines, time spent in labour was accepted by Locke, Smith, Ricardo, Mill, and Marx as the real source and measure of value, the basic cost of all forms of production and the 'natural price' underlying the monetary fluctuations in market prices. 'The real price of everything', Smith wrote, 'is the toil and trouble of acquiring it. [...] Labour was the first price, the original purchase-money that was paid for everything.'[1] The Providential mechanism of the 'invisible hand' described by Adam Smith, where free competition and monetary exchange operates as the optimising regulator of the division of labour and the distribution of wealth, was famously extended to international trade by David Ricardo (1772–1823). A stockbroker and loan agent who in 1815 retired young and very rich

[1] Cited in Ricardo, D. ([1817] 2001). *On the principles of political economy and taxation.* Kitchener, ON: Batoche, p. 9.

J. Walker, *More Heat than Life: The Tangled Roots of Ecology, Energy, and Economics*,
https://doi.org/10.1007/978-981-15-3936-7_5

from speculation on government war bonds, Ricardo subsequently turned his hand to book authoring. His *On the Principles of Political Economy and Taxation* (1817) is most famous for attempting to demonstrate theoretically, through a speculative, context-less analysis of trade dynamics in terms of the barter of labour-hours, that unrestricted international trade was to the general advantage of all nations. This was a response to the agricultural protectionism of England's landed oligarchy, manifest in the Corn Laws. As the Napoleonic wars drew to a close in 1815, existing statutes regulating the grain trade were amended to maintain the high wartime prices of grain. Imposing steep duties on imports, this was to the political and financial advantage of the landowning class, but at the cost of high prices for food for common people. The following year, Mount Tambora erupted in Indonesia, filling the upper atmosphere with ash particles which reduced the influx of sunlight to the Earth's surface. 1816 was cold and dark, a 'year without a summer'. This catastrophic climate event caused major reductions in crop yields, leading to widespread famine and bread riots across the Northern hemisphere. In 1818, Ricardo purchased a seat in Parliament, where he would pursue the case for free trade until his death. With the repeal of the Corn Laws by the Parliament of 1846, free trade became the general policy of Great Britain, and support for its doctrine almost by default the definition of what it meant to be an economist.

In this chapter we consider the nineteenth-century famine events in Ireland and India, 'natural' disasters that expose contradictions in the 'natural laws' and 'freedom' of liberal economics, and its opposition to 'government intervention'. In the following chapter I will turn to the moment in which political economy was transformed into economics, as the neoclassical economists of the 1870s claimed to erect an objective science of universal laws analogous to those revealed by physicists. Before the (marginal) revolution, political economy was taught in law departments as an applied form of moral philosophy, in recognition of its provenance in state and corporate administration. After the revolution, economics presented itself as a self-standing discipline distinct from politics, law, and philosophy, a 'social science second to none in rigour and social status'.[2] It was from this vantage point of scientific neutrality that Alfred Marshall would distance his project from those 'harsh employers and politicians, defending exclusive class privileges early in the last century [who] found it

[2] Skousen, M. (2009). *The making of modern economics*. Armonk, NY: M.E. Sharpe, p. 198.

convenient to claim the authority of political economy on their side', and who 'spoke of themselves as "economists"'.[3]

Here, our interest is in a precursory and implicitly theological conception of the *nomos* of liberal economics as simultaneously an expression of natural laws of the market and a providential unfolding of moral progress in history. I will argue that the history of liberal economics has always been a 'Whig history'—and remains so in the selective historical narrations of neoliberal writers. This is a term given to describe a particular genre of Anglophone liberalism, in which the history of the constitutional settlements that tamed Stuart absolutism and gave form to Britain's unrivalled commercial empire are portrayed as the inevitable and triumphal unfolding of a progressive moral and historical destiny. This position was exemplified in Nassau Senior's 1847 lecture as the inaugural Professor of Political Economy at Oxford, in which he described political economy as a 'moral science', a science founded on the notably un-Christian proposition that

> [...] the pursuit of wealth [...] is to the mass of mankind, the greatest source of moral improvement.[4]

From the *a priori* psychological assumption: 'that every person is desirous to obtain, with as little sacrifice as possible, as much as possible of the articles of wealth', Senior's moral science proceeded by axiomatic inference to articulate universal 'laws regulating the production and distribution of wealth, so far as they depend on the action of the human mind'.[5]

Insofar as it drew upon the tradition of British political economy which privileged self-interested labour and the free exchange of property as the origins of the common-wealth, the free-trader's vision that liberalised world trade would naturally result in an optimal international division of labour, increasing the wealth of nations, and securing a perpetual peace, had the effect of disguising an indispensable source of the material success of British 'laissez-faire' liberalism. Namely, the interventionist practice of

[3] Marshall, A. (1890). *Principles of economics*. Book 1, Ch. 4. https://www.econlib.org/library/Marshall/marP.html. Accessed 26 April 2019.

[4] Cited in Skousen (2009, p. 28).

[5] Senior is acknowledged by Austrian economists as Ludwig von Mises' (and thus Hayek's) intellectual precursor. Cited in: Centre for Economic Liberty (2013, 3 August). Nassau Senior, praxeology, and John Stuart Mill. https://centerforeconomicliberty.blogspot.com/2013/08/nassau-senior-praxeology-and-john.html. Accessed 12 May 2019.

land appropriation by organised violence, and the subjection of native peoples to a suite of political technologies for the orderly extraction of land-rent and labour. As sovereign states and corporations extracted wealth through their powers of law, legislation, and repressive policing, imposing the privatisation of lands and rivers, the displacement of subsistence farming communities, and the collection of taxes and land-rents with the force of law, the commercial export economy of the plantation landlords and new industrial capitalists was portrayed by liberal economists as a spontaneous order of natural laws. This vision, only partially secularised, was smuggled into the 'scientific' liberal economics of late nineteenth and early twentieth century, and revived in the contemporary neoliberal faith in the 'growth miracles' of the free market, whose operation is claimed at once to be both benevolent and inexorable. This was a 'moral' science by which the often disastrous effects of the commerce of the British Empire upon commoners and subject peoples would be justified as the historical outworkings of divine Providence, manifest not only in the progressive moral re-ordering of social life by the 'invisible hand' but also in the Malthusian 'natural checks' upon surplus populations of 'natural' disasters compounded by colonial rule—disasters which resulted in millions of deaths to starvation and disease, deaths in no small measure due to systemic plunder and the deprivation of livelihood.

A contrasting perspective to the liberal economists' account of globalisation as the result of the 'natural laws' of society immanent in an ahistorical, pre-juridical state of human nature is offered by the conservative German constitutional theorist Carl Schmitt (1888–1985). Schmitt played a significant public role in precipitating the crisis of the Weimar constitution's parliamentary democracy, advising conservative opponents of the welfare state and the Social Democrats from 1919 until 1933—when, perhaps reluctantly, he offered his services to the Nazi party.[6] Despite this troubling career, Schmitt's post-war history of the establishment of a Eurocentric order of international law through centuries of maritime empire-building—*The Nomos of the Earth* (1950)—has the merit of being scholarly, direct and insightful.[7] Whilst textbooks of economics translate the term from its Greek root-words simply as the laws (*nomos*) of the household (*oikos*), for Schmitt *nomos* refers to 'the fundamental process of apportioning space that is

[6]Scheuerman, W. (1997). The unholy alliance of Carl Schmitt and Freidrich A. Hayek. *Constellations*, 4(2), 172–188.

[7]Schmitt, C. ([1950] 2003). *The nomos of the Earth in the international law of the Jus Publicum Europeaum*. Trans. G. Ulmen. New York: Telos Press.

essential to every historical epoch.'[8] Whether it refers to an 'original distribution of land' forgotten in earliest antiquity, or a 'constitutive act of spatial ordering' that imposes a new imperium after invasion, land confiscation and 'settlement', for Schmitt the genealogy of all law can be traced to land appropriation: the 'primeval act of founding law', from indigenous societies to the liberal international order.[9] Schmitt's intention was to transcend a technical concept of law which defines it merely in terms of the enactments of regulations and statutes by modern states: the state, it might be said, is founded in the estate. Grounding *all* law in the spatial divisions of land and sea established through conquest and treaty, and international law in the history of Europe's territorial expansion, Schmitt's account is antithetical to that of the liberal economists, who sought to depoliticise the concept of law by presenting commerce and exclusive property as the 'natural' state of human relations, separating the intensely geopolitical acts of state sovereignty that established rule over specific territories entirely from the 'natural' economic realm. Whilst there is little to distinguish between Hayek and Schmitt's drastic critiques of parliamentary democracy, the difference in the conservative and liberal perspectives might be indicated in Hegel's words, which Schmitt cites to mark the before and after of the *nomos* of the Earth transformed by the expansiveness of the industrial revolution:

> The principle of family life is dependence on the soil, on firm land, on *terra firma*. Similarly, the natural element for industry, animating its outward movement, is the sea. Since the passion for gain involves risk, industry [...] instead of remaining rooted to the soil and the limited circle of civil life with its pleasures and desires, [...] embraces the element of flux, danger, and destruction.[10]

A Laboratory of Natural Law: The Great Famine of Ireland

As the longest established of English colonial possessions, Ireland might be thought of as a 'laboratory of modernity', in which the political and economic techniques of colonial rule were honed, from the legislative construction of social hierarchies in unequal allocations of property and political rights, through to the counter-insurgency tactics and emergency legislation deployed to repress popular uprisings. From the early invasions

[8] Schmitt (2003, p. 78).
[9] Schmitt (2003, pp. 68 & 45).
[10] Schmitt (2003, p. 49).

of Ulster in the twelfth century, soldiers, loyalists, and investors in military adventures were paid in confiscated farmland, with the 'bog Irish' forced into ever more marginal lands. In the wake of Oliver Cromwell's conquest and the passage of the 1652 Act of Settlement, the economist William Petty was hired to map and survey the estates of Ireland. On the basis of these surveys, Irish and Old English Catholic landholders deemed guilty of rebellion were made to forfeit lands to Scottish and English Protestants. With death by hanging the penalty for non-compliance, the evicted 'transplanters' were ordered to county Clare and the province of Connacht, lands chosen because they were surrounded by the sea, military camps, and garrison towns, all of which the Irish were forbidden to approach. Petty was paid handsomely in confiscated Irish land for his services.

The anti-Catholic Penal Laws enacted upon the Irish between 1691 and 1760 created a society of confessional apartheid, forbidding the Irish from practising their religion, from purchasing land, or from participating in public office. Upon the death of a Catholic landowner, his existing property by law went to his sons in equal shares, unless one of them converted to Anglicanism, in which case the Anglican son received the entire property, along with the right to wrest management from his family.[11] This policy won few conversions, and led to the continual division of Catholic-owned fields into smaller lots. Meanwhile, the English expanded and consolidated their estates.

As agricultural capitalism developed into the nineteenth century, the people of Ireland were further subjected to the imperatives of Protestant landlords to supply wool and grain to English and other overseas markets. The effect of the Corn Laws was to price wheat and barley beyond the reach of Irish tenant farmers, whose work for Anglican landowners was often compensated not in money wages, but merely for the right to feed themselves from food grown on small allotments. Much of the product of Irish soil and labour was exported, and as the population increased, the land available for subsistence agriculture became increasingly scarce. Under such cramped conditions, the potato, recently introduced from America, offered the greatest yield of food for the smallest amount of land.

It was in the context of his advocacy for free trade against the Corn Laws that Ricardo offered his theory of wages and the 'natural price' of labour. His exposition of this doctrine, however, came with caveats which exposes prejudices common to nineteenth-century British economists:

[11] Connolly, S. (1992). *Religion, law, and power: the making of Protestant Ireland 1660–1760*. Oxford: Clarendon Press.

> It is not to be understood that the natural price of labour, estimated even in food and necessaries, is absolutely fixed and constant. [...] It essentially depends on the habits and customs of the people. An English labourer would consider his wages under their natural rate, and too scanty to support a family, if they enabled him to purchase no other food than potatoes, and to live in no better habitation than a mud cabin; yet these moderate demands of nature are often deemed sufficient in countries where 'man's life is cheap', and his wants easily satisfied.[12]

Through land dispossession, the Irish became largely dependent upon a mono-crop of four or five varieties, and the failure of subsequent potato harvests to a fungal infestation led to intensifying famine between 1844 and 1849. The starvation and hazardous exodus suffered by millions of Irish is entirely misrepresented as a natural disaster. Its gruesome effects were caused and amplified by the political technologies of the English estate-holders and rising middle-class, who mythologised the just origins of their own wealth amidst the abject poverty surrounding them by equating social position to a moral hierarchy arranged according to the obscure but just dictates of Providence. In reality this wealth was acquired by the appropriation of land and labour. Behind the 'natural law' of the market was sovereign fiat backed by the King's army, and for political dissenters convicted to transportation under emergency legislation, the utter isolation, forced labour and brutal imprisonment of Van Dieman's Land, an island in Britain's gulag archipelago whose Indigenous peoples were brought to the brink of extermination within thirty years of 'settlement'.

The dour and moralising English bourgeoisie targeted in the social satire of Charles Dickens railed at the negative representation of their activities in the writings of Smith and Ricardo. Adam Smith was a moral philosopher: he disdained the pursuit of wealth for its own sake. Rather than the heroic 'wealth creator' of neoliberalism, the businessman appears as an avaricious bystander in *The Wealth of Nations* (1776). Smith offered criticism, not support, for the corruption and lawlessness of joint-stock companies, then establishing the flag of liberal Empire around the globe. With Ricardo's career as financier, the commercial class took no issue. His political economy, however, lacked their faith in the providential tendency of the emerging market society to produce a Smithian 'natural harmony'

[12] Ricardo, D. (1817). Chapter 5: On wages. *On the principles of political economy and taxation*. https://www.marxists.org/reference/subject/economics/ricardo/tax/ch05.htm. Accessed 13 April 2019.

of interests between the higher and lower social orders. Working from a distributional analysis of the profits of land ownership, for Ricardo there were inherent conflicts of interest between landlords and tenants, capitalists and labourers. Falling wages, increasing rents and rising profits all represented net transfers of wealth from the poor to the rich. As populations expanded, less fertile land would be ploughed, and food prices would rise. Capitalist profits would be reduced by the need to pay higher wages to prevent workers from starving, a dynamic captured in Ricardo's law of diminishing returns. This would ultimately benefit only the landlords, as land prices would increase with the demand for food, and rents with them. Ricardo's analysis confirmed Smith's observation that 'landlords love to reap where they never sowed'.[13]

The now familiar neoliberal ideology—with its praise of the investor, its hostility to redistributive welfare, and its claim that the foundations of the wealth, social harmony, and moral order of the household lie in the natural justice of private property and the free market—does not, according to the historian Gordon Bigelow, come to us directly from classical political economy:

> The group that bridled most against these pessimistic elements of Smith and Ricardo was the evangelicals. These were middle-class reformers who wanted to reshape Protestant doctrine. For them it was unthinkable that capitalism led to class conflict, for that would mean that God had created a world at war with itself. The evangelicals believed in a providential God, one who built a logical and orderly universe, and they saw the new industrial economy as a fulfilment of God's plan. The free market, they believed, was a perfectly designed instrument to reward good Christian behaviour and to punish and humiliate the unrepentant.[14]

The parliament of Evangelicals that enacted the Poor Laws of 1834 advocated the divine ordering of society by the market. The care of the sick, elderly, orphaned, and destitute had previously been organised by local parish churches; the Poor Laws nationalised the problem of poor relief within a bureaucratic system of state administration. Banning the payment of cash relief to the unemployed and the unemployable, the Poor Laws provided no option but enrolment in the mass 'welfare' of the workhouse, warehouses, and asylums for the rejects of industrialism. According to the

[13] Smith, A. ([1776] 1965). *The wealth of nations.* New York: Modern Library, p. 49.
[14] Bigelow, G. (2005). Let there be markets. *Harper's Magazine*, 310(1860), p. 6.

Biblical principle that 'he who does not work, neither shall he eat', the able-bodied who failed to find employment had their utilitarian value as labouring bodies redeemed in return for their gruel by such instruments as the treadmill, a brutal device for performing the labour theory of value. The misery of poverty and indebtedness was thought to be a goad to religious conversion and the reform of poverty stricken individuals, whose sins of idleness and pilfering were frequently castigated by the middle classes.

When the failure of the Irish potato harvest became known to the Tory Prime Minister John Peel, he secretly arranged for the government to purchase and import cheap cornmeal from the United States, to be distributed and sold in Ireland at wholesale prices in order to avert famine. This policy would, however, require the repeal of the Corn Laws. Against opposition from landed elites within and beyond his own party, who warned repeal would destroy the 'territorial constitution' of the British social hierarchy, Peel managed to pass the Bill of Repeal through Parliament into law in 1846, with support from Whig members and other opposition parties. Peel's parliamentary authority crumbled, however, when his Irish Coercion Bill, which sought wider emergency powers to maintain order and protect property holders in response to the crisis, was defeated in the same sitting.[15]

Lord Russell's incoming Whig administration, staffed by evangelicals, promptly halted Peel's famine relief programme on the grounds that it was an 'artificial' intervention in the free market. As shiploads of beef, butter, bacon and wheat left Irish ports under armed guard, Lord Trevelyan, the incoming administrator given responsibility for responding to the famine, denounced the programme as a 'monstrous centralisation' that would 'come into competition with our merchants and upset all their calculations'. The forcing off the land of the unproductive Irish by bailiffs, bankruptcy, and starvation would liberate the countryside for more economically rational (lucrative) uses by its Protestant owners, such as raising cattle and wool for the world market. The depopulation of the countryside would establish a flexible, low-wage urban labour market that would attract investment and trigger the rapid development of industrial manufacturing, in accordance with 'the wholesome operation of the natural laws of society': words the Colonial Secretary Lord Grey chose to celebrate the

[15] Bigelow, G. (2003). *Fiction, famine, and the rise of economics in Victorian Britain and Ireland*. Cambridge: Cambridge University Press, p. 119.

introduction of free trade in 1846.[16] Through the temporary inconvenience of famine, the mysterious workings of Providence would result in growing prosperity in the long run. The predicted boom, however, never materialised. Up to one and a half million Irish perished as a result, with another two million going into exile in the United States. Whilst various schemes of government-supported mass emigration to Canada and Australia were mooted from various quarters, the prospect of an influx of shiploads of pauperised, traumatised, and typhus-ridden Irish refugees was vigorously resisted by colonial legislators in those countries.[17] The complacency of British politicians should not be confused with incompetent indifference. 'Lassez-faire' (or 'classical') liberalism depended upon the sovereign violence of state intervention to maintain the 'natural' legal order of land appropriation, in and through the social catastrophes it wrought. As much was clear to Benjamin Jowett, the respected master of Balliol College. Recalling his Oxford colleague Nassau Senior, 'moral scientist' and economic advisor to Lord Russell's government, Jowett confessed that: 'I have always felt a certain horror of political economists since I heard one of them say that he feared the famine of 1848 in Ireland would not kill more than a million people, and that would scarcely be enough to do much good'.[18]

The Late Victorian Holocaust of British India

In an 1846 speech given to the Anti-Corn Law League on the eve of the movement's triumph, Richard Cobden outlined a prophetic liberal millennialism, a vision which married Seniors' moral science of progress to the determinism of a Newtonian social physics:

> I see in the Free Trade principle that which will act on the moral world as the principle of gravitation in the universe—drawing men together, thrusting aside the antagonisms of race, and creeds and language, and uniting us in the bonds of eternal peace. I have looked even farther. I have speculated, and probably dreamt, in the dim future—aye, a thousand years hence [...] on what the effect of the triumph of this principle may be. I believe the effect will be to change the face of the world, so as to introduce a system of

[16] Gray, P. (1999). 'Shovelling out your paupers': the British state and Irish famine migration 1846–50. *Patterns of Prejudice*, 33(4), 47–65.

[17] Bigelow (2005, p. 9).

[18] Woodham-Smith, C. (1968). *The Great Hunger*. Hamilton: London, p. 373.

> government entirely distinct from that which now prevails. I believe the desire and the motive for large and mighty empires and gigantic armies and great navies [...] will die away [...] when man becomes one family, and freely exchanges the fruits of his labour with his brother Man.[19]

To the contemporary reader, the oddness of this gravitational theory of trade might appear to be as easily dismissed as the promise of Manchester liberalism that 'freedom' will wither away the state—the gigantic armies and great navies are still with us.

Working up imported Misissippi cotton grown by African slave labour, the mechanised, low-wage textile mills of Manchester are still cited as evidence for the enhanced productivity of machine technology and the factory system of labour, emblematic of the orthodox economists' view that the combination of Adam Smith's economics and steam-engine technology led to the continuous 'physical gain' of the industrial revolution, a world of rising prosperity and freedom for all. But as Vandana Shiva reminds us, Manchester only took over India's position as the world's leading manufacturer and exporter of textiles a century after the British East India Company established a monopoly in the Indian textile trade through the systematic suppression of local weavers and merchants, a process which began when the Company built its first manufactory in Bengal in 1653. Having secured a free-trade order from the Mughal emperor in 1717 which allowed them to forcefully assert their rights to trade, the Company began a process of 'primitive accumulation', strong-arming the independent master weavers and the communities dependant on them and attacking Bengali merchant vessels at sea. More and more weavers become dependent upon company agents, who lowered their wages and bound them to contracts that did not permit them to work for others. The Company fixed prices well below those at which the cloth would sell in the public bazaar on a free sale, driving out competitors. In the wake of continual military conquests, the Company acquired monopoly control over the entire Indian weaving industry, driving down wages and increasing the export of commodities and profits overseas. Cheaper prices for calico on British markets, however, threatened to undermine local cloth-makers, who campaigned for high import tariffs upon Indian goods. As English

[19] Cobden, R. (1846, Jan 15). Free trade with all nations. Speech given to the Anti-Corn Law League at Manchester. http://www.cooperativeindividualism.org/cobdenonfreetrade.html. Accessed 13 July 2007.

textiles came to be dominated by industrial manufactories, this pressure increased, and eventually British policy was to export its own cloth and to shut down the Indian workshops.[20]

From this point on, the trade balance (e.g. the net flows of silver and gold currency) between the two countries came to favour Britain for the first time in history. The impoverishment, deskilling, and displacement of the populous Indian weaving caste was complete. As Shiva notes, this moment reflected the more general capture and restructuring of India's prosperous agricultural and artisanal communities, processes that led to its modern status as a 'third world' economy, a market for Western manufactures and a source of cheap labour and primary commodities, an object for 'development' and 'technology transfer'.

In many cases it was Royal-charter corporations such as the British East India Company that, as the corporate vanguard of empire, first subjected conquered nations to foreign domination. Only later would their lands be juridically defined as territories and dominions incorporated into the legal order of the imperial system of colonial government. Privately owned enterprises in competition with those licensed by other European powers, merchant corporations drew vast new territories in Asia, the New World and Africa into the sphere of empire. The sovereign authority of the royal charter made these first advocates of 'foreign direct investment' quasi-governmental entities, licensed to raise private armies, declare war, issue currency, make law, judge, and execute, as they aggressively expanded and enforced granted trade monopolies in tea, textiles, opium, and other exotic commodities. On the Indian subcontinent, the British East India Company established a system of corporate client rule through existing princes, rajahs, nawabs, and zamindars, whose traditional rights to claim tributes and taxes from native populations were signed over to the Company at gunpoint. By 1803, the Company's mercenary army had swelled to more than quarter of a million and much of the subcontinent was under Company rule. Native systems of subsistence farming, land tenure, and labour were overwritten by the private property relations of plantations which grew mass monocrops for export markets, reducing the availability of land for local food production.

Despite the unrestricted looting of portable wealth from temples, palaces, and households which followed new territorial conquests, despite the

[20] Shiva, V. (1997, June). How free is free India? *Resurgence*. www.resurgence.org/resurgence/articles/shiva_jun97.html. Accessed 9 May 2008.

subsequent imposition upon many millions of Indian peasants of higher land-rents than they had endured under the Mughal empire and the Hindu caste-system, and despite significant public subsidy in the form of the military protection of Company shipping and trade routes by the British Navy, the Company was periodically on the verge of insolvency, as senior Company agents built immense personal fortunes carting off treasure, defrauding the Company finances, and provoking costly wars. By such means Bengal, once among the richest provinces of the world, was eventually reduced by the Company to a deathscape of mass starvation, killing *en masse* the taxpayers at the basis of Company profit streams. In response to such crises, a long list of East India Company Acts were passed through the British parliament between 1773 and 1898 to bail out its debts, restructure its stock, appoint government administrators to its board, and reign in its excesses. The forerunner of the contemporary transnational corporation, the British East India Company was deemed too big to fail by sympathetic stock-holding officials, and its deep pockets were able to swing elections. The penultimate crisis for the Company, however, came with a widespread rebellion amongst the native soldiery of the Company's militia in 1857. As British opinion turned against it, the Company enrolled one of its senior executives in a public relations campaign. The liberal philosopher John Stuart Mill, who worked for thirty-five years in the London office, argued that the Company had 'at its own expense, and by the agency of its own civil and military servants, originally acquired for this country its magnificent empire in the East, without the smallest cost to the British Exchequer'.[21] The costs of the magnificent empire had, of course, been borne by others. In 1858 the Crown formally resumed the governmental powers of the Company, establishing the direct rule of British India through a Vice-Regal governor and a professional civil service and military.

The tragedy of the Irish famines would be paralleled in the even more horrific Indian famines of the 1870s and 1890s—which while possibly the most extreme in terms of mass mortality, were by no means the only famines endured during British rule. After the failure of successive monsoons between 1876 and 1878, attributed by contemporary meteorologists to El Nino oscillations, almost sixty million people faced chronic food

[21] Robins, N. (2012). The corporation that changed the world: how the East India Company shaped the modern multinational. *Asian Affairs*, 43(1), 12–26.

shortages.[22] Despite this, the British Raj refused to organise any effective aid or regulate the grain market. The collection of land rents from peasants, which were paid not in cash but in kind, was continued. The produce of Indian farmers continued to leave the ports of Calcutta and Bombay for external markets. As speculators began hoarding food in anticipation of sky-high prices during famine conditions, Vice-Regal administrators refused to provide any form of relief for the destitute but hard labour on public works. Labour camps were often located far from the districts experiencing famine. Those who managed to arrive alive at the camps were already exhausted, and the food on offer did little to restore health and resistance to infectious disease. The minimum diet given in Bengali prisons, one pound of rice per day, was considered adequate for continuous backbreaking labour. While no statistics were compiled for the whole of British India, by 1878 between 3.5 and 4 million people were estimated to have died of starvation and the epidemics that followed—in the Madras Presidency alone.

Famine struck India again between 1895 and 1898. Despite pressure from Queen Victoria and others to forbid hoarding and speculation, to buy as much local and imported food as possible and distribute it, the passionate commitment of the Viceroy Lord Elgin to laissez-faire ideology held firm. He repeatedly argued that state interference in the grain market was an 'extreme measure' which no emergency could ever justify.[23] While the expanding network of railways had added to the mobility of bulk goods transport, perhaps contributing to population growth in the interim years, up to thirty-three million people are thought to have died, a figure probably smaller than that of the previous famine.

The chief commissioner of Gujarat, when pressed upon the failure of the government to prevent thousands of deaths even among those who had access to rations through forced labour, attributed the body count to the endemically weak and work-shy habits of the people. The famine clearly showed that 'the Gujarati' was a 'soft man', unfit in evolutionary terms for survival under the emergent conditions of industrial labour and market exchange.[24] Here we see the influence of Herbert Spencer's evolu-

[22] Davis, M. (2001). *Late Victorian holocausts: El Nino famines and the making of the third world*. London: Verso.

[23] James, L. (1998). *Raj: The making and unmaking of British India*. London: Little, Brown & Co., p. 304.

[24] James (1998, p. 306).

tionary defence of laissez-faire liberalism, widely disseminated through the free-trade journal *The Economist* (est. 1845). For Spencer, the laws of the market were in perfect accord with the natural principles of biological evolution. These were couched in Lamarckian terms, despite his standing as the prototypical Social Darwinist. No government should attempt to contravene them by alleviating the sufferings of the weak and the vulnerable, as this would be to the detriment of healthy individuals and the progressive development of 'the social organism'.

Reports of the famines in British India had caught the attention of William Stanley Jevons, a key figure in the development of the neoclassical price theory. Jevons set to work on a hypothesis that the incidence of famine could be correlated with the recently observed phenomena of sunspot cycles on the surface of the sun.[25] Deprived of long-range data sets and modern climatology, Jevons devoted considerable effort to proving his hunch that periodic fluctuations in the intensity of solar radiance had caused the Indian monsoons to fail. If sunspot events could be shown to radically alter food production, they could be identified as the root cause of financial crises and the booms and busts of the trade cycle, as impacts to agriculture percolated through to other sectors. This speculation persuaded very few even at the time—with one satirist positing a correlation in the results of boat races between Cambridge and Oxford to sunspots. The effect of the theory was to preserve the idea that the 'free market' was a stable, self-equilibrating system that might fail temporarily due to 'exogenous shocks' from 'outside' the system (e.g. climate catastrophes, or the misguided policies of social reformers) but would quickly return to 'equilibrium' of its own accord. This loss of scientific face perhaps tarnished another of Jevons' physical correlations, *The Coal Question* (1865), in which he argued that British pre-eminence relied almost entirely upon the unrepeatable combustion of coal resources.[26] In an 1875 lecture, Jevons offered the following thoughts on famine as a natural limit to populations surplus to the free economy:

> In other (outside England) portions of the world, famines come in as one means of check; and in fact in certain stages of society it is a normal check of

[25] Jevons, W. S. (1876). The periodicity of commercial crises, and its physical explanation. *Journal of the Statistical and Social Inquiry Society of Ireland*, 7, 334.

[26] Mirowski, P. (1984). Macroeconomic instability and the 'natural' economic process in early neoclassical economics. *Journal of Economic History*, 44(2), 345–354.

> population—a famine comes to be looked upon as a kind of natural event [...] Then there is the way people live—the way the Irish live, especially in some of our large towns and in some parts of their own country makes it a priori probable that they die fast.[27]

In Jevons we can see the shift from a Providential conception of the laws of commerce as a moral system of divine origin, to secular appeals to the physical sciences that attributed the suffering of colonised people to natural causes, justifying the progressively civilising mission of liberal Empire. The effect, as far as the victims of land appropriation, famine and rent-extraction were concerned, was the same. The British could be blamed neither for their interventions or non-interventions, for their world-spanning commercial empire merely reflected the operation of impersonal laws of nature imminent in the market mechanism.

As Foucault documents in *The Order of Things* (1966), the late nineteenth century saw the comprehensive systematisation of a plethora of novel sciences, with often unclear boundaries between the human and the natural sciences.[28] While not all of these have stood the test of time, it was in this period that those that are still with us began to adopt forms much resembling our present institutions. English political economists, building the case for free trade, insisted that the 'laws' revealed by their discipline were universal and timeless, As British experience was alleged to have showed, freedom of commerce was the most efficient passage to a world of rising living standards for all. A counter-tradition of nineteenth-century economics was maintained by the German Historical School, which resisted this generalisation of British attitudes on the grounds that it erased the British Empire's historically specific trade advantages given its wide access to colonial land, labour, and resources. Germany was initially well behind Britain in the process of industrialisation, and held comparatively few colonial possessions. German economists, following Friedrich List, recommended that rather than be seduced by the individualist abstractions of Smith and Ricardo, developing countries should follow the actual example of British policy, and build up national industries behind the shelter of high tariffs until they were in a competitive position, with secure

[27] Cited in: O'Boyle, E. (2006). Classical economics and the Great Irish Famine: a study in limits. *Forum for Social Economics*, 35(2), 21–53.

[28] Foucault, M. ([1966] 2005). *The order of things: an archaeology of the human sciences*. London: Routledge.

industries and technocratic government, to benefit from trade and avoid being locked into stratified relations of unequal exchange.

This was a strategy deployed successfully by Meiji Japan, and by the United States, perhaps the most protected national economy in the world between the Civil War and the end of World War II.[29] As the centre of liberal economic orthodoxy shifted from Europe to the United States in the 1940s, and as the United States emerged from World War II as the new military and economic superpower, American economists increasingly embraced the narrative that free trade would benefit all nations. Against the developmental approach to national and international economy that informed the social liberalism of John Maynard Keynes, a new 'classical liberalism' emerged, which revived the free market economics of the nineteenth century. Increasingly, American economists shared with the early British political economists the conviction that there were *a priori* natural laws of economics that operated universally in exchange, and the claims of the neoclassicals of the late nineteenth century to have elevated the social sciences into a methodological perfection on par with the physical sciences. Yet as Melinda Cooper, Jessica Whyte, and Quinn Slobodian have shown, the imperial nineteenth-century discourse of 'moral science' was never abandoned by the neoliberals, who from the 1920s down to the present have portrayed social democratic and postcolonial independence movements as symptoms of a pervasive moral crisis threatening 'freedom' and 'Western civilisation' itself; insofar as these movements had undermined the appropriate attitude of submission to the class and racial hierarchies of the 'true liberalism' that preceded twentieth-century norms of universal civil rights and national self-determination.[30]

This chapter has attempted to place the development of liberal political economy in an appropriate historical, political, and material setting, providing a context for the development of its doctrines ignored by Whig histories seeking the genealogy of 'economic science' in the *thought* of the great economic thinkers, abstracted from their participation in events. In conclusion, one can only concur with the American philosopher Charles Sanders Peirce, who, in reflecting in the closing years of the nineteenth

[29] Chang, H. J. (2002). *Kicking away the ladder: development strategy in historical perspective*. Anthem Press.

[30] Whyte, J. (2019). *The morals of the market: human rights and the rise of neoliberalism*. London: Verso; Slobodian, Q. (2018). *Globalists: the end of empire and the birth of neoliberalism*. Cambridge MA: Harvard University Press; Cooper, M. (2017). *Family values: between neoliberalism and the new social conservatism*. New York: Zone Books.

century on what assessment future historians might make of its philosophical contributions, offered the following prophetic words; which from our present vantage point seem to anticipate both the crisis of the biosphere and our paralysis in confronting a constitutional entrenchment of the neoliberal 'moral science' in the discourse and exercise of government.

> What I say then, is that the great attention paid to economical questions during our century has induced an exaggeration of the beneficial aspects of greed [...] until there has resulted a philosophy which comes unwittingly to this: that greed is the great agent in the elevation of the human race and the evolution of the universe [...] The twentieth century, in its latter half, shall surely see the deluge-tempest burst upon the social order [...] upon a world as deep in ruin as the greed-philosophy has long plunged it [...] Soon a flash and quick peal will shake economists quite out of their complacency, too late.[31]

[31] Peirce, C. S. (1893). Evolutionary love. *The Monist*, 3(2), 176–200.

CHAPTER 6

The Fire Machine: Economics as a Social Physics of Natural Law

The Thermoindustrial Revolution as the Context for the Marginalist Revolution

Economics as we now understand the term emerged from the 'marginalist revolution' of the 1870s, as Carl Menger (Austria), William Stanley Jevons (England) and Leon Walras (France) sought to purge from political economy the irreducibly political questions of labour and land, and to constitute economics as a pure science of 'economic laws' and 'market mechanisms'. In his edifying Whig history of the Austrian and neoclassical economists' triumphs over their Marxist and Keynesian critics, Mark Skousen (MPS), a former chair of the Atlas-affiliated Foundation for Economic Education (FEE), writes of the 'colossal forces of the industrial revolution that catapulted the Western world into a new age of prosperity', and the 'massive power of capitalism as industrial might spread from Britain to Germany to the United States during the nineteenth century'.[1] No mention is made of imperial land appropriations, nor of the vast conflagrations of carbon fuel channelled through industrial pyrotechnology. Nor do we hear of the geopolitical powers of 'fossil capital'—the abiding sponsors of the neoliberal project—to secure access to energy and other strategic resources. The colossal forces and massive powers of industrial capitalism

[1] Skousen, M. (2009). *The making of modern economics: the lives and ideas of the great thinkers*. New York: M.E. Sharpe, p. 171.

J. Walker, *More Heat than Life: The Tangled Roots of Ecology, Energy, and Economics*,
https://doi.org/10.1007/978-981-15-3936-7_6

are rather attributed to the miracles of the market, the innovations of the entrepreneur, and the 'discoveries' of 'good economists'.[2]

As Mirowski demonstrates in *More Heat than Light* (1989), the history of neoclassical economics and the 'marginalist revolution' cannot be told in abstraction from the parallel development of the physics of energy in the later nineteenth century.[3] Mirowski's method and aim in this work is the deconstruction of the mathematical edifice of neoclassical economics, exposing its dubious attempts to appropriate conservation principles from statistical mechanics and energy physics. As important as this critical exposition of economists claims to have instantiated a social physics is, the book likewise contains little discussion of fossil fire as the central fact of steam power and the energy revolution. We have a different furrow to plough: the hegemony of neoliberal economic thought in the thermal crisis of the Anthropocene.

Drawing selectively on Mirowski, I will focus on an aspect of the narrative rarely considered by historians of neoclassical economics: the development of the combustion engine, which affected economic theory in two ways. First, it catalysed a grand synthesis of physical science within universal and confirmed laws of energy and entropy, an achievement which economists in their physics-envy aspired to emulate. Secondly, the ideology of Progress was transformed, from the Enlightenment faith in the progressive realisation of reason in social history, into the productivist eschatology of liberation through infinite industrial expansion, later banalised in the ideology of GDP. Both the promissory utopianism of growth economics and the theoretical justification of its laws of general equilibrium can be traced to the vast energies released with the advent of hydrocarbon-fuelled heat engines, and the physics of heat and power which deduced universal laws of nature from their operations. And yet the obvious problem that the expansion of industrial economies is achieved by the irreversible dissipation of fossil energy (heat) and natural ecosystems (life) has long been externalised from the narrative arc of Western economic history. The patent fact that the 'industrial organism' has an appetite for combustion is as absent as the actual content of physics is from economic theory.[4]

[2] Skousen (2009, p. 468).

[3] Mirowski, P. (1989). *More heat than light: economics as social physics, physics as nature's economics.* Cambridge: Cambridge University Press.

[4] Georgescu-Roegen, N. (1971). *The entropy law and the economic process.* Cambridge: Cambridge University Press.

The 'clockwork metaphysics' of early modernity were altered beyond recognition by the energy physics of heat and work developed in the nineteenth century, as the industrial revolution gained pace and solar-based sources of mechanical power—wind, water, firewood, and the labour of humans and beasts—were complemented and then displaced by the superior forces realised by steam engines burning solid hydrocarbon fuel. Thomas Savery is widely credited with building the first working heat-engine, a water-pump utilising the pressure of steam in a chamber to drive a piston and rod, patented in 1698. Savery attempted to capitalise on his patent, marketing it to mine operators in a prospectus entitled *The Miner's Friend; or, A Description of an Engine to Raise Water by Fire*. Sales, however, were few. The engine was highly inefficient and prone to explosions. In 1702, Thomas Newcomen developed the first genuinely practical steam engine, five times more powerful than its predecessor.[5] Newcomen's Atmospheric Engine was installed over the pit of a flooded coal mine to pump the water out. Steam-engine technology only really came into its own with James Watt. Invited by Adam Smith and Joseph Black to repair scientific instruments for the University of Glasgow, Watt tinkered with the resident Newcomen engine for years before marketing the radically superior Watt and Boulton engine in the revolutionary year of 1776.

Before the steam engine, mining took place in mountainous regions, or in shallow pits. As soon as the water table was reached, the pit had to be abandoned. Previously, the work of pumping out flooded mines was driven by horses, which allowed a direct comparison to be made between the work output of the machine and the animal. Since then the power of engines has been measured in 'horsepower' or metric equivalent. With the coal-fired water pump, deeper mineshafts could now be operated, and deeper still when lifts were attached to steam engines to winch miners and coal between carboniferous strata and the surface. This drastically expanded the 'vertical power' of mining operations. In expanding the scope of coal extraction, the heat engine inaugurated the exponential expansion of technomass, which despite the exclusive focus of macroeconomists on monetary aggregates is the material reality of 'economic growth'. To this moment we may also attribute the beginning of the end of the natural carbon cycle and the ominous gain in the temperature of history.

It is little remembered that the harnessing of coal fire in mining, machinery, and metallurgy was a response to a Europe-wide energy crisis

[5] Niele, F. (2005). *Energy: engine of evolution*. Amsterdam: Elsevier, p. 66.

of increasing severity—the serious and widespread shortage of wood as the great European forests that once extended unbroken from Caledonia to Russia were cleared for agriculture, construction timber and fuel. By the early eighteenth century the depletion of forests had raised the price of wood so high as to seriously retard the output of numerous industries dependent on heat: salt, glass, ceramics, soap, potash, tin, lead, iron, and other metals. Nearly all of Europe's forests within range of water transport were depleted by this time. Peasants in southern France were simply unable to heat their hovels in winter. Radical deforestation occurred in the British countryside, the result of the incredible combined appetite of all industries, but especially that of the navy's shipbuilders, and of the burgeoning ironworks developing rapidly in England. Increasingly, even the wealthiest English resorted to sulphurous coal to heat their homes, widely regarded as an inferior fuel.[6]

Due to its high sulphur content and other impurities, coal was initially unsuited to many production processes, and especially iron-making. Technical improvements in the coking of coal—cooking it in airless ovens at high temperatures to raise carbon content and remove water and other impurities—went hand in hand with the development of iron technology. Pig iron was replaced by wrought iron, and later, by the Bessemer process of rolled steel milling. Higher quality iron allowed the precise, thin castings needed for the cylinders of engines. As Niele writes: '[…] advanced coal-fired cooking in ironworks made production of Newcomen's steam engine possible, which in turn enabled deep shaft mining, thus boosting coal supply. This way the coal-based societal metabolism evolved through industrial symbiosis, or cooperative relations between several industrial processes.'[7]

The Manchester textile industries usually presented as the cutting edge of mechanisation and the factory system of labour were among the last industries to employ steam power. These industries emerged from the solar-based plantation capitalism that world systems theorists refer to as 'the triangular trade'. Mississippi cotton, grown with mass slave labour, was imported to England on wooden wind-powered ships, and processed by factory workers with wooden spinning and weaving machines which were powered by water wheels well into the nineteenth century. Economic historians for more than a century have portrayed steam-engine *technology*

[6] Ponting, C. (1971). *A green history of the world*. London: Penguin, pp. 277–282.
[7] Niele (2005, p. 66).

(e.g. the design specifications of the machines) as the 'pivot on which industry swung into the modern age' and as the 'central fact of the industrial revolution'. It is far more accurate to speak of the 'steam engine complex' as a whole.[8] Coal mines were connected to iron-works that produced locomotives, rolling stock and steamships, as well as the infrastructures such as canals, railways and bridges that connected them to dockyards and ports. Together with the mass production of the precision high-powered firearms that for a hundred years gave Europeans a nearly complete military advantage elsewhere in the world, this complex of fire-power gave Imperial Britain and other European powers access to the vast, 'untapped' resource frontiers of the New World.

In 1840, Britain imported no more than 5% of its food, however by 1900, 80% of its grain, 70% of its dairy products and 40% of its meat was imported. Importing food and raw commodities in bulk, and exporting factory goods along with surplus populations of the landless poor, Britain displaced its Malthusian crisis onto the native peoples of Australia, North America and other peripheral regions converted to its supply hinterlands, allowing the majority of England's land and labour to be devoted to industrial manufacture. The radical improvement in coal, iron and steel-based industrial technology quickly enlarged what we would might now call Britain's 'ecological footprint'. These technologies enabled greater economies of scale by increasing Britain's military and commercial dominance of world trade, as royal-chartered joint-stock corporations could ship ever greater quantities of bulk goods from ever more remote locations. As local resources are consumed first, 'economic growth' and 'globalisation'—usually treated separately, can be seen as the same self-reinforcing process arising first from local depletion and finally from the need to secure access to increasingly remote frontiers of resource appropriation.[9]

As Sieferle observed as recently as 1982, 'considering the central importance of coal as the energy basis of the Industrial Revolution, it is quite astonishing that [...] it has been almost completely ignored by economic history'.[10] The vast geological energy input to Britain's commercial-

[8] Niele (2005, p. 66).

[9] Bunker. S. & Ciccantell, P. (2005). *Globalization and the race for resources.* Baltimore: John Hopkins University Press.

[10] Sieferle, R. ([1982] 2000). *The subterranean forest: energy systems and the industrial revolution.* White Horse Press.

industrial-military complex remains by and large invisible to economists, who portray the wealth concentrated in late-nineteenth-century Britain as a result of the acceptance by the Parliament of 1846 of Ricardo's argument that easing restrictions on the import of grains would stimulate industry in Britain, allowing it to specialise in manufactures. And so down to this late hour, the global 'freedom' of commerce is justified *in itself* as the font of 'innovation' and 'growth' a boon to all humanity.

The Carnotian Revolution: Labour Power, Arbeitskraft, and the Conservation of Energy

From the late seventeenth to the mid-nineteenth century, Newton's physics of motion was the crowning jewel of physical science. The Newtonian system—with its smooth, continuous mechanics and its temperature independent and time-reversible geometry of inertia, momentum and velocity—was subsumed within the more general science of energy and entropy which coalesced from the 1840s to the 1870s. Synthesizing previously disparate theories of light, heat, force, motion, pressure, chemical affinity, sound, electro-magnetism, physiology, and the atomic structure of matter within the concepts of energy and entropy, thermodynamics was hailed as the defining achievement of scientific materialism.[11] Thermodynamics returned physics from the orbital motion of the solar system to the Earth, to the resolutely historical and intimately phenomenal world of heat and cold, water and air, food and photosynthesis, fire and pyrotechnology. It remains the only branch of physics in which life matters.

If a full genealogy of thermodynamics is beyond the scope of this inquiry, for our purposes it is noteworthy that French engineering schools generated some of its signal breakthroughs. It was this *milieu* that later produced the central figure of the neoclassical pantheon, Leon Walras, who first formulated economics as an axiomatic science of the market as an abstract system in general equilibrium. The marginalist 'subjective utility' theory of value was provided by Austrian economists and British utilitarians, who followed Nassau Senior in articulating an *a priori* science of individual greed, but went beyond the 'moral science' of political economy in claiming to have discovered universal economic laws on par with those of the physicists. In what follows, I first attempt a select history of

[11] Until it in turn was subsumed within quantum mechanics and relativity theory around the turn of the century, a development well beyond the scope of this book.

thermodynamics, followed by an account of the 'social energetics' movements inspired by its popularisers, before returning to the neoclassical economists.

In 1803, the French engineer O'Reilly called for a general theory of *applied* mechanics:

> Nearly all modern authors who have treated mathematics or physics have presented us with systems of mechanics; and nevertheless these systems very rarely advance beyond the principles of equilibrium. These principles are indispensable for the knowledge and the theory of machines; but yet they are far from furnishing us with practical notions on working machines. They are never in equilibrium. The machine must move [marcher] in order to work.[12]

Seeking an academy devoted to 'machine theory' to bridge the gap between 'pure' analytical mechanics and the applied industrial arts, Gaspard Monge, a founder of the École Polytechnique, wrote that 'one of the most useful applications of descriptive geometry [...] is the effective description of the forms and of the construction of the basic parts of machines', such as water wheels, pumps, axles, brakes and other devices for transmitting moving forces.[13]

Expanding on the overlapping problems addressed in Adam Smith's labour theory of value and James Watt's combustion engineering—the question of the optimal price efficiency of work in terms of maximum output—the physico-mathematical conception of work that entered physics through the French engineering tradition was directly economic. As Navier, a graduate of the military engineering *écoles* put it: 'the comparison of diverse machines, for the merchant and the money-lender [capitaliste], comes naturally after the quantity of work which they execute, and the price of this work'.[14]

Another military engineer of the École Polytechnique, the brilliant Nicholas Sadi Carnot, took up the study of steam engines, suspecting that it was the superior efficiency of the recently mechanised British coalmining industry and its greater capacity to produce coke and iron for armaments that had led to the defeat of Napoleon's revolutionary forces. Inspired by the relative inefficiency of the French war machine, Carnot's research into

[12] Grattan-Guiness, I. (1984). Work for the workers. *Annals of Science* (41), 1–33. See p. 8.
[13] Grattan-Guiness (1984, p. 28).
[14] Grattan-Guiness (1984, p. 8).

the operations of what he described as the *machine à feu*—literally, the 'fire machine'—was to have profound ramifications for physical theory. In his *Reflections On the Motive Power of Fire* (1825), Carnot identified the fundamental process by which the energy of heat is converted to kinetic energy, or 'work'. This slender volume, ignored until after his death, initiated the thermodynamic revolution in physics, with all its profound implications for our contemporary existential predicament on a heating planet. Here I follow Jacques Grinevald's historical trajectory of 'the Carnotian revolution', which was realised for the Earth sciences in Vernadsky's solar-energetic account of *The Biosphere* (1926), in the social sciences by Georgescu-Roegen in the 1970s, and in the political sphere in the widespread confirmation of the anthropogenic greenhouse effect in the early 1980s.[15]

Carnot introduced his epochal work with these lines:

> Everyone knows that heat can produce motion. That it possesses vast motive-power none can doubt, in these days when the steam-engine is everywhere so well known. To heat also are due the vast movements which take place on the earth. It causes the agitations of the atmosphere, the ascension of clouds, the fall of rain and of meteors, the currents of water which channel the surface of the globe, and of which man has thus far employed but a small portion. Even earthquakes and volcanic eruptions are the result of heat. From this immense reservoir we may draw the moving force necessary for our purposes. Nature, in providing us with combustibles on all sides, has given us the power to produce, at all times and in all places, heat and the impelling power which is the result of it.[16]

Carnot was the first to identify the thermodynamic principle, formalised later as the 2nd law, that in the final analysis the work of the engine was solely attributable to the universal tendency of heat to flow irreversibly towards a cold sink. The 'moving power' of the engine was solely due to the temperature differential, or thermal gradient, between the hot and cold parts of the engine. Carnot described this as the 'fall of caloric', in direct analogy to the potential energy of water in a lake becoming the kinetic energy of a waterfall that could drive a mill wheel. Carnot also

[15] Grinevald, J. (1992). De Carnot à Gaïa: histoire de l'effet de serre. *La Recherche*, 23(243), 532–538.

[16] Carnot, N.L.S. ([1825] 1897). *Reflections on the motive power of heat*. New York: Wiley & Sons.

demonstrated that heat engines never gave as good as they got: most of the heat from coal fire was 'wasted' to the cooler environment of the engine. Engineers could never achieve anything approaching perfect efficiency in the conversion of fuel to 'work': the steam engine wasted perhaps 90% of the energy of heat. If classical machine theory begins with the assumption of time and temperature independent reversible motion, Carnot's insights laid the groundwork for the rational analysis of irreversible systems, informing the work of engineers such as Rudolf Diesel in developing the internal combustion engine.[17]

In Carnot's time, fire was thought to be the visible manifestation of an invisible fluid, a weightless substance named 'caloric' by the chemist Antoine Lavoisier. Reporting on the series of chemical experiments involving combustion which had isolated the basic constituent elements of air, biomass and water (N, H, O, C) and established the method of modern chemistry, Lavoisier disposed of 'phlogiston', the older substance theory of heat, by demonstrating that the heat involved in a chemical reaction had no effect on the mass of the material before or after the reaction. He proposed 'caloric' as an alternative term encompassing heat, light, and fire. This lead to his statement of the Law of the Conservation of Matter in his *Trait'e El'ementaire de Chimie* (1789):

> Nothing is created, either in the operations of art or in those of nature, and it may be considered as a general principle that in every operation there exists an equal quantity of matter before and after the operation; that the quality and quantity of the constituents is the same, and that what happens is only changes, modifications.[18]

It was the development of the concept of energy, described in the 1st law of thermodynamics—the law of the conservation of energy—which would lead to the full realisation of the profound importance of Carnot's analysis of the relationship of heat to motion in the form of the 2nd law, the result of precursory work across many enquiries that led to the elaboration of the energy concept.

Aristotle wrote of the equilibrium of forces in an analysis of balances and levers, and used the term *energeia*, vaguely translatable to 'work' to

[17] Diesel, R. (1893). *Theory and construction of a rational heat-engine to replace the steam engine and combustion engines known today*. Berlin: Springer-Verlag.

[18] Lavoisier, A. ([1789] 1965). *Elements of chemistry*. Courier Corporation.

describe them. Comparable concepts of conserved velocity and force were developed in the mechanics of Galileo, Newton, Bernoulli, and Leibniz. In the words of Reiner Kümmel, the 'decisive steps that extended the law of energy conservation beyond the realm of mechanics, and revealed energy as the fundamental constituency of the world besides matter', were taken by the physician Julius Robert Meyer, the private scientist James Prescott Joule, and the physicist William Thomson (Lord Kelvin) in the 1840s.[19] Analysing phenomena as varied as the role of the heat released by respiration in the circulation of the blood, the raising of the temperature of water by intensely shaking it, and the heating of wires carrying electric current, they established experimentally that mechanical work could be converted into heat, and vice versa, at the same mathematically constant 'exchange rate'.

The Scottish engineer William Rankine, a key contributor to the theory of the steam engine, provided the formal definition of mechanical energy in 1845. But it was the German physiologist Herman von Helmholtz who is credited with the first precise formulation of the first law of thermodynamics. Building ambitiously on Rankine in his *Über die Erhaltung der Kraft* (*On the Conservation of Force*, 1847), Helmholtz posited a relationship between heat, light, chemical affinity, mechanical force, electro-magnetism and bio-physical phenomena by treating them all as manifestations of a continuous field pervading the entire universe, which he named *Arbeitskraft*: 'labour power'. Rigorously combining several previously separate areas of inquiry under the doctrine of energy conservation, Helmholtz united the search for the mechanical equivalence of heat with rational mechanics, doing away with earlier substance theories of heat and the vitalist notions of *vis viva* or Lebenskraft ('living force') that had been central concepts of physiology ('animal economy') and the Continental *Naturphilosophie* tradition. Initially posited as the Law of the Conservation of Force, the older language of forces would be subsumed in the greater generality of the energy concept (in which the action of 'force' is merely one of the possible phenomena attending the transformation of energy) after a 1853 paper by Rankine: '[T]he law of the *Conservation of Energy* is

[19] Kümmel, R. (2011). *The second law of economics*. Dordrecht: Springer Science+Business, pp. 30–33.

already known—viz. that the sum of all the energies of the universe, actual and potential, is unchangeable.'[20]

Mirowski defines a conservation principle as '[the] rule that some particular aspect of a phenomenon remains unaltered or invariant while the greater phenomena undergoes certain specified transformations.'[21] The law of the conservation of energy was an attempt to unify all the previously disparate conservation principles in natural science. Helmholtz and his peers showed most rigorously that the quantity of energy was fixed: energy could not be created and could not be destroyed, but was constant through all transformations. Heat, matter and motion were unified in the concept of energy and subjected to a system of mathematical equivalents. Nature, it would seem, balanced her own accounts and could not be cheated.

What then, is energy? As Richard Feynman noted in his lectures to undergraduates, physicists don't know what energy *is*: rather its existence is reliably inferred from its effects:

> There is a fact, or if you wish, a law governing all natural phenomena that are known to date. There is no known exception to this law—it is exact so far as we know. The law is called the conservation of energy. It states that there is a certain quantity, which we call 'energy,' that does not change in the manifold changes that nature undergoes. That is a most abstract idea, because it is a mathematical principle; it says there is a numerical quantity, which does not change when something happens. It is not a description of a mechanism, or anything concrete; it is a strange fact that when we calculate some number and when we finish watching nature go through her tricks and calculate the number again, it is the same. [...] It is important to realize that in physics today, we have no knowledge of what energy 'is.' We do not have a picture that energy comes in little blobs of a definite amount. [...] It is an abstract thing in that it does not tell us the mechanism or the reason for the various formulas.[22]

Energy physics renovated the machine metaphors of Western cosmology for the age of the combustion engine, with its spectacular harnessing of seemingly infinite geological powers. Nature as revealed by thermodynamics profoundly transformed the European cultural imagination, and

[20] Rankine, W. (1853). On the general law of the transformation of energy. *Proceedings of the Philosophical Society of Glasgow*, 3(5), 276–280.

[21] Mirowski (1989, p. 13).

[22] Feynman, R., Leighton, R., & Sands, M. (1989). *The Feynman lectures on physics*. Addison Wesley. See Ch. 4.1.

reshaped the entire social and natural environment in which we live and work. Contemplating Turner's paintings, Michel Serres writes: 'Matter is no longer left in the prison of a diagram. Fire dissolves it, makes it vibrate, tremble, oscillate, makes it explode into clouds.'[23] The order of the material world appeared ever more transient and ephemeral as energy underwent its myriad temporary transformations, just as long-abiding textures of social life were rent asunder by the relentless dynamism of thermoindustrial capitalism. As Marx and Engels wrote in the 1848 Communist Manifesto: '[A]ll that is solid melts into air.' Beyond all these protean transformations, only energy itself remained constant and universal.

Social Energetics and the Sciences of Work

The attempt to unify the natural sciences around the concept of energy had its corollary in the social sciences, manifest in a variety of 'social energetics' movements. This was the intellectual milieu from which neoclassical economics arose, so some background context will serve our aims. In *The Human Motor* (1992), Anson Rabinbach offers a fascinating genealogy of the ideology of productivism—now manifest in the contemporary commitment to infinite growth in and through crisis—in the social energetics movements that that shadowed the development of energy physics from the 1840s through to the early twentieth century.

> The discovery of labour power—and its subsequent elaboration in political economy, medicine, physiology, psychology and politics—was emblematic of a society that idealised the *endless productivity* of nature [my emphasis]. The Promethean power of industry (cosmic, technical and human) could be encompassed in a single productivist metaphysic in the concept of energy, united with matter, was the basis of all reality and the source of all productive power—a materialist idealism, or as I prefer to call it, *transcendental materialism* [emphasis original]. The language of labour power was more than a new way of representing work: it was a totalising framework that subordinated all social activities to production, raising the human project of labour to a universal attribute of nature. [...] The body, the steam engine, and the cosmos were connected by a single unbroken chain of energy.[24]

[23] Cited in: Rabinbach, A. (1992). *The human motor: energy, fatigue, and the origins of modernity*. University of California Press, p. 63.

[24] Rabinbach (1992, p. 52).

In his many public lectures on the subject of energy, of which he was perhaps the most important populariser, Helmholtz was fond of referring to the El Dorado of eighteenth-century inventors. The Cartesian-Newtonian image of the cosmos as a frictionless clock with a motion that never needed to be wound was parlayed back into speculation about the development of machine technology. Many engineering and scientific enthusiasts of this period, as Helmholtz noted in an 1861 address to the Royal Institution, were inspired by the dream of a 'machine which would give perpetual motion and produce any mechanical work they liked. They called this machine a perpetual mover. They thought they had an example of such a machine in the body of every animal.'[25] Helmholtz explained this dream, and its death, in economic terms:

> The solution of this problem promised enormous gains. Such a machine would have had all the advantages of steam without requiring the expenditure of fuel. Work is wealth. A machine which could produce work from nothing was as good as one which made gold. [...] That a perpetual motion could not be produced by the aid of the then known mechanical forces could be demonstrated in the last century by the aid of the mathematical mechanics which had at that time been developed. But to show also that it is not possible even if heat, chemical forces, electricity, and magnetism were made to co-operate, could not be done without a knowledge of our law in all its generality. The possibility of a perpetual motion was first finally negatived by the law of the conservation of force, and this law might also be expressed in the practical form that no perpetual motion is possible, that force cannot be produced from nothing; something must be consumed.[26]

In this lecture Helmholtz addressed the problems of physiology from the perspective of the conservation of energy, arguing that 'it was extremely probable that the law of the conservation of force holds good for living bodies' by way of an analogy between labouring humans and steam engines. Physicists frequently underplayed the important precursory discoveries of chemists. This parallel was demonstrated much earlier, in 1789, by Lavoisier:

[25] Cited in: Raitt, S. (2003). Psychic waste: Freud, Fechner and the principle of constancy. In G. Hawkins & S. Muecke (Eds.), *Culture and waste: the creation and destruction of value.* Lanham, MA: Roman and Littlefield, p. 75.

[26] von Helmholtz, H. (1863). On the conservation of force. Introduction to a series of lectures delivered at Carlsruhe in the winter of 1862–1863. https://www.bartleby.com/30/125.html. Accessed 5 May 2019.

> [...] in general, respiration is nothing but a slow combustion of carbon and hydrogen, which is entirely similar to that which occurs in a lighted lamp or candle [...] from this point of view, animals that respire are true combustible bodies that burn and consume themselves.[27]

Animal bodies and pyrotechnologies produced work from heat while dissipating high quality inputs into valueless waste. Helmholtz referred to the physiological research of Edward Smith, who studied work efficiency, diet, and stress amongst convicts forced to work treadmills in the much-feared Coldbath Fields Prison. The British penal code of 1775 defined 'hard labour' as 'labour of the hardest and most servile kind [...] such as treading in a wheel, or drawing in a capstern for turning a mill or other machine'. In most cases the treadmill was used purely for the punishment of inmates, its rotating axle unconnected to any industrial application—a miserable pedagogical device for communicating the labour theory of value.[28] Smith's studies were taken up by reformers calling for the abolition of treadmills and improvements of diet in prisons and workhouses. Helmholtz, by contrast, praised the new 'sciences of work' for showing that 'the best method of getting the greatest amount of work out of a human being is the treadmill'. Equivalent to continuously walking uphill, Smith's studies had shown that one quarter of the effort resulted in mechanical work, while the rest was lost as heat. This was more efficient than the conversion ratios achieved by steam engines, and so for Helmholtz '[T]he human body is [...] a better engine than the steam engine, only its food is more expensive than the fuel of steam engines.'

In all his writings on *Arbeitskraft*, Helmholtz made no distinction between the work of nature and the work of machines, or between the mental labour of violinists and the muscular work of labourers. Nature was indifferent to the social uses to which energy-conversion was put. Labour power was an objective and quantifiable economy of force distinct from any cultural, social or biological setting. 'The sum of all forces capable of work in the totality of nature remain eternal and unchanged throughout all variations. All change in nature amounts to this, that the labour power can change its form and locality without its quantity being changed.'[29]

[27] Cited in: Lehninger, A. Nelson, D. & Cox, M. (2005). *Lehninger principles of biochemistry*. Macmillan.

[28] Chapman, C. (1967). Edward Smith (1818–1874) physiologist, human ecologist, reformer. *Journal of the History of Medicine and Allied Sciences*, 22(1), 1–26.

[29] Rabinbach (1992, p. 61).

Here Helmholtz did not 'reduce' the organism to a thermodynamic motor, but posited their formal equivalence in the physical terms of energy conversion. In so doing, the heat engine was metaphorically transposed to the entire universe, its essence anthropomorphised as 'labour power'.

The unity of the energetic principles governing the physiology of organisms, motors and the cosmos were popularised by Helmholtz, the industrial chemistry magnate Ernst Solvay, and the Darwinian biologist Ernst Haeckel, and parlayed into a proliferation of publications on the significance of labour power and the vital importance of economising upon its use, often with detailed programmes for social reform. The defining feature of the various social energetics movements was the realisation that because energy could be wasted but not created, its direction into useful activity was paramount. Idleness, indolence, sloth, non-reproductive sex, begging, nomadism, theft, the arts, aristocratic privilege, gambling, speculation, and other forms of excessive expenditure were railed against from various quarters as the misdirection of energy into unproductive uses, a reduction in the efficient expenditure of society's energy budget, and a brake on the optimum output of industrial society.

As the principles of energy conservation were extended to the physiology of the working body on the micro-level, and extended to the macro-level of the 'health of the nation' in doctrines of national economy, the efficient unity of worker and machine became the explicit goal of the popularisers of 'productivism'. For socialist reformers campaigning for a living wage and limits on the working day, the question of optimising labour meant the reduction of exploitation. Statistical studies were compiled showing that those nations whose factories operated the longest working days and paid the lowest wages were the least productive. Deprived of a high quality diet and rest, their workers were poorly fuelled and excessively fatigued. Since the happiness of workers was co-extensive with the most rational use of labour power, higher productivity was compatible with social justice. Such studies culminated in controversial and radical French social legislation in 1903, which enforced one day per week of closure upon factories to allow workers to rest and restore their energies. Conversely, conservative liberals and industrialists accused labour activists of disrupting the efficient harmonisation of workers and machines, sapping the energies of the nation in wasteful strikes and the propagation of class conflict. Finally, both authoritarian conservatives and revolutionary socialists criticised the liberalism of American industry for its pointless inter-factory competition that wastefully diminished the potential of the productive forces through amplifying speculative cycles of boom and bust.[30]

In the early-twentieth-century United States, the dominant approaches to the ergonomics of industrial organisation are known to us as Taylorism; in every plant an engineer was trained as a 'time-and-motion' man, subordinating the movements of the worker to a rational calculus of profitability and a tight ergonomic discipline such that the speed of the production line could be intensified. Prior to the Russian revolution, Lenin criticised Frederic Taylor's system as a 'science of sweating', emblematic of 'man's enslavement by the machine'.[31] Once the Bolsheviks had seized power, American methods, advisors and engineers were imported wholesale in the massive industrialisation drives of the 1920s and 1930s.

A Marginally Energetic Revolution

The question of labour became central to economics with Adam Smith, who held that the value of the total product of a national economy was derived from a national fund of aggregate labour time. Smith's concept of labour value was inherited from Locke and the Scottish Enlightenment, in which labour was an explicitly social activity and a philosophical category of 'moral science'. On the continent, under the influence of Hegel (whose dialectics were influenced by Smith's famous discussion of the division of labour in a pin factory) and the German historical school, labour was understood inter-subjectively, as the essential human activity through which arose self-consciousness and self-recognition as intentional, social beings. Under the sway of energetics and the sciences of work, labour power was abstracted from social individuals and objectified as a quantitative input to the production process.

> By the last quarter of the nineteenth century, the social implications of the doctrine of energy were becoming apparent. In national economy, as in the work of Marx, the body was the site where the natural force of labour power was converted into energy to power the industrial dynamo. [...] Political economy was thus harmonised with the doctrine of protean energy. Energy

[30] Rabinbach (1992, p. 59).

[31] Lenin, V. (1913, Mar 13). A 'scientific' system of sweating. *Pravda*; Lenin, V. (1914, Mar 13). The Taylor system—man's enslavement by the machine. *Put Pravdy* (35).

is the universal equivalent of the natural world, as money is the universal equivalent of the world of exchange.[32]

In the earlier writings of Marx, the Hegelian theme of sensuous, social labour is evident. By the time of writing of *Capital* (1867–1894), Marx had become acquainted with Helmholtzian energetics, and now distinguished between social labour and 'labour power' (*Arbeitskraft*), the latter of which the work of humans and heat engines were but special cases. From this distinction, Marx argued that the capitalist division of labour was dehumanising: in the drive to accumulate capital by extracting surplus labour from the working class, capitalists subordinated the living to the functional requirements and soulless regularity of industrial machinery, transforming the worker into the 'automatic motor of a detail operation'.[33] Transcending the exploitative class relations of capitalism meant the redemption of the social value of labour, this would come by the progressive liberation of workers *by* machine power from the burdens of physical exhaustion and mindless, repetitive, soul-destroying work. In the mature Marx, the inevitability of a more scientific, humane, and just modernity can be detected in the teleological roar of the forces of production exploding into the future.

The social discourses of nineteenth-century scientific materialism would develop a mirrored and seemingly contradictory impetus; on the one hand, 'the insistence on the materiality or physicality of persons, representations and actions' and on the other hand 'the insistent abstraction of persons, bodies, and motions to models, numbers, charts, and diagrammatic representations'.[34] One focuses on behaviour abstracted into statistical ensembles of data, and the other on the material processes of energy consumption and dissipation. The former illustrates the construction of bodies and populations as probability distributions while the latter represents them as material objects. These different facets of what counted as scientific materialism under the reign of classical energetics were parlayed into economics. If the physiological measurement of the working body's metabolism and the Taylorist subjection of workers to production-line

[32] Rabinbach (1992, p. 61).

[33] Marx, K. & Engels, F. (1996). *Collected works: Karl Marx: Capital* (vol. 2), International Publishers, p. 469.

[34] Hayles, K. (1999). *How we became posthuman*. Chicago: University of Chicago Press, p. 99.

efficiency represents the empirical tendency of attempts to devise a social science consistent with energetics, it was in the neoclassical approach to economics that we see the alternative tendency, the insistent drive towards mathematical modelling and abstraction from the material world.

With the early neoclassicals, a bid was made to create a decisively scientific economics, a 'pure' economics of market mechanics and economic laws, devoid of politics, rhetoric, metaphor and moral suasion. As Mirowski has shown, the neoclassical model of 'the economy' directly appropriated its mathematics, models and metaphors from rational mechanics and 'proto-energetic' of versions of energy physics, prior to its completion with the statement of the entropy law in the 1860s. For our purposes, what was fundamentally new about the neoclassical approach was its specific and, it must be said, peculiar construction of 'the economy of nature'.

It is in the vital field of value theory that we can see the changing representation of human-environment relations, as moral science became political economy and political economy became economics. Touted as 'revolutionary' in standard economics textbooks, the marginal utility theory of value was developed in the 1870s simultaneously by Jevons in Britain, Menger in Austria and Walras in France. As a full account of the origins of 'pure economics' is beyond our scope, here I will present only a sketch of the work of key originating figures: Menger, Böhm-Bahwerk, Walras, and Jevons.

The 'revolution' that established the now dominant form of economic analysis is routinely presented as the becoming-scientific of economics. Jevons himself contributed to this characterisation of the intellectual moment. In 1871, as Paris revolted and was largely under control of the Communards, Jevons wrote,

> If, instead of welcoming inquiry and criticism, the admirers of a great author accept his writings as authoritative, both in their excellences and in their defects, the most serious injury is done to truth. In matters of philosophy and science, authority has ever been the great opponent of truth. A despotic calm is usually the triumph of error. In the republic of the sciences, sedition and even anarchy are beneficial in the long run to the greatest happiness of the greatest number. [...] Our English Economists have been living in a fool's paradise. The truth is with the French School, and the sooner we recognize this fact, the better it will be for the world.[35]

[35] Jevons, W. (1871). *Theory of political economy*. London: Macmillan & Co., p. 275, pp. xliv–xlv.

In claiming to be 'revolutionary' Jevons did not, however, follow through with his intellectual metaphor to the bitter end. The Paris Commune would be violently repressed, with thousands of working-class activists indiscriminately executed.

Conventional accounts refer to the 'discovery' of marginal analysis as a triumph that elevated economics into the pantheon of the sciences, by presenting a mathematical approach to the fluctuation of prices and consumption behaviour in terms of an individualist utilitarian psychology. The marginal utility theory of value was, as Jevons wrote, posited in 'direct opposition to the erroneous simplification of the science effected by Ricardo' and his labour theory of value,[36] an error which had engendered the heresies of Marx. By contrast, Jevons sought to advance a 'Theory of Economy which will reduce the main problem of this science to a mathematical form.'[37] The following Robinson Crusoe parable, written against the socialists by the Austrian finance minister Eugen von Böhm-Bawerk, outlines the starting point of marginal analysis:

> A colonial farmer, whose log hut stands by itself in the primeval forest, far away from the busy haunts of men, has just harvested five sacks of corn. These must serve him till the next autumn. Being a thrifty soul he lays his plans for the employment of these sacks over the year. One sack he absolutely requires for the sustenance of his life till the next harvest. A second he requires to supplement this bare living to the extent of keeping himself hale and vigorous. More corn than this, in the shape of bread and farinaceous food generally, he has no desire for. On the other hand, it would be very desirable to have some animal food, and he sets aside, therefore, a third sack to feed poultry. A fourth sack he destines for the making of coarse spirits. Suppose, now, that his various personal wants have been fully provided for by this apportionment of the four sacks, and that he cannot think of anything better to do with the fifth sack than feed a number of parrots, whose antics amuse him.[38]

Each sack of corn has the same labour value or cost of production, but if one of the sacks were spoiled, the colonist would not divide the remainder equally over the five uses, but let the parrots fend for themselves. The

[36] Jevons, W. (1866). Brief account of a general mathematical theory of political economy. *Journal of the Royal Statistical Society* (29), 282–287.

[37] Jevons (1866, p. 284).

[38] von Böhm-Bawerk, E. ([1890] 1959). *Capital and interest* (vol. 2, book 3). South Holland: Libertarian Press, pp. 143–145.

'utility' of the grain is different depending on how much one has. The utility of each additional unit of food declines as we approach the 'marginal' limit of how much an individual needs or can find value for in exchange. Here Böhm-Bawerk continued in the tradition of his teacher Carl Menger, deducing fundamental principles of human social existence from an imaginary 'individual' in the tabula rasa of a 'virgin forest', one ready to enter into trading relations with others in the fictitious pre-monetised 'barter economy' that Adam Smith imagined, without any anthropological warrant, to be present everywhere that money was not. Compare, for example, the attitude of Indigenous Australian fire ecologists to the carefully managed forests of their sacred and abundant Country to that of Menger's *homo economicus*, a colonist who has erased all memory of the Indigenes he has dispossessed, who perceives trees not as venerable upholders of the living cosmos but only as disposable resource units at his command, content to leave surplus communities of life to burn to the ground:

> If an inhabitant of a virgin forest has several hundred thousand trees at his disposal while he needs only some twenty a year for the full provision of his requirements for timber, he will not consider himself injured in any way in the satisfaction of his needs, if a forest fire destroys a thousand or so of the trees, provided he is still in a position to satisfy his needs as completely as before with the rest. In such circumstances, therefore, the satisfaction of none of his needs depends upon his command of any single tree, and *for this reason a tree also has no value to him.*

From the nihilistic proposition that nothing matters except what is mine, Menger advances the essential ultra-relativist subject position of marginal utility theory:

> *Value is therefore nothing inherent in goods, no property of them*, but merely the importance that we first attribute to the satisfaction of our needs. [my italics][39]

Extrapolated to all individuals and commodities presented for monetised exchange, the effect was simply to view money prices as sensitive aggregates of multiple individuals' value calculations, or in other words to

[39] Menger, C. ([1870] 1976) *Principles of economics.* Alabama: Ludwig von Mises Institute, p. 117.

collapse 'value' entirely into prices. Economics is restricted to price analysis, and this depends on positing a universal subjectivity which cannot be falsified with any extra-monetary knowledge of ecological phenomena. Beyond money, nothing matters: money rules supreme over life and matter, of which it is separate and independent.

While not so clear in the early days, the distinction between the Austrian and the neoclassical position came to be that for the Austrians, price determination was always locally subjective, subject to enormous complexity and uncertainty, which precluded predictive quantification in a general theory of value. By contrast, for the English, French, Italian and American economists, the presentation of mathematical equations resembling the conservation principles of physics was essential to their ambition to attain a scientific status, and to dispel the suspicion that political economists were mere intellectual hacks representing elite class interests. Where standard histories portray the 'marginal revolution' as a scientific innovation that opened the door to a mathematically 'pure economics', here we consider the largely forgotten metaphorical conflation of 'value' to 'energy' as the crucial development of the neoclassical science by which this was achieved.

This metaphor was explicitly asserted by Jevons in his unfinished *Principles of Economics*, a work devoted to the analysis of 'the industrial mechanism' of society: 'The notion of value [...] is to our science as that of energy is to mechanics.'[40] In his *Theory of Political Economy* (1871), Jevons elaborated Bentham's utilitarian psychology, arguing that all human activity could be reduced to a calculus of rational decisions to maximise pleasure and avoid pain. This formulation provided a continuous field, or spectrum of value as 'utility'. Agents would tend to minimise disutility (i.e. hard labour) and maximise utility (i.e. leisurely consumption of fine whiskey). Bentham went so far as to imagine a technical apparatus which could objectively measure utility values, a device he called the 'moral thermometer'.[41]

In his treatment of utility, Jevons represented the activities of the labourer as a mechanical analogue of balancing the pain of work with the pleasures of consuming commodities, for which work was exchanged at the margin. Behaviour was asserted as analogous to a 'force' in mechanics,

[40] Jevons, W. S. (1905). *The principles of economics: a fragment of a treatise on the industrial mechanism of society and other papers*. London: Macmillan, p. 241.

[41] Georgescu-Roegen (1971, p. 101).

allowing the introduction of infinitesimal calculus into the representative agents decisions to exchange 'at the margin'. This in turn meant that the prices governing and generating economic behaviour were the instances of a mechanical equilibrium between the opposing 'forces' of subjective valuations of present and future experiences of pleasure and pain, manifest in commodity production and exchange, and aggregated at the scale of national and international markets. As Jevons put it:

> The theory of Economy thus treated presents a close analogy to the science of Statistical Mechanics, and the Laws of Exchange are found to resemble the Laws of Equilibrium of a lever as determined by the principle of virtual velocities. The nature of Wealth and Value is explained by indefinitely small amounts of pleasure and pain, just as the Theory of Statics is made to rest upon the equality of indefinitely small amounts of energy.[42]

While the content of these references to physics need not concern us in detail, they are adequate to convey the ambition of the research programme. However, Jevons never satisfactorily demonstrated the coherence of the analogy (much less the implied isomorphism) of 'force/energy' to 'value', and declined demonstration with the relevant mathematics, which according to White, he simply did not understand.[43] This is, in any case, an impossible task. Money and energy are metaphorically related through human economic action and discourse, but are wholly incommensurable phenomena at the ontological level. The early drafts of Jevon's theory (written around 1862) were oblivious to the earth shaking Continental developments in energetics (as is evident in the above), but by the time of writing of the *Theory of Political Economy* (1871), Jevons had made efforts to acquaint himself with the new vocabulary, and replaced much of his discussions of utility as a mechanical 'force' with the more up to date metaphor of 'energy'.[44]

'Utility' was thus established as a kind of universal field of psychic desire in the value theory of the neoclassical paradigm. The origins of economic value henceforth were no longer to be found in the biological yield of land estates, in the silver and gold of treasuries, in the ores and fuels of its mines, nor in the quantity of 'labour time' embedded in the production of

[42] Jevons (1871, p. viii).

[43] White, M. (2004). In the lobby of the energy hotel: Jevon's formulation of the postclassical 'economic problem'. *History of political economy*, 36(2), 227–271.

[44] White (2004, p. 241).

commodities, but in the psychology of the individual consumer in relation to a continuous commodity space, where the desired goods and the wage-work required to obtain them were unified in a single field of 'utility'. This metaphor allowed the application of the calculus to 'value', represented in the metaphor as a continuous and conservative energy field. Just like money, and crucially, just like 'energy' as described in the first law of thermodynamics, 'value' was infinitely convertible into other forms, the 'real' metric and universally abiding constant in the otherwise enormous diversity and complexity of empirical phenomena.

For Margaret Schabas, it is with this step that 'the economy' was detached from any recognisably social or natural setting. As the physical reality of economic activity was neglected for the abstract indications of consumer sentiment in the traces of prices registered in 'the market', even capital was recast subjectively with the neoclassical theory of value, not as the 'savings' of the capitalist re-invested in industrial infrastructure—the means of production which themselves must be produced—but simply as 'that which is expected to yield utility in the future'.[45] Schabas interprets this as the moment when 'the economy' became 'entirely artificial'. Yet this could equally be interpreted, in light of the economists' intended conflation of value theory with the status of the law of the conservation of energy in the intellectual imagination, as the moment when the 'free economy' and the expansion of fossil-fired industry became 'entirely natural'.

Traditional histories of economics tends to accord Jevons with priority for developing a scientific theory of price, while omitting the baroque analogies underwriting this 'discovery'. Unlike the English economists, whose philosophical roots were in the psychological empiricism of Hume and Locke, the French School associated with Augustin Cournot, Jean-Baptiste Say, Leon Walras and Vilfredo Pareto pursued a rationalist approach deploying the mathematics of engineering and mechanics. It was Say who had reformulated economics by reinterpreting Adam Smith's figure of the unseen hand as the study of the 'balance' between the 'forces of supply and demand'. While no phenomenologically detailed theory of the nature of this balance had been proposed, that Say's 'laws of supply and demand' accomplished equilibrium was accepted as axiomatic by the

[45] Schabas, M. (1992). The greyhound and the mastiff: Darwinian themes in Mill and Marshall. In P. Mirowski (Ed.), *Natural images in economic thought: markets read in tooth and claw* (pp. 322–335). Cambridge: Cambridge University Press, pp. 329–330.

French, as it was later by Alfred Marshall, the British economist who presided over the first heyday of neoclassical economics as applied policy.[46]

It was Walras, at one time enrolled in the École des Mines de Paris, who was to complete the neoclassical 'revolution' with the publication of his *Elements of Pure Economics* (1874). This work presented a mathematical model of a highly idealised and simplified market operating according to principles of general equilibrium, a static and ideal system of economy modelled on the rational mechanics of the day. It contained a mathematical proof of the theoretical possibility of a general law of equilibrium that is regarded as the defining moment and superlative achievement of the neoclassical school. Schumpeter would hail this work as 'the Magna Carta of Economics'. Anointing Walras as the 'greatest of all economists', Schumpeter bestowed upon him the Popperian mantle of the Newton of the social sciences:

> His system of economic equilibrium, [...] is the only work by an economist that will stand comparison with the achievements of theoretical physics.[47]

Walras himself was not averse to promoting this interpretation of his 'discovery', concluding his *Elements* with a passage which betrays a Laplacian image of the ideal science as one far removed from the Earth, wholly abstracted from the world of labour, heat and life which we actually do (and can only) inhabit.

> Thus the system of the economic universe reveals itself in all its grandeur and complexity, a system at once vast and simple, which, for sheer beauty, resembles the astronomic universe.[48]

The Walrasian 'system'—a system in the sense of a system of mathematical equations—would come to exemplify the *nomos* of economics as a science of 'natural law', but at the expense of entirely excluding the *oikos*—the lands and resources of 'the estate'—from the sphere of the discipline's inquiry. Walras believed that only one aspect of the social reproduction of

[46] Ayres, R. (1994). *Information, entropy and progress: a new evolutionary paradigm*. New York: American Institute of Physics, p. 134.

[47] Schumpeter, J. ([1954] 1994). *History of economic analysis*. Oxford University Press, p. 795.

[48] Walras, L. ([1874] 1954). *Elements of pure economics*. Homewood, IL: Richard Irwin, p. 374.

the lifeworld was amenable to 'pure' scientific analysis: the configurations of market prices in a regime of perfectly free competition.[49]

Walras formulated economics as the study of a static balance of forces between supply and demand, directing economic enquiry into the theoretical question of the existence of a set of equilibrium prices such that all demand matches supply and 'clears' the market—that is everything offered is sold and no buyers walk away disappointed. This vision is clearly analogous to the conservation of energy framework: equilibrium is achieved because value is conserved through each exchange; machines, commodities, labour and land are all fully convertible to abstract exchange value. Among the assumptions made to get this model to 'work' are the assertion of the universality of 'the market' and its automatic operation as a frictionless machine:

> [...] *the whole world* may be looked upon as a vast general market made up of diverse special markets where social wealth is bought and sold. Our task then is to discover the laws to which these purchases and sales tend to conform automatically. To this end, we shall suppose that the market is perfectly competitive, just as in pure mechanics we suppose to start with, *that machines are perfectly frictionless.* [my emphasis][50]

The Walrasian model has been correctly labelled as 'a system devoid of human beings', and a 'form of analysis without much substance'.[51] The 'actors' in the perfectly frictionless market are algorithms, automatons without purchase on the world. In the Walrasian economy of nature, persons, machines and commodities disappear entirely, but not into money, which despite neoclassical claims to address prices, is dismissed as a mere 'veil' for the 'real' underlying motion of the psycho-social phenomena of subjective utility in commodity exchange. Walras developed the market as a 'frictionless' allocation device as an aid to abstraction, relieving economists of the task of accounting for the materiality of the economic process.

The conflation of the economy with 'the market', and the reification of the market as a frictionless, reversible machine—the antithesis of Carnot's empirical analysis of the entropic 'fire-machine'—robs theory of any traction in the analysis of the 'industrial mechanism' as the organisation of production. Like all physical objects in motion, machines are subject to

[49] Mirowski (1989, p. 220).
[50] Walras (1954, p. 84).
[51] Seligman, B. (1962). *Main currents in modern economics.* Glencoe: Free Press, p. 385.

friction. Internally, an engine's moving parts cannot avoid losing efficiency to friction, which dissipates energy in the form of waste heat, and its order through wear on the parts. Friction is defined in physics as 'resistance'. As Reuleaux's classic definition of the machine tells us, friction is integral to its ontology:

> A machine is a combination of resistant bodies so arranged that by their means the mechanical forces of nature can be compelled to do work accompanied by certain determinate motions.[52]

Motion is essential to distinguish a machine from a simple structure, and they are built for no other reason but to change the magnitude, direction or point of application of moving forces. Machines designed for cutting, drilling, boring, milling, digging, grinding, processing, mixing, polishing and transport all apply the force realised from fossil fuel combustion to surfaces in ways useful for the transformation of raw materials into artefacts, or for otherwise enhancing the 'exosomatic' architecture of the human body by various prostheses.

Walras offered a machine theory of 'the industrial mechanism of society' stripped of industrial machines. That his 'perfectly frictionless' machine required 'perfect competion' in order to work has been interpreted by his neoliberal descendants as an exhortation to remove inhibitions to the free operation of the price mechanism: for example social legislation, taxation, imperfect knowledge, transaction costs, labour unions, unwillingness to exchange at the equilibrium price, mandated price floors and ceilings, and so on. The state is enrolled to accomplish the utopian 'removal of all obstacles' to the free operation of the price mechanism.

An important predecessor of Walrasian economics was the German economist Henri Gossen. Largely ignored during his lifetime, his work directly inspired Walras by conceiving of the problem of national economy as beginning with a strictly budgeted quantity of a putatively physical ur-substance—*Kraft*—morphing into different manifestations of capital, labour, and commodities:

> Upon removal of all obstacles that interfere with not only each person's most purposive use of the money but also his choice of productive activity

[52] Reuleaux, F. (1894). *The constructor: a handbook of machine design*. Philadelphia.

> that, under the circumstances, is most advantageous to him, each person will receive a portion of the means of employment that corresponds exactly to the burden assumed by him in the productive process. Thus what socialists and communists conceive to be the highest and ultimate aim of their efforts is accomplished here with the cooperation of the forces of nature.[53]

Here the physical quantity *Kraft* is conflated with all monetised human activity, and a conservation principle is pressed into service to enforce the 'natural justice' of the market: no-one gets a free lunch. In a precise reversal of the Marxist slogan 'from each according to his ability, to each according to his needs', Gossen's economy exacts productive labour from each according to her desires and returns commodities to each in exact equivalence to her input of labour-power.

While long since disavowing the ambitions of that generation of economists regarding the cardinal measurability of 'utility' as the equivalent or analogue of 'energy', what has persisted in economic theory since its openly physicalist period is the notion of *the permanent conservation of value* in the cycles of market exchange.[54] By definition, the equilibrium model of the market requires that 'value be conserved' through each exchange in the circular flow of the economic process. For Mirowski, this essential feature of neoclassical reasoning reflects the internalised and forgotten cosmology of what he calls the 'proto-energetics movement', which flowered briefly in the decades following von Helmholtz's 1847 formulation of the first law of thermodynamics, and waned with the completion of energy physics in 1865 by Clausius' elaboration of entropy law, with all its implications of irreversibility, exhaustion, depletion, dissipation, and decay:

> The metaphor of energy-utility that was appropriated by neoclassical economics was derived from the physics of a specific historical moment, namely, the middle of the 19th century just prior to the elaboration of the second law of thermodynamics [...] In this vintage of physics, all physical phenomena are portrayed as being perfectly reversible in time; there was no room in the theory for hysteresis. In other words, nineteenth-century physical law could have no history. This stubbornly anti-historical bias of neoclassical economics has been excoriated by critics [...] In pre-entropic physics, all physical phenomena are variegated manifestations of a protean energy that

[53] Gossen (1853; cited in Mirowski, 1989, p. 216).
[54] Mirowski (1989, p. 64).

can be fully and reversibly transformed from one state to another. When this metaphor was smuggled into the context of economic theory, it dictated that all economic goods be fully and reversibly convertible into utility, and thence into all other goods in the act of trade.[55]

Against Equilibrium: Entropy and the Legitimacy of Economics

This raises an important question: if the quantity of value can neither be created nor destroyed, how is economic growth possible, and infinite growth at that? As economics is conceived of as a science of equilibrium, the mathematical treatment of 'economic growth' added to the neoclassical edifice as GDP ideology took hold in the post-WWII period is entirely an afterthought. As the physicist Robert Ayres points out, a constitutive assumption of neoclassical models from Walras to Solow is that:

> [...] all resources, commodities and services can be produced entirely from linear combinations of others in the system—the economic system is closed to fresh inputs and outputs of waste. [...] although they relied to some extent on physical analogies, the early neoclassicists (Walras, Pareto and their successors) did not fully face the fundamental contradiction of growth in the equilibrium state.[56]

As was noted in 1922 by the radiophysicist Frederick Soddy, economists effectively portray the economy as a perpetual motion machine—entirely at odds with the financial economy, where debts expand exponentially according to the logic of compound interest, and with the 'real economy' which depends on the irreversible depletion of fossil fuels.[57] The importance of this observation to our ecological crisis was first established by Nicholas Georgescu-Roegen, the first economist in good standing to import the Carnotian revolution to the social sciences:

[55] Mirowski, P. (1992). *Against mechanism: protecting economics from science.* Totowa, NJ: Rowman & Littlefield, pp. 152–153.

[56] Ayres (1994, p. 134).

[57] Soddy, F. (1922). *Cartesian economics.* London: Hendersons.

> The patent fact that between the economic process and the material environment there exists a continuous mutual influence which is history making carries no weight with the standard economist.[58]

If the sins against the first law of thermodynamics committed by general equilibrium theory were not bad enough, as Georgescu-Roegen demonstrated, economics simply ignores the crucial principle of the fully realised science of energy: the second law of thermodynamics. The second law is a notoriously difficult subject, yet we must attempt to give enough of an account that the lay reader can understand what is surely amongst the gravest blind spots of orthodox economics.

What has the entropy law to do with economics? As our ordinary experience tells us, things fall apart. Coal is reduced to ashes, cars rust, batteries go flat, machines wear out, organisms die. The first law posits an invariance in the *quantity* of energy within a system regardless of any process of transformation. Uniting the principles of the conservation of mass and the conservation of force, the first law can be simply stated in ordinary language as 'energy is conserved'. But what is meant by 'conservation'? Ultimately, this can only be explained by introducing the second law, which insists that for any change to occur, the *quality* of energy must be altered, where the quality of energy refers to the capacity of energy to be available to do 'work', that is to cause some kind of transformation or change of state. This quality of energy is referred to as available energy, free energy, or in some formulations, 'exergy'. Entropy is a measure of the *unavailability* of energy within a system to perform 'work', and within a system closed to fresh influxes of exergy, entropy will always increase. Importantly, the boundaries of the 'system' are never impermeable in reality, the system must be posited by the observer. Clausius famously framed the two laws at the scale of the universe, the total cosmological system:

> The energy of the universe is constant.
> The entropy of the universe tends to a maximum.[59]

[58] Georgescu-Roegen, N. ([1970] 1993). The entropy law and the economic problem. In H. Daly & K. Townsend (Eds.), *Valuing the earth: economics, ecology, ethics* (pp. 75–88). Cambridge, MA: MIT Press.

[59] Clausius, R. (1867). *The mechanical theory of heat*. London: van Voorst, p. 365.

By way of example, let us posit a system: the common garden-variety leaf-blower, for my money the epistemological totem *par excellence* of the unconsciousness of post-natural consumer culture. After an hour of fighting the warming breeze and blowing leaves from one end of your garden to the other, the concentrated energy (exergy) stored in the chemical bonds of petroleum is now emptied from the fuel tank, irreversibly dissipated in the combustion reaction. The high-quality energy (low entropy) stored in the fuel is converted by the 'work' of the heat engine into low-quality energy (high entropy), no longer available to perform work. If we were to sum the quantities of chemical energy dissipated as heat, mechanical work, wind and irritating noise, we would find the same number, but the quality of the energy is irrevocably degraded. There is no way to bind that energy together again as petroleum. The law of the conservation of matter is maintained: the combustion reaction is the precise reverse of the photosynthesis reaction. The carbon molecules separated from atmospheric CO_2 by marine phytoplankton in an eon long past are rebound to atmospheric oxygen to again form atmospheric CO_2. Of course photosynthetic life can capture and concentrate solar energy into new biomass, and specific geological processes over millions of years can convert carbohydrates into new hydrocarbon minerals, but only from fresh flows of solar energy transformed by the biosphere into geological strata over multi-million year timeframes. The entropy of the petrol-leafblower 'system' has irreversibly increased, while the energy has remained constant.

Formalised by Rudolf Clausius and Lord Kelvin around 1865, the second law was formulated because nobody could find a way around its rigorous constraints on heat and work, which hold good for all processes of energy conversion. No known violations exist, even at the molecular level.[60] Clausius proved that it is impossible for a self-acting machine, unaided, to transfer heat from one body at a low temperature to another having a higher temperature. Heat flows in only one direction, and it is gradients in the condition of energy that allow the possibility of mechanical work to be performed.

Energy can be concentrated, but tends toward diffusion. For example, an inflated balloon pierced by a pin will deflate until its internal pressure is equivalent to that of the atmosphere surrounding it. Drop a ball on the floor, and it will bounce and come to rest. Bring a boiling pot of soup into

[60] McClare, C. (1971). Chemical machines, Maxwell's demon and living organisms. *Journal of Theoretical Biology* (30), 1–34.

a perfectly insulated room, and the temperatures of the room and the pot will converge until they are not different. The room and the pot of soup form a closed system and once in thermal equilibrium, will not change from that state. Open the door to the winter night, and the rooms' temperature will equalise with that outside. We never observe the reverse of these processes. Soup does not heat itself, balls do not start bouncing on their own, and balloons don't self-inflate. In each of these cases energy required for changes of state is sourced externally, and ultimately from the sun.

The second law of thermodynamics is widely held to be the foundation of modern physical theory. Sir Arthur Eddington held it to be the most secure and fundamental of the laws of nature:

> The law that entropy will always increase—the second law of thermodynamics—holds, I think, the supreme position among the laws of Nature. If someone points out to you that your pet theory of the universe is in agreement with Maxwell's equations, then so much the worse for Maxwell's equations. If it is found to be contradicted by observation, well these experimentalists do bungle things sometimes. But if your theory is found to be against the second law of thermodynamics then I give you no hope: there is nothing for it but to collapse in deepest humiliation.[61]

Thus far, few economists have exhibited any such a sense of shame.

Given the assertion of the universality of the entropy law the problem was immediately posed: if everything is winding down, if order is always reduced to chaos, then how do organisms grow in size and complexity? How is the process of biological evolution possible? Some pointed to evolution to try and fault the second law, whereas Creationists persist in using the second law to falsify evolution. This problem ricocheted around the halls of science for many years until Ervin Schrödinger, in his 1944 work *What is Life?*, first set out clearly the relationship of biological life to the rigorous limits of entropy.[62] In it, he foreshadowed Ilya Prigogine's concept of 'dissipative structures', by suggesting that 'the device by which an organism maintains itself stationary at a fairly high level of orderliness (= fairly low level of entropy) really consists in sucking orderliness from its

[61] Eddington, A. (1928). *The nature of the physical world.* Cambridge: Cambridge University Press, p. 74.

[62] Schrödinger, E. ([1944] 1967). *What is life?* Cambridge: Cambridge University Press.

environment.'[63] Organisms maintain their 'orderliness' by importing what Schrödinger referred to as 'negentropy' (negative entropy—e.g. sunlight, or food) from the environment and exporting or externalising back to it waste, disorderliness and heat (entropy). Socio-economic systems, ecosystems, and the biosphere as a whole are 'dissipative structures', temporary islands of order in a sea of disorder, maintaining and increasing their 'orderliness' by dissipating inflowing energy into more chaotic forms (and thus more probable, according to the second law).

The consequences of this for the question of long term 'sustainability' are clear enough. The biosphere, having evolved a 'complex adaptive system' of species life that forms a near-perfect system for the solar-powered recycling of biogeochemical materials, is 'sustainable' for as long as the Earth is daily bathed in sunlight and photosynthesis maintains a thermal envelope suitable for life via the carbon cycle. On the other hand, global industrial technomass is radically unsustainable as long as it burns fossilised biomass faster than carbon can be 'fixed' by geological processes.

It may be objected that the import of a metaphor from one field into another need not imply an isomorphic correspondence: in the history of the sciences, metaphorical exchange has often proven useful in providing novel heuristics that have led to valuable breakthroughs. Thus the equilibrium concept could be seen to have found an autonomous meaning within economics and detached itself from the original signifier in physics. This defence might be entertained were it not the case that the rationale for equating subjective utility with energy was to avoid the moral implications of political economy and to bury the political questions raised by the labour theory of value. 'Equilibrium' was not intended to be a metaphor to conjure with; its task was to do away with metaphor by revealing the 'natural laws' governing production and exchange. But without doubt the most serious problem of the neoclassical equilibrium model is that its physical metaphor of equilibrium so badly coheres with, and indeed acts to obscure, the physical reality of the economic process. Take hydrocarbon extraction for example: what economists describe as the 'production of resources' is, from the biophysical point of view, the irreversible dissipation of resources and the catastrophic overproduction of waste.

Whereas thermodynamics is the study of (bio)physical phenomena in the approach to thermal equilibrium, economics purports to be the study of equilibrium itself. The strictest definition of equilibrium is provided by

[63] Schrödinger (1967, p. 79).

the second law of thermodynamics, where it refers to a final maximization of entropy. In the words of Ayres:

> Equilibrium is a homogenous unchanging state in which there are no gradients of any kind, including the time dimension. This implies uniformity of temperature, pressure, density, chemical composition as well as uniform gravitational and electro-magnetic fields. The equilibrium state is one in which no part of the system can be distinguished from any other part of the system.[64]

Imagine an exhausted cluster of gas particles, thinly distributed across the emptiest corner of the universe, as remote as possible from any radiant star, with a temperature of absolute zero degrees Kelvin (−273.15° Celsius). So much for *ceteris paribus*. So much for the question of 'how shall we live'?

Attempting to appropriate the scientific aura of the most prestigious and confirmed of the natural sciences of his time, Walras' concept of general equilibrium was in the final analysis a purely mathematical imaginary of the monetary logic of market order, stripped of any empirical institutional, industrial or biophysical setting. It is precisely this feature of the neoclassical system which gives it the virtue of having no necessary political content. Thus the Walrasian model is a logical machine which lends itself to the imagination of all manner of social engineering projects. Milton Friedman once remarked that 'we curtsey to Marshall, but we dance with Walras.'[65] And yet the political project of market freedom that Léon Walras' believed his 'scientific' economics supported was the one that Friedman dedicated his lifelong project as a neoliberal protagonist to suppressing.

Walras expressly identified his project with social democracy, in the spirit of progressive reformers who advocated the neutralisation of the political and economic powers of the monopoly capitalist by the taxation of unearned rents, which resolved, in the analysis of Henry George, to the 'despotic dominion' of government enforced monopolies in the 'free gifts of nature' represented by exclusive private property in 'land': whether urban real estate, railway corridors, ocean fisheries, grazing grounds, or mines. According to Renato Cirillo, Walras:

[64] Ayres (1992, p. 3).
[65] Ayres (1992, p. 134).

> […] strongly believed that without an equitable distribution of wealth there could be no social justice. Though he defended the right of private property, he considered that land was a special case and that it belonged to all the community. His social reform involved the nationalization of land, the abolition of taxation on wages, the curbing of monopoly power, and the promotion of a strong cooperative movement. He insisted that the only way the working class could regain their freedom was by becoming property owners. The influence of Henry George on Walras' thinking is obvious. They both shared the same humanitarian ideals, and both believed in a capitalist system working side by side with the social reforms they advocated.[66]

Whilst claiming 'scientific' descent from Walras, Friedman advocated for precisely the reverse reforms, praising private wealth-holders and plutocrats, and opposing labour unions and any institutional foothold for their democratic project of social justice. Friedman was one of the architects of the Internal Revenue Service's system of withholding of personal income taxes. In *Capitalism and Freedom* (1962), Friedman advocated the abolition of corporate taxation and the privatisation of all lands held in public trust, such as national parks and nature reserves. As a participant in the founding research projects of Chicago economics and law—the Free Market Study and the Chicago Antitrust Project—his brand of neoclassical analysis was applied to critique and then replace normative legal concepts of justice designed to uphold free and fair market competition, justifying the progressive dismantling of the enforcement apparatus of the Sherman Antitrust Act (1890), an anti-monopoly statute won by popular protest which in 1911 authorised the US federal government in to break up the gigantic Standard Oil corporation.

Given the intensifying climate catastrophe arising from our managed dependence on fossil fuel infrastructure, and the well-documented role of Standard Oil's successor ExxonMobil in aggressively promoting science-denialism and undermining climate policy via its sponsorship of the Atlas Network, the public interest case for the trust-busting and nationalisation of fossil-fuel corporations is far more compelling than it was in 1911. Contra Friedman, who claimed that 'horses are more polluting than automobiles,' nothing less than planetary survival is at stake. That such a programme of public defence presently seems almost unimaginable can largely be attributed to the success of the Hayekian project to immunise world

[66] Cirillo, R. (1984). Léon Walras and social justice. *American Journal of Economics and Sociology*, 43(1), 53–60.

economic order from democracy, through the capture of state power by a cadre of authoritarian liberals in the service of 'pollutocrats'. Perhaps the final words on the biospheric nihilism of economics as the social science of thermoindustrial society could be given to FEE, a 'business propaganda' outfit that argues against any legislative programme to limit global heating, on the grounds that 'the free market [will always] outperform government intervention, regardless of the fragility of Earth's ecosystems'.[67]

[67] Callahan, G. (2007, Oct 1), How a free society could solve global warming. Foundation for Economic Education: New York. https://fee.org/articles/how-a-free-society-could-solve-global-warming. Viewed 16 April 2019. I am indebted to Matt Johnson for this reference.

CHAPTER 7

Economics as Agnotology: Unlimited Growth and the Limits of Knowledge

Writing in 1919, long before permanent growth had become the defining condition of the 'strong economy', the Austrian economist Joseph Schumpeter defined imperialism as what you have 'when a state evinces a purposeless propensity to expansion by force, beyond all definable limits'.[1] States cannot easily be distinguished from the strategic enterprises that they are enmeshed with, and whose need for legal order and security they serve. By Schumpeter's definition, all governments that orient their policy according to rising GDP growth are irrational in their purposeless expansion, and imperialist in their application of the sovereign powers of the state to the project of unlimited 'creative destruction'.

As the twentieth century drew to a close, the sociologist Ulrich Beck noted the general absence of serious treatments of planetary ecological hazards in the social sciences of the day:

> [...] the following question has been criminally rejected by the social sciences: what does the threat of self-annihilation mean to society, its institutions, its understanding of progress and itself; to the legal, scientific and economic system; to politics and culture?[2]

[1] Cited in: Ferro, M. (1997). *Colonization: a global history*. London: Routledge, p. 13.
[2] Beck, U. (1995). *Ecological politics in an age of risk*. Cambridge: Polity Press, p. 82.

J. Walker, *More Heat than Life: The Tangled Roots of Ecology, Energy, and Economics*,
https://doi.org/10.1007/978-981-15-3936-7_7

A compelling way to frame an answer to the question of the 'apocalypse blindness' of the economic sciences appears in an obscure text by Vladimir Nuri, which aptly describes the integral contradiction of modern economics:

> [...] any rational model of the economy absolutely *must* consider the *equilibrium* state one that does *not* involve growth. A century and a half of the 'dismal science' may be based on an evasion or defiance of that principle. It is quite possible that economics is based on a mass collective rationalization in much the same way that US citizens subscribed to the vision of their 'manifest destiny' during the era of the expansion to the west. Relative to serious worldwide dangers of environmental degradation (e.g. global warming, pollution, deforestation, *etc.*), at the dawn of the twenty-first century the rationalization is taking on the signs of mass psychological delusions of grandeur. [emphasis original][3]

For many decades, the social good has been held by economists to be indistinguishable from 'the growth of the economy' and best achieved by free markets. If one lives in a 'first world' democracy, it is more than likely that the governing parties more or less agree upon the neoliberal policy programme—free markets, privatisation, deregulation, minimalist social policy, foreign investment, and export-led growth. If one lives in a 'developing' country, or one that has suffered a systemic financial crisis, it is likely that the neoliberal policy programme was imposed upon the country by governments fulfilling the conditions of external agencies managing external public debt, restricting the scope of parliamentary opposition to 'structural adjustment'. When we take into account the escalating rates of deforestation and hydrocarbon extraction thus enabled, it is hard to imagine any communities of human and extra-human life not directly affected by the policy decisions made in the closed forums where finance ministers and corporate lawyers forge further rounds of trade and investment liberalisation. And yet as late as the 1990s, when protest modified its position, the IMF maintained that 'macroeconomics has nothing to do with the

[3] Nuri, V. (2001). Fractional reserve banking as economic parasitism: a scientific, mathematical, and historical exposé, critique and manifesto (No. 0203005): University Library of Munich.

environment'.[4] It would seem that economists have acquired a professional authority out of all proportion to the epistemic warrant of their science.

Another approach to the 'apocalypse blindness' of economic science in the era of ecological crisis is to analyse it as a form of *agnotology*. Counterposed to *epistemology*—the procedures by which ignorance is reduced, knowledge produced, and truth validated by a deliberative or scientific community—the term agnotology was coined by Proctor and Schiebinger to analyse the intentional production of 'culturally induced ignorance or doubt'.[5] As I have argued in earlier chapters regarding the 'classical liberalism' of the nineteenth century that neoliberals claim to represent, the elimination of 'land' via the *a priori* method of economists from Nassau Senior to Friedrich Hayek is an essential and abiding feature of economic agnotology, one which diverts attention from the land appropriations necessary to the operation of extractive industry. As we have seen in the case of the Irish and Indian famines, 'methodological individualism' allows economists to ignore the 'state of nature' on the ground and to deduce natural laws of the market with which to naturalise politically constructed economic arrangements, to which it will then be claimed that 'there is no alternative'. The agnotological functions of economic discourse are integral to its history as a governmental rationality. This is evident when we attend to the forms of knowledge it routinely excludes, either through the intellectual complacence of a dominant culture, or intentionally through propaganda devoted to 'producing ignorance and doubt in the general populace for specific political motives'.[6] The elimination of resource dissipation from neoclassical models of 'production' allowed mid-twentieth-century economists to explain infinite 'equilibrium-path growth' with a mysterious 'technological residual', at a time when the geopolitical fortunes of American empire increasingly turned on the control of oilfields and mines beyond its national territory. As we approach the present, the fortification of a global 'neoliberal constitution' through a thicket of international treaties guaranteeing the rights of transnational 'investors' over those of citizens coincides with a highly organised

[4] Daly, H. (1996). *Beyond growth: the economics of sustainable development*. Boston, MA: Beacon Press, p. 144.

[5] Proctor, R., & Schiebinger, L. (Eds.) (2008). *Agnotology: the making and unmaking of ignorance*. Stanford University Press.

[6] Mirowski, P. (2013). *Never let a serious crisis go to waste: how neoliberalism survived the financial meltdown*. London: Verso, p. 227.

campaign conducted through the fossil-sponsored Atlas Network to propagate science denial and mass-market 'climate skepticism'.[7]

Mirowski has characterised neoliberalism as a 'double-truth' doctrine, consisting of a public narrative retailed to the demos celebrating the freedom and prosperity that will emerge from the rollback of the 'red tape' and 'green tape' of 'big government', and a more candid discourse deployed in private sanctums, in the elite spaces where strategic interventions to fortify the discipline of 'free' markets are contemplated and conducted.[8] By way of example, let us consider the views of a senior economist and consummate political insider of institutions devoted to re-making the 'positive laws' of states in the service of the 'natural laws' of the market. In a 1997 interview given during his tenure as US Deputy Treasury Secretary, Summers argued that

> The laws of economics are more like the laws of physics than many people once supposed. You can't wish them away and they don't change because of politics.[9]

The idea that differing shares in the common-wealth are allocated and determined by an eternally fixed natural law, from the plutocrats at the top to the slum-dwellers at the bottom, is hard to argue with when framed this way, for what fool would advocate for a policy which contravenes the laws of physics? The 'double truth doctrine' is nicely illustrated by comparing Summer's public statements on the 'laws of economics' to comments made privately in the position of Chief Economist at the World Bank, which were revealed in a leaked internal email:

> December 12, 1991
> To: Distribution
> From: Lawrence H. Summers.
> Subject: 'Dirty Industries'

[7] On 'neoliberal constitutionalism' see: Gill, S. (2002). Constitutionalizing inequality and the clash of globalizations. *International Studies Review*, 4(2), 47–65; Purdy, J. (2014). Neoliberal constitutionalism: Lochnerism for a new economy. *Law and Contemporary Problems*, 77, 195–213; Schneidermann, D. (2013). *Constitutionalizing economic globalisation: investment rules and democracy's promise.* Cambridge, UK: Cambridge University Press.

[8] Mirowski, P. (2009). Postface: defining neoliberalism. In P. Mirowski & D. Plehwe (Eds.), *The road from Mont Pelerin: the making of the neoliberal thought collective collective* (pp. 418–451). Cambridge MA: Harvard University Press, p. 426.

[9] Heilemann, J. (1997, Jan 7). The integrationists vs. the separatists. *Wired*.

Just between you and me, shouldn't the World Bank be encouraging MORE migration of the dirty industries to the LDC's [Less Developed Countries]? I can think of three reasons:

1. The measurements of the costs of health impairing pollution depend on the foregone earnings from increased morbidity and mortality [...] from this point of view health impairing pollution should be done in the country with the lowest wages.
2. The costs of pollution are likely to be nonlinear as the initial increments of pollution probably have very low cost. I've always thought that underpopulated countries in Africa are vastly UNDERpolluted [...].
3. The demand for a clean environment for health reasons is likely to have very high income elasticity. Concern over an agent that causes a one in a million change in the odds of prostate cancer is obviously going to be much higher in a country where people survive to get prostate cancer than in a country where under 5 mortality is 200 per thousand.

   The problem with the arguments against all of these proposals [...] (intrinsic rights to certain goods, moral reasons, social concerns, lack of adequate markets, etc.) could be turned around and used more or less effectively against every Bank proposal for liberalization.[10]

The memo became public in February 1992, as world leaders and NGOs prepared for the UN World Conference on Environment and Development at Rio de Janeiro, a conference that resulted in the Convention on Biological Diversity and the Framework Convention on Climate Change, treaties intended to limit the destruction of ecosystems and to initiate coordinated efforts to slow the global acceleration of greenhouse gas emissions. In an open letter responding to Summers, Jose Lutzenberger, Brazil's then-Secretary of the Environment wrote:

> Your reasoning is perfectly logical but totally insane [...] Your thoughts [are] a concrete example of the unbelievable alienation, reductionist

[10] Vallette, J. (1999, June 15). Larry Summers' war against the Earth. *Counterpunch*.

thinking, social ruthlessness and the arrogant ignorance of many conventional 'economists' concerning the nature of the world we live in.[11]

Summers claimed that the memo was a satire. In any case, Summers had merely explicitly stated in the terminology of the dominant discourse a long-institutionalised process of unequal exchange between the centres and peripheries of the 'world system'. In public, Summers appealed to economics as a science of inexorable laws of nature, independent of fanciful human wishes. Away from the cameras, Summers deployed his executive authority at a global governance institution to advocate for the active acceleration of environmental degradation in poor countries, a process already well underway as a result of the imposition of 'austerity measures' upon the Bank's debtor nations. Later renamed Structural Adjustment Programs, these were entirely *artificial* policies intentionally developed and actively administered. Requiring governments to attract foreign investment by offering business the most laissez-faire tax and labour conditions, they led to the restriction of civil rights and the rollback of environmental regulation. The strategy of deregulation, privatisation, foreign direct investment, and 'export led growth' imposed through client government, odious dictators and the Washington Consensus ensured a steady flow of fuels, minerals, logs, cash crops, and low-wage manufactures to Western markets. It should be noted that the top-down imposition of financial, fiscal, and economic structures, even with specific preferences for certain kinds of industry, completely reverses the order of causation in the economic theory mobilised to justify the policy recommendations. Economics is allegedly built on the 'microeconomic foundations' of the quotidian desires of sovereign consumers, whose free and rational choices result in the spontaneous emergence of macroeconomic forms of organisation.

The analytic fiction of general equilibrium theory was that it held constant arguably the most important factors—population, capital, technology, and land—to derive a self-balancing market. In other words, equilibrium was 'existence proofed' by excluding the existence of all material phenomena, including 'growth'. This problem has not bothered economists very much. Take for example the encomiums of Alan Greenspan during his tenure as governor of the US Federal Reserve, amidst the financial boom that preceded the financial crisis of 2008:

[11] Vallette (1999).

> [globalisation] has altered the economic frameworks of both advanced and developing nations in ways that are difficult to fully comprehend. Nonetheless, the largely unregulated global markets, with some notable exceptions, appear to move smoothly from one state of equilibrium to another. Adam Smith's invisible hand remains at work on a global scale.[12]

Appealing to a different concept of 'equilibrium', ecological economists have argued that the only solution to the biospheric crisis—the increasingly irreversible violation of the 'balance of nature'—is a transition from growth to some kind of steady state. For Herman Daly, the most pressing legitimate task of economic science is to discover the optimal scale of the economy (as a total industrial metabolism) that can be shown to be sustainable in the long term from the point of view of the Earth sciences, and then hold population and capital constant at those levels, or even plan for 'de-growth' if we are already beyond the limits of planetary survival.[13] While such a policy seems politically impossible, surely this is the only rational 'biophysical' interpretation of the equilibrium concept, recognising the 'mass-balance' condition of the first law of thermodynamics, the logic of 'predator—prey' balance in community ecology (in which a population of consumers is limited by the size of its food supply), and the relative thermal stability of the Earth's heat balance as accomplished by the photosynthetic carbon cycle.

The Post-WWII Economics of Infinitude

Prior to World War II, economists were largely preoccupied with bringing the wild booms and bust of the trade cycle under stabilisation. Reflecting a myopia opposite to the farsightedness conferred upon the rationally maximising 'representative agent', economists have rarely concerned themselves with the long-term consequences or horizons of growth. Prior to World War II, even as sophisticated a thinker as John Maynard Keynes tended to think of economic growth as a short-term phenomenon, neither enduring nor important. He expected economic growth in the industrial countries to cease within two or three generations at most, arriving at a plateau of consumption (called 'bliss' in that period's

[12] Greenspan, A. (2005, Mar 10). Globalization. Speech to the Council on Foreign Relations, New York.

[13] Daly, H. (Ed.) (1993). *Valuing the Earth: economics, ecology, ethics.* MIT Press.

literature) as the declining marginal utility of consumption approached zero.[14] Despite the normal ubiquity of the growth fetish in our day, maintaining permanent growth became the central concern of applied economics only in the post-war period.

A key instrument of social technology was developed in this era and applied to society at large, an informatic tool with which the productivist state continues to organise its planning and policy decisions. Described by Alan Greenspan as 'one of the great inventions of the twentieth century', this is the system of national income statistics devised by Simon Kuznets in the 1930s at the Federal Bureau of Economic Research, a response to the crisis of the Great Depression.[15] The development of national accounts would be given further importance by the planning needs of the total war economy of World War II. The need to rationalise society for the maximum output of war materiel was the impetus behind the development of production estimates (e.g. Gross Domestic Product). By the mid-1940s, the accounts had evolved into a consolidated set of income and production statistics, a model which generated an 'an integrated birds-eye view of the economy'. Though Kuznets himself refused to accept GDP as a measure of welfare or progress, it has been taken as exactly that, to the point where it is the ground zero of political measurement. The problems of GDP, which measures only the monetary value of annual sales of goods and services inside a nation's borders, are well known. For example, GDP may rise with increases in the wealth of the already wealthy, whilst the majority experience falling incomes, living standards, life expectancies and the erosion of civil and political rights. An oil spill that destroys coastal ecosystems will add to GDP due to expenditure on cleanup efforts. Irreversible extractions of old-growth timber or mineral assets add to GDP. What is more interesting to our narrative is how the pseudo-energetic ontology of general equilibrium analysis accommodated the newfound vocation of economics as a science of exponential 'growth' and 'development' in the post-war era, particularly as the radical re-calibration of neoclassical doctrines by the Chicago law and economics movement coincides with the emergence of the modern environmental movement.

[14] Ayres, R. (1994). *Information, entropy and progress.* New York: American Institute of Physics, p. 137.

[15] Landsfeld, S. (2000). GDP: one of the great inventions of the 20th Century. *2000 Survey of Current Business*, U.S. Bureau of Economic Analysis.

As we have seen, the Walrasian model of the economy, focussed upon defining mathematical 'proofs' for the existence of an invisible balancing hand, is fundamentally static. The basic model was to be much upgraded by John von Neumann in his classic multi-sector growth model (published in English in 1945). Despite its influence upon the economics academy and its increased complexity, the model suffered from much the same drawbacks. Growth occurred only homothetically—that is, smoothly, continuously and with all sectors expanding at exactly at the same rate. Moreover, the model ruled out technological change, resource depletion, and waste accumulation, as it specified a closed system in which there were a fixed number of processes and all sectoral outputs were produced by combinations of inputs from other sectors.[16] Later growth models were more flexible in their portrayal of technology and open to empirical time-series data from capital markets. Nevertheless, they repeat the same sins against ecology and the laws of thermodynamics by maintaining the assumption that nature is infinitely expandable, infinitely convertible, infinitely substitutable, infinitely pollutable, infinitely depletable, or what amounts to the same, more or less irrelevant to the phenomena of economic growth. In his earlier growth models, Solow excluded natural resources entirely. Commodities were produced by varying combinations of capital and labour. In later models, some natural resources were required, but the amount of 'growth' that could be accomplished with some initial quantity was again unlimited. Herman Daly has compared this to a chef, who in order to increase the number of cakes generated by a favoured recipe, does not order more eggs, butter and flour, but hires a kitchen hand, stirs the mixing bowl faster and increases the size of the baking tray.[17]

The economist Mark Blaug had this to say about the methodological limits of growth theory:

> Consider [...] the preoccupation since 1945 of some of the best brains in modern economies with the esoterica of growth theory, when even practitioners of the art admit that modern growth theory is not yet capable of casting any light on actual economies growing over time. The essence of modern growth theory is simple old-style stationary state analysis in which an element of compound growth is introduced by adding factor-augmenting

[16] Ayres (1994, p. 146).

[17] Daly, H. (1999). *Ecological economics and the ecology of economics: essays in criticism.* Cheltenham UK: Edward Elgar, pp. 86–87.

> technical change and exogenous increases in labour supply to an otherwise static, one-period general equilibrium model of the economy. In view of the enormous difficulty in handling anything but steady-state growth (equiproportionate increase in all the relevant economic variables), the literature has been almost solely taken up with arid brain-twisters about 'golden-rules' of capital accumulation.[18]

One can only wonder why hydrocarbon combustion was never entertained as have something to do with the growth of actual economies over time.

In 1932, Lionel Robbins (MPS) gave a definition of the 'nature of economic science' which became the textbook classic: 'economics is concerned with that aspect of behaviour that which arises from the scarcity of means to achieve given ends. [...] What the ends are is rather unimportant to the economist—he is more concerned with how a person chooses between scarce alternatives.'[19] The logic of economics as a deductive science of individual choice unfolds from the primordial fact of scarce time, means and resources. In the American neoclassical growth literature, however, a new assumption was added to the traditional list of perfect foresight, perfect information, and perfect competition:

> [...] the objective of the 'representative agent' (Oiko Nomos) is to maximize utility over an infinite horizon. Let us suppose that the utility function at any time t is hedonistically defined as a positive function of consumption per capita at time t, $U(c_t)$. Given an infinite horizon and continuous time, consumption will thus be infinite and continuous.[20]

As the Great Acceleration of the age of oil got underway, scarcity itself was quietly abolished as the defining feature of the economic problem. The 'given ends' were now endless, along with the means to those ends.

Whatever may have preoccupied the mathematicians of academic growth theory, as Richard Lane has shown, the consolidation of the post-war US consensus on economic growth (and the geopolitical strategies which followed from this) owed much to a series of widely publicised

[18] Blaug, M. (1980). *The methodology of economics*. Cambridge UK: Cambridge University Press, p. 254.

[19] Robbins, L. (1932). *An essay on the nature and significance of economic science*. New York: New York University Press.

[20] Fonseca, G. (2006). Neoclassical optimal growth: golden rules and the Ramsey exercise. History of Economic Thought Website. http://cepa.newschool.edu/het/essays/growth/optimal.htm. Accessed 13 April 2019.

reports compiled by the US President's Materials Policy Commission, under the title *Resources for Freedom* (1952).[21] Commonly known by the name of its Chairman, William Paley, director of the CBS television network, the Paley Commission was convened to respond to widespread public concerns regarding the vast quantities of American resources depleted and dissipated in the conflagrations of World War II. While the United States enjoyed continental sovereignty over a mineral resource base second only to the Soviet Union, it was nevertheless of great concern that US oil reserves had been significantly drawn down by the war, to the point where the United States would soon become a net importer of petroleum. The report made similar findings across a range of strategic minerals, suggesting that the United States had outgrown its resource base, at a time when economic growth, as the Commissioners noted in the opening lines of the Report, that had come to define the post-war American prospect:

> [...] we share the belief of the American people in the principle of growth. Granting that we cannot find any absolute reason for this belief we admit that to our Western minds it seems preferable to any opposite, which to us implies stagnation and decay.[22]

In its projections of resource demand forward to the 1970s, the key conceptual innovation of the report was its reconfiguration of the question of the absolute scarcity represented by quantitative measures of known resources into one of a merely relative scarcity, in which the key factors governing resource availability were relative prices. Higher prices for scarce resources would trigger new technological substitutions, extractive processes, or investments in previously diseconomic ventures. Scarcity was no longer a geological question, but one of the balance sheet of a multinational business inventory. Thus arrived the American orthodoxy that '[i]t is a mistaken notion [...] that on a given day the world will find that the last ounce or foot of a given resource has been used up. At a cost there is always more.'[23]

From the outset, the Commission had the public relations goals of US corporations in mind. Popularised in a 1954 national broadcast hosted by

[21] Lane, R. (2019). The American Anthropocene: economic scarcity and growth during the great acceleration. *Geoforum* (99), 11–21; President's Materials Policy Commission (1952). *Resources for freedom*, Vols. 1–5. Washington, DC: US Government Printing Office.

[22] PMPC (1952; cited in Lane, 2019, p. 13).

[23] Lane (2019, p. 16).

Ed Murrow on the Paley's CBS television network, the essential message of *Resources for Freedom* was that 'the free economy' was destined to grow toward an open future, beyond local resource limits. As Lane reports, when it came to the resources policy of the US government, rather than recommending a policy of national conservation, efficiency measures, or recycling, the Commission urged:

> [...] the expansion of overseas extractive industries under the banner of the 'least cost principle' as a means to govern material resources [...] and ensure future economic growth. This expansion was to be undertaken by: 'unfettered private enterprise, free from government controls' and regulated to the greatest extent by the 'spur of the profit motive,' 'the competitive market structure,' and 'the price system'.[24]

Lane describes this as a policy for 'the global expansion of the "American environment" [via] the planet-spanning supply chains of US state-supported corporate actors'.[25] In this respect it is noteworthy that much of the work of the Paley report in compiling industry statistics and projecting resource demand and price dynamics of strategic minerals through to 1975 was performed by 'a promising young economist': Arnold Harberger (MPS) of the Chicago School.[26] Harberger would later be a leading architect of the neoliberal experiments in 'shock therapy' conducted by General Pinochet's military dictatorship in Chile (1973–1990), following a US-sponsored coup d'état provoked by the 1971 nationalisation by Salvador Allende's Popular Unity government of Chile's US-operated copper mines.

Widely cited as the text that inaugurated the modern environmental movements' science-based critique of the 'growth miracles' of industrial technoscience, Rachel Carson's *Silent Spring* (1962) was published at a time when W.W. Rostow's *Stages of Economic Growth: a non-Communist Manifesto* (1960) had become a standard text of the development economics genre.[27] A political insider long connected with military intelligence, in 1961 Rostow was appointed director of the Policy Planning Staff

[24] Lane (2019, p. 17).

[25] Lane (2019, p. 12).

[26] Cooper, R. (1975). Resource needs revisited. *Brookings Papers on Economic Activity* (1), 238–245.

[27] Rostow, W. (1960). *The stages of economic growth: a non-communist manifesto.* Cambridge: Cambridge University Press.

of the US State Department by President Kennedy. In *Stages*, Rostow employed an organismic metaphor for industrialisation, arguing that all national economies can reproduce the pattern of industrialisation experienced by Western Europe and North America. 'The economy' is presented ontogenetically, that is, as going through several pre-determined 'stages of growth' resembling an organism's genetically programmed pathway from embryo to maturity. Presenting the problem of national development as beginning with a *tabula rasa*, Rostow attributed 'third world' poverty not to a history of colonial appropriation, but to the ideological obstruction of 'traditional' culture to the 'Newtonian' view of nature. 'Traditional' socio-economic organisation is for Rostow confined by pre-modern ignorance and values to 'limited production functions'.[28] The realisation that nature is knowable according to 'a few simple laws', he argues, allows the systemic manipulation and transformation of nature in the direction of progress. Thus he associates 'Newtonianism' with 'unlimited production functions'. Rostow saw the ultimate goal of national development as the achievement of an 'age of high mass consumption', an economy of consumer abundance similar to that realised in the United States in the 1950s. Rostow argued that this historical process could be engineered by policies favourable to capitalist industry. Where Rostow departs from his implicit metaphor of the industrial 'organism' is that reaching the final 'mature' stage of economic growth does not mean an end to growth, nor a transition to the maintenance of the body's integrity in the face of inevitable senescence and death. In the mature Rostovian economy, 'growth is the normal condition'.[29] Against the zero-sum model of dependency theorists such as Raúl Prebisch, Fernando Cardoso, and Immanuel Wallerstein, who argued that concentrations of wealth in the 'advanced' nations more or less reflected concentrations of poverty in the underdeveloped world mediated by historically unequal trade relations, modernisation theorists championed perpetual growth as the historical destiny available to all nations willing to embrace a 'Newtonian' view of social relations and nature.

Two years after Rostow's essay, and a decade before the *Limits to Growth* report, another inquiry into the burgeoning mineral resource demands of US industry by Barnett and Morse returned the following reassuring verdict on the relationship between *Scarcity and Growth* (1963):

[28] Rostow (1960, p. 4).
[29] Rostow (1960, p. 10).

> Advances in fundamental science have made it possible to take advantage of the uniformity of matter/energy, a uniformity that makes it feasible *without preassignable limit* to escape the quantitative constraints imposed by the character of the earths' crust. [my emphasis][30]

While this does nothing to reassure us of the unlimited availability of fresh water, fertile soil, old growth forests, or fish, at least in the mining sector it was claimed that modern technoscience had effectively abolished scarcity. No doubt bolstered by the early 1960s faith in the 'peaceful atom', there are obvious parallels here with the neoclassical model, with its proto-energetic assumptions of the unlimited convertibility of capital, labour and resources. Remember that, strictly speaking, the metaphorical appropriation of the conservation principle of the first law of thermodynamics implies there can be no creation in production, only conversion. For the assumption of balance to work, the value substance must remain constant, neither created nor destroyed through all its economic transformations. With Rostow, Barnett and Morse, the application of 'technology' to the 'uniformity of matter/energy' now licensed a creationist infinitude. Not only is 'the economy' a perpetual motion machine, it is also the Philosopher's Stone of the alchemists: even if the Earth were made entirely of a fixed quantity of lead there would be no limit to the amount of gold that 'fundamental science' could produce from it.

Rostow's view of the malleability of nature exposed to technoscience was reflected in his view of the malleability of societies confronted with externally imposed forces of historical change. According to J.K. Galbraith, another of the 'best and the brightest' assembled by the Kennedy Administration, Rostow in his advisory role to war planner Robert McNamara advocated the systematic cluster bombing of the Vietnamese countryside on the view that this would lead to mass migration to towns and cities. Reminiscent of the approach of the British to the Irish famine, the mass evacuation of 'low-value' subsistence farmers forced from their land by hunger and terror would provide a cheap, flexible, urban labour force, which would attract investment and trigger the initial stages of industrial modernisation, without the 'erroneous' diversion through

[30] Barnett, H. & Morse, C. (1963). *Scarcity and growth*. Baltimore: John Hopkins Press, p. 11.

Marxist-Leninism.[31] When the obscene policy of carpet-bombing was applied to rural Cambodia, where nearly 2.8 million tons of ordnance were dropped (the Allies are thought to have used 2 million tons in WWII) it yielded only the auto-genocidal year-zero 'communism' of an agrarian society utterly traumatised into the absolute rejection of modernity.[32] Yet it is usually only the Marxists who are remembered as the utopian historical determinists of the period. McNamara later served a dignified stint as the chair of the World Bank, presiding over the first decades of 'development aid', issuing the initial debts to which many nations remained in thrall generations later. These loans were made in lieu of the 'savings' (capital investment in industry) that orthodox economists saw as the key driver of economic growth, in order to place the nation a stage or two higher on Rostow's ladder of development, to the point where growth become a self-reinforcing process such that growth rates would exceed the interest charges on the development loan and allow it to be repaid. This alarming contradiction between the imperatives of 'free market' growth economics and violent social engineering would be less so were it safely confined to the dustbin of history.

Following the fall of the Berlin Wall, Francis Fukuyama, a staff member of the Middle East divison of Policy Planning at US Department of State, famously declared the 'end of history'. Later appointed professor of public policy at George Mason University, Fukuyama foresaw the impending globalisation of an unchallengeable order of 'liberal democratic capitalism' which would bring universal peace and prosperity. This vision rested upon a now familiar ontology of human-environment relationships: for Fukuyama, the origins of wealth were not in the intentional labour which transforms land and resources into sustenance and artefacts, but in the spontaneous order of the market and the 'knowledge' of capitalist technoscience:

> [...] modern natural science establishes a uniform horizon of economic production possibilities. Technology makes possible the *limitless* accumulation of wealth, and thus the satisfaction of an *ever-expanding* set of human

[31] Garrett, K. (2006, Jan 15). Interview with J.K. Galbraith. Background briefing. Sydney: Australian Broadcasting Corporation.

[32] Owen, T., & Kiernan, B. (2007, May 12). Bombs over Cambodia: new light on US air war. *Third World Traveller*.

> desires. [...] the logic of modern natural science would seem to dictate a universal evolution in the direction of capitalism. [my emphasis][33]

At the core of Fukuyama's argument was the notion that liberalism was the 'endpoint of mankind's ideological evolution'—an ideological system free from internal contradictions. Yet his version of 'modern natural science' is innocent of the Earth sciences and contradicts the laws of thermodynamics, to the point of rejecting, as so many economists are willing to do, 'the simple fact that the Earth is finite.' Fukuyama's view of 'science' is informed by his Hegelian idealism, an eschatology in which 'consciousness will ultimately remake the world in its own image'.[34] Wealth is the result of boundless technological wish-fulfilment. If technology allows 'limitless accumulation', then the obvious question is why is everyone not yet a billionaire? According to the Policy Planners of the State Department, the billions of poor people in the global South remain outside the post-historical 'age of high mass consumption' because they persist with the wrong 'ideas'.

Beyond liberal economics there are but few discourses where infinitude is routinely contemplated—astronomy, pure mathematics and Christian eschatology come to mind—all of which have little to do with the immediate facts of life for mortal, embodied creatures confronted with the question of 'how are we to live' on the surface of the Earth. This is perhaps less than surprising: as I have argued, economics has long held elective affinities with the former group of arcane pursuits, and is perhaps more accurately described as an anti-scientific political theology than a moral science of *oikonomia*.

Since Smith, it might be said, we have been living in the Age of Economics. The fundamental flaw of all economics derived from the classical focus on the dynamics of exchange is its neglect of the depletion of resources in production, and the limited resilience of ecosystems to over-harvest, pollution and radical shifts in the chemistry and temperature of sky and sea. It is remarkable that the extreme optimism exemplified in Engels' claim that 'the productivity of land can be infinitely increased by the application of capital, labour and science'[35] still remains the default position of official economic theory and practice.

[33] Fukuyama, F. (1992). *The end of history and the last man*. Avon Books, p. xv.
[34] Fukuyama (1992, p. xvii).
[35] Cited in: Ponting, C. (1991). *A green history of the world*. London: Penguin, p. 158.

The continuing polarisation of wealth and poverty within and between nations through four decades of neoliberal globalisation suggests that the widening discrepancy in the 'exchange value' of human life may be correlated with the acceleration of economic growth. Since Rostow, these disparities have been presented as temporary features of the historical process of debt-financed economic development—wealth would one day trickle down—because the promised destiny of all nations that submitted to the 'system of natural liberty' and a 'Newtonian' policy of industrialisation was a 'state of high mass consumption, where 'growth is the normal condition', and 'compound interest becomes built [...] into habits and institutional structure.'[36] The fetishistic view of machines as intrinsically 'fertile' in economic discourse is connected to the fetishism of money: after all, both industrial infrastructure and financial wealth are ambiguously conflated as 'capital', in turn defined as 'that which will yield more capital in the future'.

It might appear that money invested in an interest-bearing bank account can increase infinitely; after all it is only digital information. When anchored to a physical index such as the gold standard, however, the logic of compound interest yields utterly implausible outcomes. According to the calculations of Heinrich Haussman, one Deutsch *pfennig* (trading at about half a US cent at the time) invested from the year 0 AD at 5% compound interest, would by 1990 have yielded a volume of gold bullion 134 billion times the weight of the Earth.[37] The need to repay debt out of future earnings is fine if growth is perpetual, but to the extent that growth depends on irreversible depletion, today's credit binge risks emptying the bank of 'natural capital' when Mother Nature, the insurer of last resort, refuses to cover all these claims.

Such are the illogical contradictions at the heart of the central metaphor of market economics: <energy = utility = value = money>. It would of course be naive to suggest that the damage of this ubiquitous, subterranean cultural doctrine can be undone by insisting that money is symbolic and therefore illusory, and that energy is material and therefore real. The troubling implications of the following comment made a century ago by Frederick Soddy are still with us today:

[36] Rostow (1960, pp. 7–12).

[37] Monbiot, G. (2004). *The age of consent*. London: Flamingo, p. 239.

> Energy, someone may say, is a mere abstraction, a mere term, not a real thing. As you will. In this, as in many other respects, it is like an abstraction that nobody would deny reality, and that abstraction is wealth. Wealth is the power of purchasing, as energy is the power of working. I cannot show you energy, only its effects. […] Abstraction or not energy is as real as wealth—I am not sure that they are not two aspects of the same thing.[38]

The questions raised by this observation have only been touched on here, and our history of the relationship of economics to physical nature has not yet to dealt directly with nature as biological life. In the next half of the book we will discover that the question of energy was also central to the disciplinary history of ecology as it matured as a science linked to the announcement of industrial apocalypse in the early 1970s. Here, it is timely to consider a third aspect of Soddy's dualism of energy and wealth: the relation of both to *power*, both in the sense of political authority and access to hydrocarbon energy.

Looking back, it seems that as prophesies of ecological apocalypse loomed in the early 1970s, an existential space opened up where capitalism as such was open to question. At the 1972 United Nations conference in Stockholm, it was widely argued that far from carrying a 'technological cross' for the underdeveloped countries, developed nations were building their wealth upon the impoverishment of the Third World. Since the 1987 Brundtland Report, the consensus view has been that there is no contradiction between economic growth and global sustainability.[39] The question of infinite growth has been culturally quarantined from discussions of our ecological future. Take for example Nicholas Stern's (2006) report on the economics of climate change, seen as a major breakthrough to the conventional circles of the business and banking professions. Describing global warming as 'the biggest market failure the world has ever seen' was shocking language in the age of the neoliberal consensus, but Sterns' revelation that runaway climate change might 'damage economic growth' seems to have clinched the argument. 'Tackling climate change is the pro-growth strategy for the longer term' says the report, 'and it can be done in such a way that does not cap the aspirations for growth of poor or rich countries'.[40] The solution is to unleash the power of the market to correct

[38] Soddy, F. (1920). *Science and life.* New York: Dutton, pp. 27–28.
[39] Hornborg, A. (2001). *The power of the machine.* Altamira Press, p. 21.
[40] Stern, N. (2006). *Stern review on the economics of climate change.* HM Treasury.

the market's failures, by creating new financial markets trading in carbon credits, carbon offsets, and pollution permits, with the aim of 'pricing in' carbon emissions and financing investments in renewable energy. This is a policy which has widely failed to generate a carbon price high enough to actually lower emissions.[41] In the Stern review, even the threats of a radically destabilised climate are mobilized to secure our faith in unceasing growth and the miracles of the market. Even as the Earth's biosphere 'shrinks' amidst rising heat, the permanent growth economy remains a stable object only loosely connected to its 'environment'.

Constructing a 'Climate of Opinion' Against Climate Action: The Atlas Network and the Engineering of Ignorance

The radicalism of the 1970s was embedded in a hope that human beings could act rationally in solidarity to avert apocalypse, and progress intellectually, politically, and morally. Since then, progress has been reduced to promises that the free market will bring more economic growth. The following comments, made on the cusp of the neoliberal counter-revolution in a 1972 address to the American Economic Association by its then president J.K. Galbraith, would be quite unthinkable in that forum today:

> The most commonplace features of neoclassical and neo-Keynesian economics are the assumptions by which power, and therewith political content, is removed from the subject [...] If the state is the executive committee of the great corporation and the planning system, it is partly because neoclassical economics is its instrument for neutralising suspicion that this is so. I have spoken of the emancipation of the state from economic interest. For the economist there can be no doubt where this begins. It is with the emancipation of economic belief.[42]

A similar problem was posed from the neoliberal perspective of James Buchanan (MPS) of the Virginia school, who along with fellow MPS members offered his expertise to the most senior officials of the Pinochet

[41] Spash, N. (2010). The brave new world of carbon trading. *New Political Economy*, 12(2), 169–195.

[42] Galbraith, J. K. (1972). Power and the useful economist. Presidential address to the American Economic Association.

dictatorship.[43] Buchanan's project in *constitutional* political economy was quietly bankrolled by the oil billionaire Charles Koch (MPS).[44] The aim of this investment, if Chile's neoliberal Constitution of 1980 can be taken as exemplary, was the permanent fortification of the powers of private capital in law and legislation and the radical restriction of popular sovereignty and democratic participation in the legislative processes of the liberal state. Yet as Buchanan himself once candidly wondered,

> The classical liberal, in the role of social engineer, may of course recommend institutional *laissez-faire* as a preferred policy stance. But why, and under what conditions, should members of the citizenry or of some ultimate political authority, accept this advice more readily than that offered by any other social engineer?[45]

One response would be to silence the citizenry with the state-terror apparatus of an Operation Condor. Another answer is hinted at in the following words of Hayek, the most eloquent expositor of the Austrian view that it is impossible for any rational policy of government to solve the problems of economic injustice it seeks to address. Hayek's lifelong political efforts to immunise economic 'freedom' from 'unlimited democracy' begins with an insistence on the ineradicable ignorance of the demos.

> Probably it is true enough that the great majority are rarely capable of thinking independently, that on most questions they will accept views which they find ready-made, and that they will be equally content if born or coaxed into one set of beliefs or another. In any society, freedom of thought will be of direct significance only for a small minority.[46]

The problem of mass communication under (allegedly) ontological conditions of general ignorance is at the root of Hayek's account of price formation as the distributed computation of a radically decentralised yet omniscient market machinery:

[43] Farrant, A. (2019). What should Knightian economists do? James M. Buchanan's 1980 visit to Chile. *Southern Economic Journal*, 85(3), 691–714.

[44] MacLean, N. (2017). *Democracy in chains: The deep history of the radical right's stealth plan for America*. Penguin.

[45] Cited in Mirowski (2013, p. 73).

[46] Cited in Mirowski (2013, p. 79).

> It is more than a metaphor to describe the price system as a kind of machinery for registering change, or a system of telecommunications which enables individual producers, to watch merely the movement of a few pointers, as an engineer might watch the hands of a few dials.[47]

Hayek's insistence that egalitarian democracy would bring ruin was never likely to win widespread approval by a democratic citizenry on the basis of its reflection upon Hayek's corpus of 'scientific' publications. Since the commoners are incapable of exercising the responsibilities of self-rule acquired in the democratic revolutions that had, from the Austrian perspective, erroneously constitutionalised popular sovereignty, the problem is to make sure that the 'ready-made views' which they will be content to adopt are the correct ones, those most amenable to the advice of the neoliberal social engineer. Thus the Leninist vanguard work of the *libertarian internationale* of MPS intellectuals and political insiders seeking to capture the policy planning powers of the state must be complemented with a permanent programme of mass communication to counter and undermine the sources of the 'wrong' sets of beliefs—public universities, public servants, public broadcasters, public scientific institutes—and to coax, confuse or intimidate the citizenry into accepting their submission to a market order in which all public knowledge, assets and services are, in the long game, to be fully privatised. In public, Hayek constantly reiterated the message that 'the free market' emerged 'naturally' and 'spontaneously' as an undesigned epistemic machinery for condensing and distributing the 'knowledge' conveyed in prices. In private, Hayek understood that protecting the market mechanism from excess democracy would require the engineering of consent, through the intentional construction of an agnotological political machine mass-marketing 'business propaganda'.

The task of setting up this parallel machinery for mass communication (and the recruitment and training of neoliberal activists) was taken up by the English businessman Anthony Fisher, a devotee of Hayek. In 1955, Fisher founded the Institute for Economic Affairs (IEA), which later launched the Thatcher revolution from the far right of the Conservative Party. In a letter to Fisher after her 1979 election victory, Margaret Thatcher wrote that the IEA had created 'the climate of opinion which made our victory possible.' In the 1970s, Fisher went on to set up the

[47] Hayek, F. (1945). The use of knowledge in society. *American Economic Review*, 4(35), 519–530.

Manhattan Institute, the Pacific Research Institute and the Fraser Institute. In 1981, Fisher established the Atlas Foundation for Economic Research (later renamed the Atlas Network) to coordinate an expanding worldwide network of 'thinktanks' dedicated to the production of the right kind of collective ignorance, through the oxymoron of the for-profit 'non-profit' research institute.

Radically pro-corporate, anti-labour, and anti-environmentalist business propaganda would of course be immediately suspect if issued in the name of the corporate sponsors themselves. And likewise, as Fisher recognised, if it was clear that the campaigns of a phalanx of arms-length 'public policy research institutes' to influence policy-making were planned and centrally coordinated. Hence Fisher's original funding pitch for the Atlas Foundation, presented to billionaire donors and the CEOs of powerful corporations, emphasised the importance of non-transparency: 'To influence public opinion, it is necessary to avoid any suggestion of vested interest or intent to indoctrinate.'[48] Each new Atlas outfit must appear to operate independently of the others, and claim to be completely uninfluenced by its strictly undisclosed list of donors.

Flooding the public sphere and corporate media with constantly repackaged opinion pieces, 'independent' research papers, cost-benefit analyses, strategic litigation, 'astroturfed' NGOs, targeted Facebook memes, and confected shock-jock outrage, these agenda-setting policy PR campaigns present to the citizen as a diverse array of opinion on a wide set of policy issues, such that it would be reasonable to accept some of these views as one's own, or at least that they were widely held by fellow citizens. Another function of the network is to provide draft legislation to aligned politicians: as Madsen Pirie of the Adam Smith Institute once put it, '[w]e propose things which people regard as being on the edge of lunacy. The next thing you know they're on the edge of policy.'[49] Whilst its existence remains little known amongst political commentators, the Atlas Network has by such means 'reshaped political power in country after country', swinging public opinion and elections rightward in the direction of an increasingly authoritarian liberalism. According to the investigative journalist Lee Fang, the Atlas Network operates as 'a quiet extension of

[48] Fang, L. (2017, Aug 9). Sphere of influence: how American libertarians are remaking Latin American politics. *The Intercept*.

[49] Rusbridger, A. (1987, Dec 22). The Adam Smith Institute's sense and nonsense. *The Guardian*.

U.S. foreign policy, with Atlas-associated think tanks receiving quiet funding from the State Department and the National Endowment for Democracy'.[50] Perhaps nowhere is this more important than the sphere of energy and climate policy, and especially in those nations whose unexploited hydrocarbon reserves are coveted by the US-based corporations that have financed the neoliberal project from its early days.

For more than sixty years the senior executives of the US oil majors have been well informed of the science linking the combustion of hydrocarbons to an increasingly catastrophic planetary heating trend. At a 1959 conference on the theme of 'Energy and Man', a celebration of 100 years of the US oil industry, Robert Dunlop, a future president of the American Petroleum Institute (API), shared the stage with the H-bomb physicist Edward Teller, who advised oil executives and policy planners present that

> carbon dioxide […] transmits visible light, but it absorbs the infrared radiation which is emitted from the earth. Its presence in the atmosphere causes a greenhouse effect in that it will allow the solar rays to enter, but it will to some extent impede the radiation from the earth into outer space.[51]

Teller warned that if atmospheric concentrations of CO_2 rose in accordance with the then current trend through to the end of the twentieth century, the icecaps would melt, and 'submerge New York. All the coastal cities would be covered'. In a 1965 keynote to the API, its president Frank Ikard briefed fellow oil executives on the recent report of US President Lyndon Johnson's scientific advisors on CO_2 and global warming.[52] Ikard accepted its seriousness without criticism, recognising that the internal combustion engine would have to be phased out before the year 2000 or the heat balance of the Earth would rise beyond the reach of policy-makers: 'the substance of the report is that there is still time to save the world's peoples from the catastrophic consequence of pollution, but time is running out'.[53]

Anticipating rising demands upon governments to restrict fossil fuel use in response to climate change, since the 1980s fossil-fuel corporations,

[50] Fang (2017, Aug 9).

[51] Franta, B. (2018). Early oil industry knowledge of CO2 and global warming. *Nature Climate Change*, 8(12), 1024.

[52] Revelle, R., et al. (1965). Atmospheric carbon dioxide. In President's Scientific Advisory Committee, Restoring the quality of our environment. Washington: White House.

[53] Franta (2018, p. 1024).

media barons such as Rupert Murdoch, chambers of commerce, and neoliberal politicians have collaborated with MPS members and the Atlas Network to mass manufacture ignorance, confusion, and doubt regarding the scientific truth of global heating, to encourage hostility toward scientists, environmentalists and advocates of renewable energy, and to undermine and permanently delay adequate policy responses to catastrophic climate change at the domestic and international level.[54] It is of course true that the influence of the Atlas Network is limited in nation-states with already-existing authoritarian governments and extensive fossil-fuel resources (such as China and Russia), and that certain democratic jurisdictions with relatively robust public spheres have managed, with wide public support, to initiate relatively ambitious policies of investment in large-scale wind and solar, moratoriums on hydrocarbon extraction, and timetables to phase out coal-fired electricity and sales of vehicles powered by internal combustion (such as Costa Rica, New Zealand, Scotland, Germany, and other European countries). Nevertheless, the Atlas strategy to turn public opinion against the pursuit of responsible climate policies, most importantly in the United States, has been successful in undermining the international cooperation required to reverse the rising trend of global GHG emissions. Increasingly, the Atlas Network cultivates the far extremes of right-wing politics in pursuit of this goal. Atlas-linked think-tanks played a major role in channelling 'dark money' funds into the campaign for Brexit, after which the Tory MP Jacob Rees-Mogg hopes to see the elimination of UK environmental laws, along with those of the EU.[55] The climate-denialist Heartland Institute reportedly now collaborates with the ethno-nationalist Alternative für Deutschland,[56] a strategy presumably aimed at exploiting discontent with the neoliberal austerity policies of European monetary authorities to destabilise the European Union and undermine the climate policies pursued by the European Parliament through its member states.

[54] Oreskes, N., & Conway, E. (2011). *Merchants of doubt: how a handful of scientists obscured the truth on issues from tobacco smoke to global warming*. New York: Bloomsbury; Brulle, R. (2014). Institutionalizing delay: foundation funding and the creation of US climate change counter-movements. *Climate Change* (122), 681–694; Kaiser, D. & Wasserman, L. (2016, Dec 8). The Rockefeller Family Fund vs. Exxon. *New York Review of Books*.

[55] Stone, J. (2016, Dec 6). Britain could slash environmental and safety standards 'a very long way' after Brexit, Tory MP Jacob Rees-Mogg says. *The Independent*.

[56] Baynes, C. (2019, May 15). Germany's far-right AfD takes aim at Greta Thunberg as it denies climate emergency. *The Independent*.

The effect of nearly four decades of this organised obstruction of climate and energy policy is that we now face an increasingly likely escalation of planetary heating towards a terrifying 'hothouse Earth' scenario, which may well render our only home uninhabitable by century's end, completing the sixth mass extinction event in the Earth's unique history as a life-bearing planet. This must surely qualify as the darkest imaginable victory of the agnotological arts, going well beyond Ulrich Beck's account of the 'self-annihilation' of society through the criminal neglect of a complacent social science. From the perspective of the young people born to the twenty-first century, who will unjustly suffer the direst consequences of the policy planning of an ageing elite in the service of fossil plutocrats, it is not enough to speak of criminal neglect. It can only be appropriate to speak of the most criminal intent.

In conclusion, one can only concur with Greta Thunberg's recent comments to the UN Climate Action Summit:

> People are dying. Entire ecosystems are collapsing. We are in the beginning of a mass extinction, and all you can talk about is money and fairy tales of eternal economic growth… How dare you! [57]

[57] National Public Radio (2019, Sept 23). Transcript: Greta Thunberg's speech at the UN Climate Action Summit.

PART III

Ecology as Social Physics

CHAPTER 8

The Age of Ecology

'Anyone Will Profit from Learning How the Biosphere Functions as an Economic System'[1]

On the 22nd of June, 1969, the Cuyahoga River flowing through the industrial district of Cleveland, Ohio caught fire. It is unclear what particular cocktail of petrochemical pollution caused the fire, but locals reported that sparks from a passing train ignited an oil slick on the river. Seven years after Rachel Carson's *Silent Spring* (1962) had punctured enthusiasm for 'better living through chemistry' by demonstrating that US farm chemicals were ubiquitous, present even in the bloodstream of Arctic nomads, this event galvanised public outrage at the scandalous liberties taken by the chemical and automotive industries, at a time when student activists were targeting Dow and Monsanto for their role in supplying napalm and Agent Orange for the war against the peasants of Indochina. *Time* magazine described the river like this:

> Chocolate brown, oily, bubbling with sub-surface gases, it oozes rather than flows. [...] The Federal Water Pollution Control Administration dryly notes: 'the lower Cuyahoga has no visible life, not even low forms such as leeches

[1] Bambach, R. (2005). Ecology as an economy: review of 'Nature: an economic history' by G.J. Vermeij, 2004. *American Scientist*, 93(2), 185–187. See p. 185.

J. Walker, *More Heat than Life: The Tangled Roots of Ecology, Energy, and Economics*,
https://doi.org/10.1007/978-981-15-3936-7_8

> and sludge worms that usually thrive on wastes. It is also, literally, a fire hazard.[2]

Lake Erie, into which the Cuyahoga drains, was from the 1960s also dying in full public view in a state of near ecosystem collapse. In addition to petrochemical toxic pollution, this was due to an extreme accumulation of phosphorus in the lake. Phosphorus is a basic nutrient for biological growth, a building block of life. The initial source was an increasing urban outflow of human excreta, which like all animal waste contains phosphorus. With the domestication, commercialisation and expansion of wartime chemical industries during the 1950s, sewerage also carried a sudden abundance of phosphate detergents. Mixed with agricultural runoff, feedlot waste and chemical fertilisers, these generated vast blooms of algae in the lake, which removed much of the dissolved oxygen in the water. The consequences of this were a shift from the remarkable water clarity expected of a Great Lake to a general murkiness, a substantial collapse of aquatic biodiversity and fish populations, and extensive piles of reeking algae rotting on the shoreline where residents had prime real estate. A minor cataclysm, the wholly unnatural Cuyahoga River fire drew its apocalyptic glare from the Promethean furnaces burning at the hidden foundation of Progress.

The mastery of fire was key to the first 'globalisation': the radical Pleistocene expansion in the northward and southward range of human groups from equatorial regions until the extremities of every landmass bar Antarctica was populated by fire users, first Tasmania, then Tierra del Fuego, and finally Aotearoa/New Zealand. Its domestication features prominently in myths all over the world as the moment when humans entered into a competitive rivalry with the gods and became properly human, cultural, beings; which is to make, as Levi-Strauss suggested, a distinction between the raw and the cooked.

Burning rivers, however, signify a historically new intensity to the human ecology of fire: the increasing volume and speed of fossil fuel combustion and the chemical sophistication that the metallurgical mastery of heat had taught. It was not the first time that the Cuyahoga River had caught on fire, but in 1969 it came to symbolise for the United States the seriousness of the unregulated dumping of industrial pollutants into the

[2] Cited in: Craige, B-J. (2002). *Eugene Odum: ecosystem ecologist & environmentalist.* Athens: University of Georgia Press, p. xii.

soil, lakes, rivers, atmosphere, aquifers and the sea, and the previously unnoticed accumulation of a battery of synthetic chemicals that had come into existence and use since World War II. It was this unnatural conflagration, along with the death of Lake Erie and several spectacular oil spills, that catalysed an increasing sense that a human-made environmental crisis was imminent. In the same year as the portent of the burning river, *Time* magazine wrote that pollution would 'soon replace the Viet Nam war as the nation's major issue of protest' and called 1969 'the year of ecology'. The article profiled 'the new Jeremiahs'—the leading ecologists of the day—who explained 'that all nature is interconnected and that any intervention has far-reaching effects'. In 1970, *Newsweek* upped the ante and announced the dawn of 'the Age of Ecology', in which ecologists would teach society how to transform its relationship to 'the web of life'.[3] The same year, President Richard Nixon established the Environmental Protection Authority, to the amazement of his opponents in the radical movements and the corporate supporters of the Republican Party. The new environmental legislation marked the internalisation of the environmental crisis within the regulatory apparatus of the American state, a state centrally concerned since WWII with facilitating economic growth. The nation-state thus acquired a subsidiary remit to ward off the environmental degradation and 'side-effects' of economic growth, the securing of which had become (and still remains) the foundational task of governments.

The social regulation of pollution is not merely the concern of environmental agencies. This concern, perhaps a cultural universal, is replicated in the fundamental taboo structures of the Judaeo-Christian tradition.[4] Mary Douglas defined pollution simply as 'matter out of place', and noted that its proliferation warned of an impending crisis in the social order, manifest in the violation of taboo and threatening the desolation of cosmic order. Elaborating on Douglas in *The Powers of Horror* (1982), Julia Kristeva writes of taboo and sacredness in patriarchal religion, arguing that waste, the undifferentiated, the slimy—'what disturbs system, order, identity'—are marked out for exclusion from the sacred. The abject is excluded by

[3] Craige (2002, p. 81).

[4] Douglas, M. (1970). *Purity and danger: an analysis of concepts of pollution and taboo.* New York: Routledge.

taboo for its dangerous capacity to overwhelm society through pollution.[5] Uncontrolled pollution, the contradictory mixture of identities that should be separate (fire and water in this case), and the contravention of the sacred balance between the order of society and the order of nature are tell-tale signs of imminent apocalypse, recognisable in numerous cosmologies.

But what exactly is meant by apocalyptic? Several candidate definitions include: 'A belief that all history has a single irreversible conclusion [...] A teleological framework for the understanding of evil [...] An attempt to usher in a new era by redefining the rules of the redemptive process [...] A sense that each passing moment stands in significant relation to a beginning and an end [...].'[6] These definitions have been offered by Malcolm Bull, in an expansive account of the secularisation of the apocalyptic genre in progress-oriented modern political theory. Bull ploughs through Hegel, the Weimar Social Democrats, Benjamin, Heidegger and Rorty, but avoids what I would argue must be the most relevant candidate for analysis of modern politics and the apocalyptic tradition: the conflict between infinite growth advocates and the modern environmental movement.

The philosopher of history Hans Blumenberg differentiated the temporal structure of Progress ideology, with its steady progress toward a perpetually deferred goal, from Christian apocalyptic, arguing that 'eschatology speaks of an event breaking into history, an event that transcends and is heterogenous to it'.[7] Political theory, political economy and social movement ideologies have generally nourished themselves upon a view of human history as elevated and distinct from natural history. What I would suggest is that the global ecological crisis—the unprecedented degradation of the biosphere and associated multiplication of existential hazards—is a total 'event breaking into history' that has just this eschatological structure. Since the 1970s, environmentalists have feared a very real apocalypse—'an event breaking into history' not through divine revelation but by an increasingly detailed and sophisticated account of the mounting destruction of nature in the annals of natural science—and have established an urgent political quest for the ecological salvation of nature and society from the wasteful, polluting, and destructive logic of the

[5] Cited in: Bull, M. (1999). *Seeing things hidden: apocalypse, vision and totality*. London: Verso, p. 61.

[6] Bull (1999, p. 47).

[7] Blumenberg, H. (1983). *The legitimacy of the modern age*. Cambridge: MIT Press, p. 30.

'millennium' of exponential industrialisation. Indeed when considering the organization of neoliberal actors in the service of powerful fossil-fuel interests to defeat any policy to deflect from runaway planetary overheating, it seems we are dealing with 'economic growth' as a cult of collective suicide—although of course the greater collective of life beyond the 'commanding heights' was never 'free to choose' this fate from among other possible destinies.

The Invisible Vernadskyan Revolution

As *Newsweek* declared the 'age of ecology' in 1970, an existential moment had arrived in the Western cosmopolitical imagination. For the first time, ecologists were offered an opportunity to communicate their science through the mass media to a wide audience, one that had benefitted from the increased opportunities for higher education of the post-war welfare state. Thus the term 'ecology' entered into common parlance for the first time. With it came arguably the first generation to develop an ecological consciousness of the Earth as whole, a movement of thought that would be concentrated in the form of the world-revealing photographs of the Earth in a single frame, beamed back to her from afar by the spacecraft of the Apollo programme. Perhaps nothing brought home 'the simple fact that the Earth is finite' more than the picture of a solitary blue orb, suspended alone in the vast reaches of outer space, perhaps the only home for life in the Universe.

The grandest aim of environmentalism in the 1970s, it might be said, was to establish an 'age of ecology' which would correct the apocalypse-blindness of the 'age of economics' and allow us to restore and maintain the 'balance of nature'. But which ecology, and which concept of balance, equilibrium, or harmony? A passing acquaintance with the history of ecology reveals that, like other branches of knowledge that address complex totalities, it has never achieved anything like an internal consensus as regards its paradigmatic methods and models, and its borders with other branches of science are often porous.

The first scientist to fully recognise solar-driven photosynthesis and biological life's evolutionary trajectory as the profoundest force in the transformation of the Earth's inanimate geochemistry was the brilliant Russian mineralogist, pedologist, and biochemist Vladimir Vernadsky (1863–1945). Developing the concept of 'the biosphere'—a term proposed fifty years earlier by Edward Suess—Vernadsky first presented empirical demonstrations of his thesis in *La Geochimie* (1924), a work published in French which contained the first use of the term 'the carbon cycle'.[8] This was followed by his great work, the visionary systematic synthesis of *Biosfera* (1926), published first in his native Russian, and in French translation in 1929.[9] These works superseded an older presumption of evolutionary history, that life had evolved on Earth within the thermal and geochemical limits imposed by a relatively stable lithosphere, hydrosphere, and atmosphere.

As Jacques Grinevald argues, it is one of the great tragedies of intellectual history that Vernadsky's revolutionary scientific paradigm remained long confined behind the iron curtains of Cold War geopolitics.[10] Vernadsky was almost entirely unknown in the West until references to his work began to appear in English, following an epoch-making article in *Scientific American* on 'The Biosphere' (1970) by G.E. Hutchinson.[11] Intellectual traffic between Western scientists and those of the Soviet Union—where Vernadsky's standing as the originator of biosphere theory had long been honoured—increased through international environmental science collaborations in the mid- to late 1970s, especially via the climatologist Mikhail Budyko's early work on global warming and the climatic aspects of the 'evolution of the biosphere'.[12] Since, mainstream scientists have come increasingly to recognise with Vernadsky that the non-living geochemical environment of the Earth's vast and complex biotic community has for billions of years been 'regulated by life for life'.[13] Whilst partial, xeroxed English translations of Vernadsky's *Biosphere* had circulated amongst a handful of specialists beforehand, the 'biosphere' concept came to public attention through the ambitious failures of the Biosphere 2 experiment. Conducted privately between 1987 and 1991 by Space Biosphere Ventures Inc, a counter-culture project sponsored by the heir of a Texas oil fortune without official research funding or supervision, this involved the enclosure of a team of 'bionauts' within a sealed greenhouse in the Arizona desert for more than a year. Containing a minimalist replica

[8] Vernadsky, V. (1924). La géochimie. Paris: Alcan.

[9] Vernadsky, V. ([1926] 1998). *The biosphere* (trans. D. Langmuir). New York: Copernicus.

[10] Grinevald, J. (1998). Introduction: the invisibility of the Vernadskyan revolution. In V. Vernadsky, *The biosphere* (pp. 20–32). New York: Copernicus.

[11] Hutchinson, G. E. (1970). The biosphere. *Scientific American*, 223(3): 45–53.

[12] Budyko, M. (1986). *The evolution of the biosphere*. Dordrecht: Reidel.

[13] Margulis, L., & Sagan, D. (1997). *Microcosmos: four billion years of microbial evolution*. Berkeley: University of California Press, p. 94.

of the Earth's biosphere, this was envisioned as a Noah's Ark-like testbed for the solar-powered, pollution-free eco-engineering techniques that would be needed for autonomous human life-support in anticipated scenarios of long-range spacefaring and space colonisation.[14]

Only in 1998 did an authoritative English edition of *The Biosphere* first appear. Even now, Vernadsky's name remains relatively unknown amongst otherwise scientifically-literate Anglophone authors. We can only speculate on the extent to which our planetary future might have been otherwise, had the profound import of Vernadsky's almost century-old achievement promptly entered into the mainstream scientific canon, say in the 1930s, and become part of the general knowledge of an educated world citizen—say just before the post-war American 'growth ideology' took hold. Perhaps the anticipatory imaginaries of the mechanistic Western knowledge enterprise might have been shifted early in an ecological direction by Vernadsky's basic message that terrestrial life forms are 'children of the sun', and his anti-reductionist insight that: 'creatures on Earth are the fruit of extended complex processes [...] and are an essential part of [the biosphere's] harmonious cosmic mechanism.'[15]

Towards a History of the Constitutional Metaphors of Systems Ecology

The particular brand of ecological science that catalysed the politics of the Western 'ecology movement' of the 1970s was the systems ecology developed most prominently in the mid-twentieth-century United States around the concept of 'the ecosystem'. Unlike Vernadsky's empirical recognition of the biosphere as an irreducibly planetary entity, the 'ecosystem' was an abstract concept, which could be applied at any scale to which the observer could identify a 'system': say to a pond, a lake, or a watershed. First proposed in 1935 by the British ecologist Arthur Tansley, the ecosystem was the central organising concept in E.P. Odum's bestselling textbook, *Fundamentals of Ecology* (1971).[16] The systems ecology

[14] On the history of attempts to construct minimally functioning artificial biospheres *in vitro*, see: Granjou, C., & Walker, J. (2016). Promises that matter: reconfiguring ecology in the ecotrons. *Science & Technology Studies*, 29(3), 49–67; Walker, J., & Granjou, C. (2017). MELiSSA the minimal biosphere: human life, waste and refuge in deep space. *Futures*, 92: 59–69.

[15] Vernadsky (1998, p. 44).

[16] Odum, E.P. (1971). *Fundamentals of ecology* (3rd edition). Philadelphia: W.B. Saunders

popularised in the metaphor of 'the coming spaceship Earth'—the title of a famous 1968 essay by the economist Ken Boulding—would present a critical challenge to the orthodox economists' concepts of growth and equilibrium, prompting a series of reconfigurations of these concepts in response to the risks of escalating ecological erosion. At a time when burgeoning US oil demand meant increasing dependence on secure access to foreign oilfields, and comprehensive and far-reaching domestic environmental legislation was under negotiation, the terms of conventional economic management were up for renegotiation. By 1971, senior government and industry figures were seeking strategic policy responses to an

> [...] acute awareness of the apparent conflict that has been emerging between two societal objectives that are both of prime importance: providing energy to meet the needs of future economic growth and protecting the quality of the natural environment.[17]

The demands for environmental protection first realised in Western democracies in the form of the science-based environmental laws of the 1970s brought an expanded role for ecologists within the regulatory apparatus of civil government. Thus was introduced a new lexicon of scientific terms into the political discourse of civil society. In a 1978 report, Gordon Edwards observed that the ecology movement had introduced a vital new metaphor for the analysis of social institutions:

> With the rise of the environmental movement in the late 60's and early 70's, a new metaphor was introduced into the public consciousness—the metaphor of the ecosystem. Infinitely more complicated than any man-made machine, a living ecological system is a marvelously well-organized interactive system that is self-regulating, resilient, and irreplaceable. It represents the culmination of billions of years of evolution. [...] Here is another way of organizing complexity that is not machine-like. [...] This image [...] may succeed in transforming our society as thoroughly as the metaphor of the machine has done in the past.[18]

[17] Schurr, S. (1972). *Energy, economic growth, and the environment.* Baltimore: Johns Hopkins University Press, p. vii; Cited in: Lane, R. (2014). Resources For the Future, resources for growth: the making of the 1975 growth ban. In B. Stephan & R. Lane (Eds.), *The politics of carbon markets.* London: Routledge.

[18] Edwards, G. (1978, June). A metaphorical framework: the ecosystem versus the machine. Part 1: CCNR Final Submission to the Ontario Royal Commission on Electric Power Planning.

In 1971, E.P. Odum declared that ecology was rapidly becoming the 'indispensable' branch of science, the one 'most relevant to the everyday life of every man, woman and child'. The necessity of an ecological grounding for the human sciences was clear, for an age of limits was at hand: '[f]or the first time in his short history man is faced with ultimate rather than local limitations'.[19] No longer an obscure branch of biology, ecology during the 1970s lent its scientific credibility to a social movement that argued for a revolutionary transvaluation of values, where the order of society would be consciously adjusted according to what ecology, a superior account of the order of nature, had revealed. Ecology as an 'indispensable' knowledge paradigm would transform the rationalities of government by overcoming the exclusion from conventional economics of the foundational realities of Earthly life as understood by scientific materialism: namely, the inseparable phenomena of life (biology) and heat (thermodynamics). Deploying the ecosystem concept, environmental thinkers placed the industrial machine as a metaphor, ideological fetish and infrastructure of social life under sustained examination, a critique which implicated the entire mechanistic paradigm from which economics had derived its faith in the powers of technoscience to deliver endless growth. From a truly scientific account of the planetary *oikos* would come a new *nomos* of the Earth.

With the benefit of hindsight, it is clear that Odum's optimism that an Age of Ecology was superseding the Age of the Economics was tragically misplaced. As a contribution to an explanation of the weakness of ecological thought in relation to the economics of the 'free economy' and an ascendant neoliberal counter-revolution, the following chapters present a selective history of a broad sweep of ecological thought, stressing its historical links to parallel developments in political and economic theory, from the early modern accounts of the 'economy of nature' up to the point of collision between the two sciences around the fundamental questions of energy and power in the American crisis of the 1970s. In much the same way as the previous chapters attempted to historicise the emergence of 'the economy', my objective here is to unpack the development of the 'ecosystem' through the complex intellectual pathways by which, for a brief moment, systems ecology catalysed an existential critique of the economics of perpetual growth in equilibrium conditions.

[19] Odum (1971, pp. 3–4).

Perhaps one reason for the failure of the ecological critique to revitalise *oikonomia* and offer powerful and transformative answers to the question of 'how shall we live' is that the image of the ecosystem, described by Edwards as 'a way of organizing complexity that is not machine-like', was in the final analysis anything but. Systems ecology developed from a history of metaphorical and conceptual traffic across systems theory, colonial resource management, energy physics, military engineering, information science, and computer technology. Systems ecology has been widely criticised as a machine theory of nature, with a reductionist tendency to focus on quantifiable energy flows. As Voigt notes, 'the main concern is with the material-energetic aspects of interactions; the actual species involved are only of interest insofar as their specific features are relevant to the transformation of matter and energy.'[20] The formal resemblance of this version of the 'balance of nature' to neoclassical theories of general equilibrium in which 'individuals' transform an invariant value substance through variegated forms of capital, labour, and 'commodities' is suggestive.

In short, the metaphor of the ecosystem did not arise fully formed *ex nihilo*: ecology has multiple roots, and shares a long, parallel intellectual history with economics, a history of mutual influence and interactions. As a science built upon the metaphor 'oeconomy of nature', ecology owes often unacknowledged conceptual debts to the doctrines of political economy. This insight is not without consequences for historians seeking to assess the biopolitical confrontation between environmentalism and neoliberalism, a contest which shaped the world we have lived in from the 1970s to the post-natural present, a contest which has made our future.

[20] Voigt, A. (2011). The rise of systems theory in ecology. In A. Schwarz & K. Jax (Eds.) *Ecology revisited: reflecting on concepts, advancing science* (pp. 183–194). New York: Springer, p. 189.

CHAPTER 9

Oeconomy of Nature: The Balance of Nature and the Struggle for Existence

The Supreme Oeconomist in the Perpetual Order of Natural History

The term 'biology' is thought to have first appeared when the Swedish scientist Linnaeus (Carl von Linné) used its Latin form *biologi* in his *Bibliotheca botanica* (1736). Despite our familiarity with biological entities such as molecules, genes, cells, tissues, organs, organisms, species, populations, and ecosystems, these categories, which form the conceptual matrix of the modern life sciences, only began to assume their modern form in the nineteenth century. Prior to this, says Foucault, 'all that existed where living beings, which were viewed through a grid of knowledge constituted by natural history.'[1] Following the secularisation of the great metanarrative of Christian millennialism by the progressive philosophies of the Enlightenment, the diverse providential and creationist narratives of natural history yielded to a recognisably secular scientific materialism only in the later nineteenth century. The disorienting, decentering effects of the discovery of deep time by geologists such as Hutton and Lyell, and the incorporation of this temporal perspective into Darwinian biology destabilised all previous discourse on human value by portraying humans as animals subject to the violence and vagaries of a contingent, inhuman

[1] Foucault, M. (1970). *The order of things: an archaeology of the human sciences*. London: Pantheon, p. 127.

J. Walker, *More Heat than Life: The Tangled Roots of Ecology, Energy, and Economics*,
https://doi.org/10.1007/978-981-15-3936-7_9

history. This loss of meaning in history was resolved for some by depicting humans as the outcome of a providential process, thus saving the ideology of progress by embedding it in 'evolution'. Given that the emerging thermo-industrial economy provided the material evidence for Progress, it is perhaps not surprising that the biological organism *Homo Sapiens* was soon relocated within a thoroughly modern economy of nature.

It was the German zoologist, Darwinist and populariser of energetics Ernst Haeckel who in 1866 first outlined the contours of the discipline he termed 'Öekologie', in a work entitled *General Morphology of the Organism.*[2] This technical term was largely unknown outside of biology departments until coming into common vernacular a century after it was coined. The 1966 Webster's dictionary defined ecology as the study of the 'totality or patterns of relations between organisms and their environment'. Popular recognition of ecology as a science was inseparable from the increasingly politicised evidence of systemic environmental deterioration, and from announcements of an unfolding global ecological crisis. As Worster once observed, '[The] science of ecology has had a popular impact unlike that of any other academic field of research.'[3] It is precisely because ecology at its most ambitious addresses the 'totality' of our cosmopolitical existence that it has acquired such importance. Just as traditional religions and political ideologies from Marxism to neoliberalism have done, as it percolated through a wide range of cultural and social movements, ecology offered an ontology of the order of society within the order of nature, thus placing the subject in reference to the whole of space and time, posing the ultimate parameters of ethical meaning and social interaction.

Theorising the 'totality or patterns of relations between organisms and their environment' is hardly a modern concern. The subject matter of ecology has of course always been of direct interest to human beings. One does not have to invoke stereotypes of the Noble Savage, or the *homo economicus* lurking within the Optimal Forager of functionalist anthropology to state the obvious, that detailed empirical knowledge of how best to hunt and gather, when to find medical or edible plants, and how to live well in the present whilst not undermining future abundance is

[2] Haeckel, E. (1866). *Generelle Morphologie der Organismen. Allgemeine Grundzüge der organischen Formen-Wissenschaft, mechanisch begründet durch die von C. Darwin reformirte Descendenz-Theorie, etc.* (vol. 2). Berlin: G. Reimer.

[3] Worster, D. (1993). *The wealth of nature: environmental history and the ecological imagination.* New York: Oxford University Press, p. 156.

survival-relevant information of primary importance for pre-industrial and pre-capitalist societies. Detailed knowledge of the interactions between different species and their seasonal distribution across different landscapes was and is the basic economic concern of all human societies which depend closely on direct interactions with local lands, as opposed to the abstraction from land of the money-oriented activities of market society. The idea of ecology, of a comprehensive way of looking at the living earth and the interrelatedness of all its organisms, is thus much older than the term we use to describe it, and takes a much wider diversity of cultural forms than the Eurocentric knowledge tradition from which it emerged as a recognised science.

Despite its nominal emergence from Darwinian biology, which stresses a deep history of contingency, qualitative change, and adaptation, according to Frank Egerton the foundational theoretical concept of ecology is the abiding idea of the 'balance of nature'.[4] In a 1973 article, Egerton traced a genealogy of this notion in natural history from the pre-Socratic philosophers through to debates amongst the professional ecologists of his day, at a time when the equilibrium concepts of systems ecology were being challenged by novel developments in second-order cybernetics, non-linear thermodynamics and theories of chaotic dynamics in complex adaptive systems. Egerton concluded that at no time had this idea of 'balance' in ecology been given critical attention sufficient to elevate it to something approaching a 'law of nature': clarification of the balance-of-nature concept would involve expelling the hidden providential and teleological foundations of the metaphor stretching back to Pythagoras, whose abstract and general explanations of natural phenomena reflected the assumption that the cosmos was constant and harmonious. According to the philosopher of science Emile Meyerson, the very definition of positive science in the West has come to be bound up with the 'discovery' and elucidation of invariance principles, assumed to exist *a priori*, of which the first law of thermodynamics—the principle of the conservation of matter/energy—was only the latest version.[5] In contrast to the divine Creation *ex nihilo* of the Abrahmanic religions, conservation principles were recognised in Greek philosophy at least as far back as the third-century BC, when Epicurus wrote,

[4] Egerton, F. (1973). Changing concepts of the balance of nature. *Quarterly Review of Biology* (48), 323–350.

[5] Meyerson, E. ([1908] 1930). *Identity and reality*. London: Allen & Unwin.

> Nothing is created out of that which does not exist, for if it were, everything would be created out of everything with no need of seeds. And again, if that which disappears were destroyed into that which did not exist, all things would have perished since that into which they were dissolved would not exist.[6]

Most cosmologies, Egerton argues, have some notion of the balance of nature. Egerton claims that the earliest Western source supporting the idea of an ecological 'balance of nature' is to be found in Herodotus, who argued that Divine Providence had ensured that the world was not overrun by malevolent creatures such as serpents and lions by assigning them a much lower level of fertility than benign species such as hares. Although detectable in the histories of Herodotus, the biological observations of Plato, and the physiological writings of Aristotle, the tendency to focus on the individual traits of organisms among early natural historians meant that an explicit formulation of the concept of balance was not attempted. Later writers attending to questions regarding the interrelatedness and constancy of animal and plant populations drew upon these earlier accounts, which suggested that the balance of nature was maintained by the differing reproductive capacities of species, traits that ensured the survival of each species in a pre-ordained position for each creature in the natural order. This evidence of natural balance was mobilised to provide support for theological positions, first Stoic and later Christian, that asserted the existence of Divine Providence and the benevolence of the Creator.[7]

Egerton describes this complex of teleological ideas as 'providential ecology'. Noting that discussion of the balance of nature largely disappears from the literature of the medieval period, he attributes this to the universal acceptance of the idea of an omni-surveillant God, continually supervising all the workings of nature. Such a belief required no elaborate theory of in-built mechanisms of balance. It wasn't until the seventeenth century, with the schismatic eruptions of Protestantism and early modern science, that the balance-of-nature concept resurfaces. Again, the balance of nature was rarely itself the direct object of inquiry: rather it was invoked in arguments that attempted to safeguard both theology and empirical

[6] Cited in: Hardin, G. (1990, Nov 10). Perpetual growth: the next dragon facing biology teachers. Address to the National Association of Biology Teachers, Houston, Texas, p. 73.

[7] Egerton (1973, p. 330).

inquiry into nature from fears that they were antithetical and hostile forms of knowledge. Given that the early naturalists were often men of the cloth, as were most who had received a higher education, there was ample reason to defend the rational study of Creation from the charge that science led to atheism. An example may be found in a 1677 work by Sir Matthew Hale, *The Primitive Origination of Mankind*, which through a compressed set of metaphors, attempts an explicit account of the balance of nature as a concert of both physical and animate natural forces.

> That the vicissitudes of Generation and Corruption are by a kind of standing Law in Nature fixed in things, and the Notions and Qualities of Natural things are so ordered, to keep always that great Wheel in circulation; and therein the Access and Recesses of the Sun, the Influxes of the Heat thereof and of the other Heavenly Bodies, and the mutual and restless Agitation of those two great Engins in Nature, Heat and Cold, are the great Instruments of keeping on foot the Rotation and Circle of Generations and Corruptions, especially of Animals and Vegetables of all sorts. [...] That yet these Motions of Generations and Corruptions, and of the conducibles thereunto, are so wisely and admirably ordered and contemporated, and so continually managed and ordered by the wise Providence of the Rector of all things, that things are kept in a certain due stay and equability; and though the Motions of Generation and Corruptions and the Instruments and Engins thereof are in a continual course, neither the Excess of Generations does oppress and over-charge the World, nor the defect thereof and Corruptions doth put a Period to the *Species* of things, nor work a total Dissolution in Nature.[8]

Already we see machine metaphors at work in the construction of an argument for a Providential equilibrium, a feature we have already observed at work in the formative debates of early physical theory and resplendent in the rhetorical arsenal of economics. Maintained here from the days of Herodotus is the permanence and stability of populations, and a conception of species as fixed, immutable and secured from the threat of extinction. The modernity of the text is evident in the recasting of the Creator as supreme Engineer.

The eighteenth century saw a rapid expansion of the scientific knowledge base, with extensive work published cataloguing the range and distribution of organisms throughout a globe made increasingly accessible to Europeans by maritime technology and conquest. This increasing body of

[8] Cited in Egerton (1973, p. 336).

evidence exposed contradictions and sparked conflicts in many areas of knowledge, although the balance-of-nature concept did not come up as a significant subject for debate and was largely preserved intact. It was in this period, with the increasing influence of Enlightenment humanism that the theological context of natural history faded somewhat, although the general belief in Providence meant that the balance of nature remained a complacent background assumption of scientific minds.[9]

Surprisingly few comprehensive histories of ecology have been written given the profound political importance of the discipline. One of the fewer still that deal directly with the relationship of ecological thought to economic thought is bequeathed to us in the form of Donald Worster's *Nature's Economy*, first published by the Sierra Club in 1977.[10] Worster says that the most important early formulation of the ecological point of view can be attributed to Carl von Linné (1707–1778) and his 'oeconomy of nature'. Known by the Latin name under which he published his prolific scientific works, Linnaeus is most famous for his *Systemae Naturae*, a logical system of taxonomy that revolutionised the classification of plants. The system of binomial classification that he devised relied on simple principles which were avidly taken up by amateur botanists: one simply counted the pistils and stems of a flower and noted their position and its location in the neatly categorised system of nature was revealed. Prior to this, biological work had been greatly obstructed by the absence of universal taxonomic principles. Linnaeus' talent for organisation brought a coherence to the work of natural history which was previously unknown, and although the tedious work of organising all species into a single plan often produced erroneous results, it was this very process of organisation that rendered anomalies and contradictions visible, thus paving the way for more coherent general theories, of which the theory of evolution was to be the most important result.[11]

Linnaeus' work *The Oeconomy of Nature* (Latin 1749, English 1775) offered the first systematic and recognisably modern exposition of the ecological world picture, albeit one that never departed from the theological and teleological reasoning of providential proto-ecology. For Linnaeus, the 'oeconomy of nature' appeared as an eternally fixed cycle of geological and biological interactions, with the 'perpetual succession' of life and

[9] Egerton (1973, p. 333).
[10] Worster, D. (1977). *Nature's economy: the roots of ecology*. San Francisco: Sierra Club.
[11] Worster (1977, p. 33).

death expressing the rationality and harmony of a universe suffused with divine intention. All creatures were carefully and wisely designed to fit into an intricate hierarchy providing its means of shelter, its 'clothing', its source of food and its relations to other natural kinds in the 'food web':

> [...] the tree-louse lives upon plants. The fly called musca aphidivora lives upon the tree-louse. The hornet and wasp fly upon the Musca aphidivora. The dragon fly upon the hornet and wasp fly. The spider on the dragon fly. The small birds on the spider. And lastly, the hawk kind on the small birds. [...] In like manner the monoculus delights in putrid waters, the knat eats the monoculus, the frog eats the knat, the pike eats the frog, the sea calf eats the pike.[12]

Both the fertility and appetite of each organism is geared to the maintenance of a 'just proportion' of each species. This image was applied to 'animal economy', the study of the physiology of the individual organism. As each species is designed to interlink with other select organisms in the hierarchical cycle of eating, excreting, reproducing, and dying, each has thus been designated an 'allotted place'—its *nomos* in the *oikos*—which is both a location in which to dwell, and a functional role to play in the grand scheme of things. Each species provides support to others even as it goes about earning its own living.[13] 'By the Oeconomy of Nature,' wrote Linnaeus, 'we understand the all-wise disposition of the Creator in relation to natural things, by which they are fitted to produce general ends, and reciprocal uses.' Thus arranged by Linnaean ecology, living beings are 'so connected, so chained together, that they all aim at the same end, and to this end a vast number of intermediate ends are subservient'.[14]

To what end or purpose did the Supreme Oeconomist direct the 'oeconomy of nature'? The study of ecology, the scientific substitute for the older phrase, is for Worster 'in its very origins a political and economic as well as Christian view of nature; the earth was perceived as a world which must somehow be managed for maximum output'.[15] The search for general ends, for an overriding purpose guiding human agency in nature, was the central impetus to the 'economical' or 'ecological' approach to

[12] Cited in: Egerton, F. (2019). History of ecological sciences, Part 63: Biosphere ecology. *Bulletin of the Ecological Society of America*, 00(0), e01568.

[13] Worster (1977, pp. 34–35).

[14] Cited in Worster (1977, p. 37).

[15] Cited in Worster (1977, p. 37).

natural history that preceded the late-nineteenth-century scientific materialism which synthesised Lyell's geology, Darwin's biology, and Carnot's thermodynamics into the Earth sciences. Providence guaranteed natural abundance; Progress meant the maximisation of the abundant Earths' potential to serve human ends. The Linnean model of ecology mingled the religious terminology of 'dispensations' with a social metaphor the political organisation of resources for the maximisation of production. This tendency to borrow heavily from political economy—absorbing its values along with its metaphors—is a definitive pattern in the history of ecology.

In its commitment to a Christian and neo-Platonic ontology of species, the Linnean system denied the possibility of the post-Edenic creation of novel species and the possibility of extinction prior to the End of Days. As much of the world remained to be explored by Western science in the eighteenth century, his disciples maintained that fossils were the remnants of animals and plants that must exist as living creatures elsewhere on the globe—a form of agnotology similar to that of the nineteenth-century Christian palaeontologists who discounted the more credible temporality of Native American accounts of the deep antiquity of dinosaur bones.[16] As this explanation became less plausible is colonial frontiers advanced to encompass the Earth, it was claimed that that fossils represented antediluvian organisms that had perished in the Flood of Noah. The post-Genesis natural history of ecological organisation remained static in its equilibrium, without birth, change or death at the species level. The economy of nature was neatly bracketed between the time of Creation and the End of Days. Such a schema could even be approved of by the Evangelicals, fundamentalists who followed Archbishop Ussher in maintaining, on the basis of the patriarchal genealogies of the Old Testament, that the age of the Earth could be no greater than 5650 years. Published in 1650, Ussher's chronology followed quite literally the six-day creationist model of millennial thought, in which 'a day in the eyes of God was as 1000 years in the lives of men'. Thus, the creation of the world took place at noon on the 23rd of October 4004 BC, and the End of the World was slated for 23 October 1997, exactly 2000 years after the birth of Christ, and 6000 from

[16] Mayor, A. (2008). Suppression of Indigenous fossil knowledge. In R. Proctor & L. Schiebinger (Eds.), *Agnotology: the making and unmaking of ignorance* (pp. 163–182). Stanford, CA: Stanford University Press.

the day of creation.[17] The Apocalypse would be followed by the Millennium, the thousand years of blissful life that anchored Christian hope in the justice and meaningfulness of history. With the creation of a new heaven and a new earth, the Redeemer would usher in a new 'dispensation' in which the lion lays down with the lamb and the war in nature ceases, an age without tears, bodily corruption, or death. Eschatology, economy and ecology remained coherent subjects in relation to one another.

So what was the proper role and relationship of human beings to other forms of life within the Linnean economy of nature? For the most part, its ethics were entirely resonant with the increasingly ruthless utilitarianism of mercantile capitalism. The absolute distinction between humans and non-human life remained in place, complicated by shifting identities of civilised, barbarian and savage nations. Possessed of language, reason, and purpose, Christians dwelt in a world created by God for their use. Every last piece of animate and inanimate nature had been created for the exclusive use of God's chosen people. Although many Linneans hesitated to embrace the full program of Cartesian mechanism, its view of non-human creatures as biological machines with neither sentience nor value resonates with the Protestant view of nature as having moral relevance only as a backdrop to the individual in relation to his Creator, and the commission to 'go forth and multiply' as delegates of the Lord's sovereign dominion.

So far we have avoided the question of predation, violence, and 'war to the death' in nature, the economy of death by which creatures fill their bellies at the expense of other lives, an economy of sovereign violence in which humanity was clearly increasing its dominance. It was clear from the processes of colonisation underway in the Americas that the deforestation, ploughing of prairies, and ranching of cattle that followed the violent dispossession of Indigenous custodians resulted in the replacement of hundreds of native plants and animal species with the handful of domesticated and weedy species that Europeans had brought with them. This process was enhanced by the open season on the shooting of native animals such as mountain lions, bears, wolves, and coyotes, described as 'varmints', vermin better dead than alive. 'The uniform scope of human industry' noted William Smellie enthusiastically, has been to 'diminish the noxious animals, and to augment that of the useful animals.'[18] Nevertheless,

[17] Eco, U., Gould, S.J., Carriere, J-C., & Delumeau, J-C. (1999). *Conversations about the end of time.* Penguin, p. 11.

[18] Cited in Worster (1977, p. 353).

eighteenth-century naturalists found it impossible to conceive that a benevolent God would have built a cosmos in which human activity could diminish the abundance and multiplicity of Creation. As the Linnean naturalist and Lutheran pastor John Bruckner argued in his *Philosophical Survey of the Animal Creation* (1768):

> It is, I say, five thousand years at least that one part of the living substance has waged continual war with the other, yet we do not find that this Law of Nature has to this day occasioned the extinction of any one species. Nay, we may add, it is this which has preserved them in that State of perpetual youth and vigour in which we behold them.[19]

Again, a divine balance of nature metaphor is here called upon to justify the permanent and providential nature of ecological diversity and creationist productivity. The religious conceptions of order and reproduction in the pre-scientific ecology of the eighteenth century thus operated to justify the aspirations of an ever-more-aggressive process of capital accumulation, European imperialism and the transformation of landscapes.

The Deep Time of Light, Heat and Coal

Through the three-stage dispensational model of historical time in Christian eschatology, in which the present Age of Faith was bracketed between the Genesis of the Old Testament and the Apocalypse of the Book of Revelations, the proto-ecologists carried the crucial metaphor of balance in a largely unexamined state through to the early nineteenth century. The shallow time depth of this cyclical model, in tune with the prevailing cultural experience of time, was to be threatened and then overthrown in the nineteenth century by the revolutionary discovery of deep time, as the rationalisation of geological research gave context to the study of what we now refer to as the 'fossil record'. An early figure in the development of geological theories of the deep-time origin of mineral strata and their uniform transformation in the 'plutonic' and 'igneous' heat of a solar-driven Earth was James Hutton (1726–1797). Hutton's *Theory of the Earth* (1795), a speculative and ambitious synthesis of the science of his day, offered an account of the Earth as an 'organised body' which attributed to coal and life major roles in sustaining and stabilising

[19] Cited in Worster (1977, p. 37).

the Earth as a 'machinery' in which 'all its different parts are adapted, in form, in quality, and in quantity, to a certain end [...].'[20] The interactions of light, life, fire, water, biogenic material, and the geological transformation of these materials by the internal heat of the Earth were presented in a synthesis that contained an early account of what we would now describe, in the wake of the chemical revolution, as the regulation of the hydrogen cycle and the oxygen cycle by the photosynthesis-driven carbon cycle.[21] Recognised as the 'father of geology', Hutton can also be seen as an early precursor of Vernadsky and other theorists of the carbon cycle. His natural philosophy was nevertheless framed in implicitly theological terms, suggesting a kind of 'intelligent design' account of the wise disposition of the Supreme Oeconomist immanent in the orderly balance of nature:

> The proper purpose of philosophy is to see the general order that is established among the different species of events, by which the whole of nature, and the wisdom of the system, is to be perceived.[22]
>
> Thus the wonderful constitution of light and heat, [...] may be traced through many processes in the wise oeconomy of nature, or in the system of this world, where ends and means are the proper subjects of our science.[23]
>
> [...] in studying the system of nature, we observe, that every thing is in action for some purpose; that opposite powers are continually balancing each other, or alternately prevailing; and that the general end view is to contribute every thing requisite for the necessities for the conveniences of animal life [...].[24]

Hutton's work would be adapted and modernised by Charles Lyell in his *Principles of Geology* (1844). Lyell's uniformitarian geology of deep time would in turn provide the background for the evolutionary theory of the

[20] Hutton, J. (1795). *Theory of the earth.* 2 vols. London: Cadell, Junior and Davies, & Edinburgh: Willam Creech; Cited in: Galvez, M. & Gaillardet, J. (2012). Historical constraints on the origins of the carbon cycle concept. *Comptes Rendu Geoscience* (344), 549–567. See p. 563.

[21] Galvez and Gaillardet (2012).

[22] Hutton, J. (1792). A chymical dissertation concerning phlogiston, of the principle of fire. Part II in *Dissertations on different subjects in natural philosophy.* Edinburgh, p. 62; Cited in: Allchin, J. (1997). James Hutton and coal. Cadernos IG/UNICAMP 7, 167–183.

[23] Hutton, J. (1794). *Dissertation on the philosophy of light, heat, and fire.* Edinburgh, pp. 305–306; (Cited in Allchin, 1997, p. 11).

[24] Hutton (1792, in Allchin, 1997, p. 11).

origin of species through natural selection developed simultaneously by Charles Darwin and Alfred Russel Wallace. These mutually reinforcing developments in the geochemical and life sciences enabled a picture of vast eras in which radically diverse biological species and communities came ever so gradually into being, adapting to changing geological conditions as landforms rose from the seas or were submerged back into them, and disappearing into the abyss of extinction over eons of time. The insights of Wallace and Darwin's contingent evolutionary biology, drawn in part from their readings of the political economy of their day, would have enormous implications for abiding accounts of the 'natural laws' of the economy of nature. Finally disposing of the theological creationism of natural history, it also presented a profound challenge to the atemporal, reversible determinism of the Cartesian-Newtonian-Laplacean 'system of the world'.

The Struggle for Existence: The Malthusian Economy of Darwinian Evolution

Until the nineteenth century the progress of natural history was the domain of dedicated amateurs, the respectable pursuit of country divines who painstakingly collected and documented the interactions of the richly varied life of the fields and forests around them in the Arcadian manner of Gilbert White's *Natural History of Selbourne* (1789). Noting Whites' comment that 'Nature is a great economist' who 'converts the recreation of one animal to the support of another', Worster describes this work as an antecedent of the ecological world view, an early exemplar of the field study method of the working naturalist, and for all its providential Christian framing, a predecessor of later nature writers' reverential evocations of the harmonious and orderly interactions of multi-species life worlds; such as those of Rachel Carson, who would give affective form to the consciousness of the modern environmental movement of the Earth's life lost to industrialisation. White's *Natural History* went all but unnoticed in his own time. It was only in the nineteenth century, as the political and economic forces of the commercial revolution uprooted the heaths, moors, hedgerows, and forests of the countryside through the rationalisation of agriculture, and as industrial manufactories relocated an increasing portion of the population to smoky and polluted cities where the din of machinery replaced the quiet music of insects, birdsong, and brooks, that White's work developed a wide and devoted readership:

> For many the mere experience of reading Gilbert White had become a drug to deaden the senses against the landscape of ruin in Manchester, Birmingham and Pittsburgh, as well as a point of renewed contact with a rural nature shrinking into vague memory.[25]

Drawing less upon local observation than the field studies of explorer-naturalists working in far reaches of world empire, the Darwinian biology of the later nineteenth century would dispense entirely with the theological language of the economy of nature and embed the life sciences firmly in scientific materialism. Biology became increasingly professionalised along with the other natural sciences, and after 1845, when William Whewell gave the name 'scientist' to those trained in empirical and positive method, it was increasingly harder to square scientific findings with the harmonious natural order of Christian cosmogony. In parallel, the delineation of 'society' as a political, national, and scientific question generated an array of human sciences whose biopolitical object was the rational organisation of industrialising societies according to principles of social order and change revealed by political economists.

Among historians of science, a widely discussed case of mutual influence between the natural and the social sciences is the influence upon Darwin of Thomas Robert Malthus. The first British holder of a professorship in Political Economy (at the East India Company College, 1805–1834), Malthus is famous for the bleak bioeconomics of his *Essay On the Principle of Population* (1798). In it, Malthus argued that due to the 'fruitfulness of marriages,' the tendency of the population to grow geometrically (1, 2, 4, 8, etc.) was blocked by the limited possibilities for increasing the availability of arable land, which Malthus asserted could at best increase arithmetically (1, 2, 3, 4; this is itself is an infinite series.) Marx, among others, complained of a lack of evidence for these claims. At the time, there was no accurate population data for Britain: a bill for a regular census had been defeated in 1753 in Parliament on the grounds that it was 'totally subversive of the last remains of English liberty'.[26] In 1801, in response to the concerns regarding population that Malthus's book had catalysed amongst the landed and literate, an Act was passed for a comprehensive census to be undertaken.

[25] Worster (1977, p. 16).

[26] Egerton, F. (2012). *Roots of ecology: antiquity to Haeckel.* Berkeley: University of California Press, p. 117.

Malthus lived during a period of rapid population growth, the enclosure of common land, the proletarianisation of commoners, and the wild throes of an ill-disciplined, undemocratic market society that suffered frequent periods of panicked speculation, unemployment, underproduction, near starvation and the breakdown of law and order along the lines of class discipline. His 'moral science' and political interventions reflect these pressing concerns, being the prescriptions drawn from his speculative demography and a bleak insistence on an 'iron law of wages'. For Malthus, wages would naturally converge to the minimum required to sustain the bare life of the worker. Should wages increase beyond bare subsistence, labourers would increase the 'fruitfulness' of their marriages, resulting in increased competition for scarce employment which would drive wages down, increasing hunger, misery and disease. Increasing population meant increasing demand for food, which would always outstrip supply. For Malthus there was no bread for the unemployed at Mother Nature's table: through famine she would impose 'equilibrium' by carrying economically surplus populations out of existence.

Here was a wholly different concept of adjustment to equilibrium in the economy of nature, at bleak odds with the divine natural harmonies and sympathetic mutualisms of a Gilbert White or an Adam Smith. For Malthus, it was simply logical that the number of individuals existing at any one time must be closely balanced with the amount of food available to feed them. Therefore, he posited the existence of a continuous, forceful array of natural 'checks' to maintain the balance of biological forces at some median level that could be sustained over the longest period of time. 'Positive' checks were accidental death, infant mortality, epidemics, starvation and war. 'Preventative' checks included the 'savage' and 'unnatural practices' of infanticide and abortion. In the moral civilisation, they took the form of late marriage and preferably abstinence. Contraception was unthinkable to Malthus, a rural parson who believed that the marital duty of sexual union was the production of offspring to the extent that God would give (and the household budget could support). With utter conviction, Malthus declared that 'all these checks [...] may be resolved to misery and vice.'[27]

Prime Minister William Pitt's proposal to increase the poor rate provided under the parish poor relief system of the 'old' Poor Law of

[27] Malthus, T. ([1798] 1970). *An essay on the principle of population*. Penguin: Hammondsworth.

1601—which levied funds from local land holders—would only increase the population (temporarily) beyond the means available for its support, whilst spreading the misery of the destitute labouring classes to the landowners taxed to provide for them. According to Malthus, if a poor labourer were given five shillings and not the 18 pence he usually received, this extra money would do nothing to increase the production of food. Rather, more money chasing the same quantity of bread would inflate prices to the point where five shillings was no longer enough to survive upon. Expenditure beyond the mere survival of the working class was considered a waste of national product to common vices such as drinking, gambling and irresponsible fornication. Expenditure could only be productive as the rational investments of an 'austere' commercial class.

In 1800 Pitt withdrew his proposal, under pressure from a Malthusian backlash from Parliament. The Reform Act of 1832 extended the franchise by lowering the assessed value of property holdings required to qualify for voting rights, sweeping a middle-class Whig administration versed in Malthusian moral science to power. In 1834, a Poor Law Amendment Act passed the House of Lords, the result of a commission of inquiry including Evangelicals and the economist Nassau Senior. Banning payment of 'outdoor' cash relief, this instituted the solution to the poor question in the disciplinary confinement of the workhouse. Families were to be strictly separated into dormitories for men and women, and girls and boys—to prevent further unwanted offspring.[28] Conditions were to be so miserable that only those on the verge of starvation and unable to find any form of employment would be willing to enrol in them. Working class resistance to the 1832 Poor Law and the workhouse would inform the mass movements of the Great Charter (first published in 1838) for universal male suffrage, to be fought for and eventually won by the revolutionary threat of the general strike. Notably, Malthus also opposed the repeal of the Corn Laws: his defence of the landed rentier stood in contrast to the analysis of Ricardo, whose analysis of rent disclosed the legislative fortification of landed property rights as a means of appropriating labour and wealth from the property-less and the industrial entrepreneur. Beyond England, we need not look far to see the influence of Malthus on the

[28] For an insightful analysis of the interventionist sexual and racial welfare policies pursued since the 1960s by American religious conservatives and neoliberals allied against the redistributive welfare of the New Deal, see: Cooper, M. (2017). ***Family values: between neoliberalism and the new social conservatism***. New York: Zone Books.

policy of the Whig administration in relation to the Irish Famine, and through his training of student 'writers' for the East India Company, upon the callous attitudes of British colonial administrators to catastrophic famines in India.[29]

The invisible hand of the market, the disciplinary division of labour, and the implacable checks and balances imposed upon the fecundity of biological life by starvation were conceptually entangled in nineteenth-century political economy. Along with diverse works of natural and geological history, these streams of social thought contributed to the revolutionary reconceptualisation of the 'balance of nature' proposed in the evolutionary theory of Charles Darwin and his less-remembered co-discoverer Alfred Russel Wallace, later in the nineteenth century.

Darwin's diaries and notebooks of 1838 record his reading of *On Population*. While we know little of what Darwin thought of applied Malthusian political economy, his extrapolation of Malthus' political economy of life and death to the problem of ecological change and speciation led to his breakthrough concept of natural selection, the 'mechanism' underlying his theory of species evolution. Until Darwin read Malthus, he had never seriously considered how it was the number of animals and plants stabilised and held themselves in relative proportions to one another. Malthus gave detailed discussions of vegetable and animate life, in which he asked why it was that the world was not entirely overrun with insects and frogs, given the vast number of eggs generated by every female. A vast overproduction of individual offspring was everywhere to be seen in nature, he asserted. Whilst arguably still influenced by his earlier creationism, the dismal science taught Darwin to think deeply about the ways in which the increase of animals must be exactly limited to the number that can actually live. This balance of nature evidently arose out of competition to the point of annihilation, or what he called the 'warring' of species. Every species in Europe, he mused, 'must lose individuals every year to cold, hawks, etc'—even the death of single hawk must effect the life chances of other life. Darwin reasoned that this 'war' would favour the strong over the weak. If only two pups of a litter of seven survived to breed, chances were that the strongest and swiftest would predominate. In this way, over long periods of time death performed the creative function of 'fixing' favourable adaptations and erasing detrimental traits. Driven along by this struggle, species would adapt to their environment or die

[29] As discussed in Chap. 5.

out.[30] In an early expression of what was to become the theory of natural selection, the organising concept of the *Origin of Species* (1859), Darwin wrote in his notebook:

> One may say there is a force like a hundred thousand wedges trying [to] force every kind of adapted structure in the oeconomy of nature, or rather forming gaps by thrusting out weaker ones. The final cause of all this wedgings [*sic*] must be to sort out proper structure & adapt it to change.[31]

Leaving aside the question of Darwin's opinions regarding the political debates of his day, there is little doubt that the bioeconomics of Malthus served as a catalyst for the modernisation of biology embodied in the Darwinian principle of natural selection. Darwin witnessed at first hand various moments of the violent treatment of 'inferior' races on the colonial frontiers of Australia and Paraguay during his journey on the *HMS Beagle*, a trip in which he records his early troubled conclusions as to the centrality of violence within the economy of nature. The contribution of the dismal science to ecology was to elevate the 'warring of species' for food, territory and offspring to its basic principle. Armed with the bio-economics of Malthusian population theory, Darwin accomplished the final removal of 'dispensational' terminology from the study of the economy of nature, placing biology upon a scientific footing.

Such is the received view of the origin of the *Origin*. Yet another author likely influenced Darwin far more than he acknowledged: Alfred Russel Wallace (1823–1913), a worthy competitor also working on the theory of speciation by natural selection. There is an air of scandal regarding the question of priority for the theory of evolution in recent work on the relationship between the two scientists. Roy Davies has assembled convincing evidence that Darwin was quite cavalier with attribution and citation, particularly with regard to Wallace.[32] Davies concludes that Wallace, a modest personality with much greater field experience as a biogeographer than

[30] Browne, J. (1995). *Charles Darwin. Voyaging: vol. 1 of a biography.* London: Pimlico, pp. 387–388.

[31] Cited in Browne (1995, pp. 387–388).

[32] Davies R. (2008). *The Darwin conspiracy: origins of a scientific crime.* London: Golden Square.

Darwin, 'has a stronger claim to the theory of evolution than commonly realized, and was shabbily treated by Darwin'.[33]

The class dynamics of the day are evident in their respective careers and the perhaps undue deference shown by Wallace toward Darwin in later years. Unlike Darwin, who was born to a wealthy family of Kentish landowners well-connected to the intellectual circles of high society, Wallace was of working class origin, born to a large family with insecure income in a poor village in Wales. After early studies in theology, Darwin's developing interest in natural history overtook his flirtation with a career as a divine, and a comfortable path to natural history was laid out for him at Cambridge. Never persuaded by theology, Wallace learned several trades as a young man, and pursued scientific learning out of work hours at the London Mechanics Institute. Influenced early by Owenite socialism, Wallace later pursued adventurous field forays in the rainforests of Amazonia and South East Asia, assembling collections of novel species to send back to European researchers, and making major contributions to biogeography. Darwin by contrast, settled into married life, and after the *Beagle* voyage remained in England, a country gentleman of high society. Influenced by Humboldt to consider the non-utilitarian aspects of evolutionary emergence (such as the elaborate colouration and sexual display of birds of paradise), Wallace was less reductionist and less 'Malthusian' than Darwin in his approach to evolution. Informed by experience on the colonial frontier, Wallace remained committed to social justice for the oppressed masses his whole life, and returning to England in later life, became involved in the politics of land reform.[34] By contrast, Darwin's later career was embroiled by public controversy, as he reluctantly attempted to clarify the implications of his work for accounts of human origins in *The Descent of Man* (1871). Darwin's thesis provoked impassioned public defences of Creationism and Christian morality on one side, but the implications of evolutionism for the social sciences would increasingly take the form of the elitist naturalisations of class and racial hierarchies offered in 'social Darwinist' interpretations of a biological war of all against all, for example by the eugenicist Francis Galton and the ultra-libertarian Herbert Spencer.

[33] Lloyd, D., Wimpenny, J., & Venables, A. (2010). Alfred Russel Wallace deserves better. *Journal of Bioscience*, 35(3), 339–349.

[34] Lloyd, Wimpenny and Venables (2010).

An 1855 article composed by Wallace in Sarawak implicitly outlined much of what would become the accepted theory of biological evolution. This was widely ignored by academicians, although both Darwin and his colleague Charles Lyell read it, and grasped its importance: Lyell urged Darwin to overcome his private hesitations and quickly bring his work to publication.[35] Recovering from a malarial episode in 1858 which had triggered new insights in his evolutionary theory, Wallace wrote them up in a manuscript, which he sent by post from the island of Ternate to Darwin, asking for his thoughts, and to send the paper on to Lyell for possible publication. Wallace approached scientific inquiry as a social and collaborative enterprise, 'little suspecting that Darwin was not so much a collector of ideas on evolution but a hoarder who was unlikely to give anything back'.[36] It was this event that moved Darwin, on the advice of Lyell and Joseph Hooker, to present an earlier private letter to the Harvard botanist Asa Gray outlining his ideas to the Linnaean society. Without Wallace's knowledge, this was presented and then published along with Wallace's manuscript in a paper containing the writing of both authors, which attributed the theory of natural selection to both but implied Darwin's independent originality and priority by presenting his letter before Wallace's manuscript, and crediting Darwin as first author.[37] In the following year, Darwin ushered the *Origin of Species* (1859) into print, to widespread scientific acclaim and public controversy, a work in which Wallace remained uncited.

In a largely overlooked passage from his underappreciated 1858 Ternate essay, Wallace invoked a signal metaphor to explain the law of evolution: a machine metaphor which, as I have argued, has since the days of Adam Smith also covertly underwritten the liberal economists' imagination of a natural law of equilibrium governing a spontaneous market order:

> We have also here an acting cause to account for that balance so often observed in nature [...] The action of this principle is exactly like that of the centrifugal governor of the steam engine, which checks and corrects any

[35] Wallace, A.R. (1855). On the law which has regulated the introduction of new species. *Annals & Magazine of Natural History*, 16(93), 184–196.

[36] Lloyd, Wimpenny and Venables (2010, p. 346).

[37] Darwin, C. & Wallace. A.R. (1858). On the tendency of species to form varieties; and on the perpetuation of varieties and species by natural means of selection. *Journal of the proceedings of the Linnean Society of London*, 3(9), 45–62; (Lloyd, Wimpenny & Venables, 2010).

> irregularities almost before they become evident; and in like manner no unbalanced deficiency in the animal kingdom can ever reach any conspicuous magnitude, because it would make itself felt at the very first step, by rendering existence difficult and extinction almost sure soon to follow.[38]

Commenting on this passage in the 1970s, Gregory Bateson hailed Wallace as an early theorist of ecological cybernetics, wondering what might have been had Wallace's paradigm not been obscured by the shadow of Darwin, who left underdeveloped the ecological implications of evolutionary theory. Developing this thought, the historian of biology Charles Smith contrasts Wallace's biogeographical approach to the reductionist attention to individualist, genetic and molecular evolution that preoccupied twentieth-century professional neo-Darwinian biologists, to the exclusion of supra-species level environment-evolution dynamics of crucial importance to the applied questions of conservation biology in a time of anthropogenic environmental transformation and gathering extinctions. For Smith, Wallace's evolutionary theory was implicitly also a theory of biogeochemical cycling, long before this approach emerged in the solar-energetic systems ecology and Gaia hypothesis of the 1970s.[39]

Marx and the Evolution of Natural Technology

As several historians have observed, neo-Darwinian biology and liberal economic theory are of similar provenance.[40] As has been discussed in previous chapters, classical liberalism was mathematically formalised into the 'science' of neoclassical economics through half-baked analogies with equilibrium energetics and statistical mechanics. Molecular biologist Mae Wan Ho notes the relevance of this history for the attempts of neoliberals such as Charles Murray (MPS) to demonstrate the biological basis (and 'naturalness') of racial inequality in IQ statistics, or of Gary Becker (MPS) to claim a grounding for Chicago school economics in the crude neo-Darwinian hybrid of sociobiology: 'Neo-Darwinian theory, expressed mathematically in population and biometrical genetics, is based on the same equilibrium, mechanistic assumptions and has an even closer link to

[38] Darwin and Wallace (1858, p. 62).

[39] Smith, C. (2004). Wallace's unfinished business: The 'other man' in evolutionary theory. *Complexity*, 10(2), 25–32.

[40] Hodgson, G. (1993). *Economics and evolution: bringing life into economics*. Cambridge: Polity Press; Ormerod, P. (1994). *The death of economics*. Boston: Faber & Faber.

statistics.'[41] It was not, however, only biological racists and liberal economists who sought to integrate natural selection and organic development into a social physics of thermoindustrial society.

It is well known that Friedrich Engels declared at Marx's funeral that just as Darwin discovered the law of development of organic nature, so Marx had discovered the law of development of human history. It is also well known that Marx chided Darwin for reproducing English bourgeois social relations in nature, with his emphasis on individual competition and inheritance, to the occlusion of symbiosis and mutual aid. It is perhaps less well known that he considered Darwin's theory of natural selection as 'the class struggle writ large.' As Pancaldi has shown in an intriguing study, an under-explored avenue of conceptual and metaphorical exchange between the natural and social sciences is the influence of Darwinian ecology upon the economic thought of Marx.[42] In a note added to the second German edition of *Capital* (1872), Marx described Darwin's study as an inquiry into 'the history of natural technology, i.e. the formation of the organs of plants and animals, which serve as the instruments of production for sustaining their life.' This unusual characterisation of Darwin's theory was a crucial first step in Marx's original proposal to develop a 'critical history of technology', a 'history of the productive organs of man in society, of organs [machines, inventions etc.] that are the material basis of every particular organisation of society.' Here we see an early intimation of the critical project of science and technology studies.

Marx's project to develop a political economy grounded in the material environment of society had much to gain from the rising prestige of Darwin. To Marx's credit he emphasised the social relations profoundly embedded in machinery, whereas the classical tradition merely abstracted technology as a generic and abstract form of capital investment. In an 1867 letter to Engels concerning a forthcoming review essay of *Capital* that Engels was writing for the journal *Der Beobachter*, Marx instructed him to emphasize the achievement of the work in demonstrating that 'cooperation, division of labour, the use of machines and the connected social relations develop according to natural laws'. Where the neoclassical

[41] Ho, M. W. (1999). Are economic systems like organisms? In P. Koslowski P. (Ed.), *Sociobiology and bioeconomics* (pp. 237–258). Berlin: Springer.

[42] Pancaldi, G. (1994). The technology of nature: Marx's thoughts on Darwin. In I. Cohen (Ed.), *The natural and the social sciences: some critical and historical perspectives* (pp. 257–273). Dordrecht: Kluwer.

'revolutionaries' of 1871 appropriated the invariance principles of proto-energetics to produce a law-bound, rational, and harmonious society where floating prices produced the best of all possible worlds, Marx's analysis of the business cycle emphasised chronic disequilibrium, with speculative investment leading to overproduction, underemployment, sectoral collapses, and polarising intensifications of wealth and poverty. These systemic moments of crisis resembled what we would now, in the language of complexity theory, call 'bifurcation points' or 'tipping points': revolutionary moments where the socio-economic system might irreversibly cross a threshold into a new state. In the letter to Engels, Marx even went so far as to dictate parts of the review, suggesting that he write:

> When he [Marx] shows that, from an economic point of view, present society is pregnant with a new and higher form, he is just showing from a social point of view the same process of transformation established by Darwin in natural history. The liberal theory of progress concurs on this and it is the author's merit to reveal a hidden progress precisely where modern economic relations display discouraging immediate consequences.[43]

Of course, where Marx enrols the law of evolution in a progressive anticipation of social development through moments of crisis, Darwin sought merely to explain speciation in terms of adaptation to existing conditions, which need not at all imply progress over time. Evolution as conceived by Darwin is non-teleological. Although Darwin was criticised by Marx and his successors for projecting the social relations of Manchesterism and its laissez-faire 'struggle for existence' into the natural world at large, Darwin was nevertheless indispensable to the progressive movement for providing a theory of development and an economy of nature that was resoundingly post-theological. The Darwin-Wallace thesis did not require an external God to supervise the economy of nature, and as disturbing as it might have been to contemplate for Darwin, there was only one possible conclusion: there was no such God.

When Ernst Haeckel defined ecology as 'the body of knowledge concerning the economy of nature' he was not merely reflecting a general enthusiasm among late-nineteenth-century intellectuals for political economy, or throwing metaphors around for the sake of literary flourish, but repeating the phrase given to the comprehensive study of biological

[43] Marx, K., & Engels, F. (1957–1968). *Werke* (vol. 31). Berlin: Dietz, p. 404.

interactions in its earliest modern forms. For Haeckel, the science of the economy of nature meant:

> [...] the investigation of the total relations of the animal both to its organic and to it in organic environment; including and above all, its friendly and inimical relation with those animals with which it comes directly or indirectly into contact—in a word, ecology is the study of all the complex relationships referred to by Darwin as the conditions of the struggle for existence.[44]

The new term 'ecology' was intended to seal the economy of nature off from its history of dispensational historicism and invest the new science firmly in the scientific materialism of the late nineteenth century. Haeckel did not himself, however, systematically pursue the science of ecology that he pointed to in outline: it would fall to others to attempt to realise this vision.

[44] Cited in: Ricklefs, R. (1997). *The economy of nature: a textbook in basic ecology*. New York: W.W. Freeman & Co, p. 1.

CHAPTER 10

Superorganism: American Ecology and National Development

A New Science of the New World

Like other disciplines, ecology did not emerge spontaneously as a unified science with a coherent set of concepts and methodologies, but has long struggled to define itself as a science amongst the sciences, to establish internal consistency amongst competing schools and paradigms, and to justify its institutional contribution to society. Ecology emerged from multiple roots across different scientific cultures—the German tradition of Humboldtian phytogeography, Russian soil science, plant physiology and geochemistry, the Zurich-Montpellier school of 'plant sociology', Anglophone botany and zoology to name a few—with geology and climatology becoming important sites of disciplinary exchange in the twentieth century. Despite its Eurocentric genealogy, in important respects ecology is a science of the New World, a discipline that arose in the wake of colonial land appropriations and the rapid exploitation of the ecological abundance long fostered by Indigenous nations. From the early twentieth century the United States would become the chief site of innovation, discipline-building, and consolidation of the new science of ecology, led by Henry Cowles, Victor Shelford, Frederic Clements and numerous other ecologists. Defining what ecology was supposed to do and who should do it was a major area of contention in these formative years, as Beeman has written of that moment:

J. Walker, *More Heat than Life: The Tangled Roots of Ecology, Energy, and Economics*,
https://doi.org/10.1007/978-981-15-3936-7_10

> By the onset of the World War I British scientists formed an ecological society, and in 1915–16 the Americans followed suit by founding the Ecological Society of America, (with the journal *Ecology* appearing a few years later in 1920). Though the field found some unity between those studying the ecology of plants and animals, the discipline of ecology was still divided and ripe for criticism in the post-World War I years, as it lacked theoretical and academic bureaucratic unity.[1]

In this chapter I offer a narrative of a constitutive moment in the history of ecology as a coherent science: the discipline-building work of the American vegetation ecologist Frederic Clements (1874–1945). Clement's research on the remnant grasslands of the Great Plains catalysed important debates over the core theoretical concepts of ecology as it sought to rise above its heritage in the field studies of natural historians and become a basic science, at a time when the United States had well and truly closed the continental frontier (although the westward drive to incorporate new territories continued across the Pacific). With its choking dust storms and mass abandonment of farms, the 1930s disaster of the Dust Bowl compounded the social suffering of the Great Depression that had followed the financial collapse of 1929, bringing into stark focus the irrationalities of speculative capitalism and the mythologies of frontier individualism, highlighting the potentially catastrophic consequences of indifference to social injustice and ecological erosion. This highly visible and unnatural national disaster drew public attention to the importance of the nascent science. Indeed, as Hannah Holleman notes, the Dust Bowl has become something of a historical touchstone for scholars, an event from which lessons are drawn and which prefigures the coming 'dust-bowlification' of once prosperous agricultural regions around the world as soil losses are compounded as land surfaces heat up in accord with the thermal trajectory of the age of mass combustion. Yet it's worth remembering at the outset that while the Dust Bowl is commonly framed in regional or national terms, as a singular event, it was but one instance of the globalising crisis of soil and society that accompanied the destruction of native subsistence economies, as ever larger regions were incorporated into the 'free economy' of international commodity trade, and as cheap land and labour were

[1] Beeman, R. (1995). 'A green and permanent land': agriculture in the age of ecology, 1935–1985. PhD thesis, Iowa State University.

pressed into plantation monocultures and export cash-cropping to support expanding metropolitan centres.[2]

As professional biologists, the training of ecologists was frequently remote from the social sciences, and yet the ecological paradigm that first achieved scientific and institutional recognition was organised around social metaphors. Propounded by Clements in *Research Methods in Ecology* (1905), the first American ecology textbook, and elaborated in numerous publications, the core organising concept of 'ecology's first paradigm' was a research object variously named the 'successional climax community', the complex organism, or 'superorganism'.[3]

> The crux of this concept was that single species populations in nature are integrated into well-defined, organic entities, and key subsidiary aspects were that temporal succession in a sere is utterly deterministic, analogous to development of an individual, and leads inevitably to one of a few climax communities. The relationship between the stylized, integrated superorganism and the deterministic successional development producing it is organic and fundamental [...].[4]

The communitarian metaphors favoured by early-twentieth-century ecologists possessed an intuitive resonance with the progressive politics that came into the ascendant with the New Deal coalition, and its institutional responses to the Dust Bowl crisis amidst the Great Depression. The tension between the goal of advancing the status of ecology as a 'pure science' on the one hand, and the justification of ecology as an eminently useful form of applied knowledge for land use and policy decisions posed problems for its practitioners: namely, an ambiguity as to whether ecology ought to be a purely descriptive or nomothetic science of untrammelled nature, or an applied, prescriptive science evaluating human economic interactions with the 'biotic community' and advising on the forms of economic activity most amenable to long-term productivity and stability.

Kingsland locates these themes within wider historical debates about progress and national development, arguing that the formation of ecology

[2] Holleman, H. (2017). De-naturalizing ecological disaster: colonialism, racism and the global Dust Bowl of the 1930s. *Journal of Peasant Studies*, 44(1), 234–260.

[3] Clements, F. (1905). *Research methods in ecology*. Lincoln, Nebraska: University Publishing Co.; Simberloff, D. (1980). A succession of paradigms in ecology: essentialism to materialism and probabilism. *Synthese*, 43(1), 3–39.

[4] (Simberloff, 1980, p. 13).

as a distinct discipline in the United States occurred in an ideological context suffused by the national mythos of American exceptionalism. A variant of liberal millennialism converted into the national ideology of the United States, American exceptionalism might be described as an historical faith that the abundant natural resources, republican institutions and democratic ingenuity of Americans would allow them to escape the historical quagmire of class-conflict ridden Europe and lead the way to a future based on liberal humanism and scientific modernity. Even Hegel, the dialectical philosopher of historical forces, exempted America from his laws of history. As 'the country of the future' he saw in it the possibility of a clean break from 'the historical arsenal of Old Europe.' As progressive Europeans emigrated across the Atlantic in disgust at the repression of the revolutions of 1848, America came to be seen as the 'true Europe', where the democratic spirit of the Enlightenment would realise a new *nomos* of freedom amidst a 'new' nature, as settlers expanded westward across a continent free of the old land and social divisions of the ancient regime and the reactionary states that maintained them.[5] This new 'law of the land' was preceded by a violent dispossession of Indigenous communities and the unravelling of their long-abiding *oikonomia*.

Ecology has always had to grapple with fundamental questions regarding the relationship of Nature to Society, of *logos* to *nomos*. Indeed, its efforts to isolate general laws and principles that could explain ecological order and abundance in the absence of human disturbances were inspired by the accelerating disappearance of species rich environments through the increasing reach of modern agricultural, industrial and urban development. Ecology's tendency to construct a pure nature, in equilibrium and external to social relations, was largely the result of early efforts to document and reconstruct the pre-industrial order of nature in order to predict and control future transformations of the environment, in the interests of disaster-aversion through scientifically rational progress.

The extent to which ecology yields knowledge of relevance to vital problems of moral, social, economic and political importance has always remained an urgent question, despite the efforts of some ecologists to distance themselves from the radical import of environmental politics and confine themselves to purely technical concerns in order to delineate the boundaries within which the discipline can be said to have a purely

[5] Schmitt, C. ([1950] 2003). *The nomos of the Earth in the international law of the Jus Publicum Europaeum*. New York: Telos Press, p. 291.

scientific character. Among those whose definition of 'science' is Comtean—the reduction of explanation to the rigorous causal 'laws' of physics and chemistry—ecology has often not been considered a true or 'hard' science, insofar as it arises from specific natural histories and the ad hoc character of local field work, or deals with ambitiously large and complex theoretical 'objects', or tends toward holism and organicism, inviting flirtation with anti-mechanistic, vitalist, or mystic philosophies. As such, it is unsurprising that like their opposite numbers in the economic social sciences, ecologists pursuing the systematic unity of ecology as a science of 'natural law' and causal mechanisms have often been accused of 'physics envy'.[6] Against the views of such critics of ecology who have at times pressured its practitioners, it must be acknowledged that the questions the early ecologists identified as comprising the research agenda of their field as they carved out a disciplinary territory from diverse fields such as evolutionary biology, natural history, biogeography, physiology, and climatology were and remain some of the most important questions science has been asked to answer. As Kingsland rightly argues,

> Ecology cannot have fixed boundaries and ecologists must not shy away from ambitious goals. They grapple with some of the hardest questions in science and cannot afford to feel oppressed by the claim that theirs is a 'soft' science. [...] We are indebted to [the early ecologists] for beginning a research tradition that has grown increasingly important for addressing the problems that beset our overcrowded world.[7]

These questions arose largely on the frontiers of the expanding settler societies of the nineteenth century, in Southern Africa, Australasia, Argentina and particularly the United States. Europe had long been a densely populated and profoundly humanised landscape. The long frontier of 'wilderness' encountered by European colonists and settlers provided much of the early impetus for the development of ecology. Ecology might indeed be thought of as a science of the frontier, always looking back to a time before the 'post-natural' landscape to construct a reference point against which to assess the present and anticipate possible futures. The global expansion of the 'free economy' in the colonial era, amplified

[6] Cohen, J. (1971). Mathematics as metaphor: a book review of 'Dynamical system theory in biology' by R. Rosen. *Science* (172), 674–675.

[7] Kingsland, S. (2005). *The evolution of American ecology 1890–2000*. Baltimore: John Hopkins University Press, p. 6.

by the expansion of long-distance movement of people and bulk goods by ship and trans-continental rail, was of a piece with the wholesale transformation of continents and biomes through mass immigration, rapid deforestation, systematic defaunation of native animals, monocrop agriculture, the introduction of vast herds of cattle and sheep, and the release of invasive non-native species. Simply Progress itself in the colonial imagination, this process came to be seen by some prophetic writers as rank plunder, potentially threatening to long-term development. A pivotal study in environmental history is *Man and Nature: The Earth as Modified by Human Action*, published in 1874 by George Perkins Marsh,[8] perhaps the first effort towards the long term, global stocktaking of human interaction with the Earth continued in the present day by the publications of the IUCN Red List, the UN Millennium Ecosystem Assessment, the Intergovernmental Panel on Climate Change and numerous other monumental efforts to document and quantify anthropogenic environmental change on a planetary scale. Marsh's systematic attempts to grapple with the effects of rapid colonial land appropriation—the extinction of large mammals, the loss of old-growth forests, the proliferation of weeds, soil erosion, desertification—presented such actions as transgressions of the natural order with grave future consequences.

> The ravages committed by man subvert the relations and destroy the balance which nature had established between her organised and her inorganic creations; and she avenges herself upon the intruder, by letting loose upon her defaced provinces destructive energies hitherto kept in check by organic forces destined to be his best auxiliaries, but which he has unwisely dispersed and driven from the field of action.[9]

The driving insight of Marsh's treatise was that the balance of nature would not be automatically self-correcting at the scale of exploitation now underway, to the point where the Earth's reproductive capacity might be so far reduced as to render it incapable of supporting human life. Human intervention was now necessary to restore what human intervention threatened to destroy.[10] Careful planning was required to re-construct and restore these vital balances. Modern civilisation had to rise to the challenges

[8] Marsh, G. P. ([1874] 1965). *Man and nature: the Earth as modified by human action*. Cambridge, MA: Harvard University Press.

[9] Marsh (1965, p. 42).

[10] Kingsland (2005, p. 8).

created by its current blindness to the loss of nature's resilience. Marsh was pessimistic about the prospects of such a challenge in Europe, whose deeply engrained pattern of land degradation produced flows of millions of impoverished immigrants to the temperate regions of North and South America, Australia and New Zealand—lands the historian Alfred Crosby has described as the 'neo-Europes'.[11] It was in these frontier locations that Marsh saw the prospect of a truer accord with a relatively untarnished Nature, although he noted with trepidation the 'horizontal' nomadism of the North American 'settlers', who tended to scour each new frontier location for immediate economic gain and then move westward, rather than settling down to establish mature and long-lasting 'vertical' relationships with the soil.

In Marsh's global proto-ecology we see already forming the two impulses of ecology as a source of narrative about the future of industrial modernity: fear that its destructive expansion would lead to the apocalyptic demise of civilisation, and calls to avert such a future with a science capable of anticipating the consequences of human activity and ameliorating them with projects of preservation, re-afforestation, and the scientific management of land use.[12] In the first instance is an acknowledgement that the 'balance of nature' had been dangerously disrupted by (white, technologically superior) humans, and in the second, the conclusion that a new balance between nature and (white, scientifically superior) civilisation must be struck. These themes are exemplified in the work of Frederic Clements: whose career reached its zenith in the unfolding catastrophe of the Dust Bowl in the 1930s. Whilst Clement's 'superorganism' approach would be superseded by ecosystems ecology from the mid-twentieth century, Clements and his colleagues firmly established ecology as a science on American soil, thus contributing to the reformist culture of the ecology movement that emerged in the 1960s, and to the institution-building that culminated in the revolutionary federal environmental laws of the 1970s.

[11] Crosby, A. (1986). *Ecological imperialism: the biological expansion of Europe, 900–1900.* New York: Cambridge University Press.

[12] Kingsland (2005, p. 7).

The Development of Clements' Ecology

As scientists that seek to describe the systematic interdependence of organisms, it would appear only 'natural' for ecologists to be drawn to organicist philosophical positions. Organicism as theory posits that certain complex wholes exhibit the characteristics of organisms, which differ from mere mechanisms or statistical aggregates because of the dependence of the existence of the parts on their integration within the whole. A butterfly's wing, for example, is and only remains as such if united to a whole, living body. And butterflies are in turn existentially dependent on integration within wider ecological 'communities' and the evolutionary history of the biosphere as a whole. Organicists assume the organism possesses qualities (e.g. life) that elude mathematical formalisation and physico-chemical analysis, and that these qualities result from the integrated functioning of the whole; in other words, the whole is more than the sum of the parts. One cannot explain the elaborate colouration, nest architecture and mating rituals of the blue satin bowerbird from knowledge of the subatomic particles of which it is composed. The organic metaphor implies not only that the parts of the whole are internally unified, but that the whole has a life cycle, a course of development, senescence and reproduction in the manner of typical living organisms. Organicism has long been applied to social institutions: from Hobbes' analysis of the commonwealth as a 'body politic' through to Hayek's evolutionary theory of the market as an emergent, undesigned 'spontaneous order' too complex to understand or rationalise—a thesis which represses its heritage in the ultra-liberalism of Herbert Spencer's account of 'the social organism'. There is also an element of organicism in Hegelian or Marxist inspired theories of historical progress, insofar as society develops, by analogy to the organism, from infantile or embryonic forms through a succession of stages to more mature and complete forms.

In Clement's application of the organic metaphor to organise ecology, the interdependent interactions between different species in the 'economy of nature' pointed to the existence of a larger whole, an entity variously termed the 'climax community', the 'complex organism', and the 'superorganism'. Whilst such theoretical concepts proved unable to withstand empirical scrutiny, Clement's paradigm nevertheless advanced ecological science in its attempts to answer the problems that distinguish ecology from biology. After all, individual microbes, plants, insects and animals cannot be completely understood as discrete objects sealed off from their environments. As Gregory Bateson once put it, 'the basic unit of survival is the organism *plus* its environment'.[13]

In the early years of ecology practitioners tended to be divided into plant ecologists and animal ecologists. A further division of approaches was discerned between autecologists and synecologists, with the former focussing on the population dynamics of a single species and the latter dealing with the relationships in the multi-species 'community'. Observations about how different plant communities change over time have a long history that predates formal science and the formation of ecology as distinct research program on American soil. Key precursors of Clementsian ecology include Alexander von Humboldt's work on plant geography, Henry David Thoreau, who wrote on forest succession after logging, and Eugene Warming, who produced the first ecology textbook which included formulations of the 'laws of succession'. Stephen Forbes is another figure often cited in the literature, whose 1887 lecture 'The Lake as Microcosm' forged a synthesis between Darwinism and early balance-of-nature concepts. Forbes remarked on the tendency of populations to return to an equilibrium value. 'Perhaps no phenomena of life is more remarkable than the steady balance of organic nature which holds each species within the limits of a uniform average number, year after year.'[14]

The Chicago ecologist Henry Cowles (1869–1839) is generally credited with providing the first modern studies in plant ecology, in which he gave a detailed study of the successive changes in vegetation associated with the migration of sand dunes on the shores of Lake Michigan. This study introduced the metaphor of the community 'climax formation' into the study of inter-species interaction. Indeed, much late-nineteenth- and early-twentieth-century research in ecology emphasised the 'social' aspect of biological life: the communal relationships and mutual interdependencies among distinct assemblages of plants and other organisms. In his *Pflanzensociologie* (1928), Josius Braun-Blanquet of the Montpellier-Zurich school developed a floristic typology of community types or 'plant associations', using concepts such as sociability, gregariousness, and constancy to refer to attributes of inter-species relationships.[15] Charles Elton's

[13] Bateson, G. (1972). *Steps to an ecology of mind.* New York: Ballantine.

[14] Cited in: Egerton (1973, p. 343).

[15] Braun-Blanquet, J. (1928). *Plant sociology* (trans. G. Fuller & H. Conrad). New York: McGraw Hill.

important *Animal Ecology* (1927) was concerned with 'the sociology and economics of animals'.[16] Thus it seemed only natural to some that ecological research should contribute to the sociology and economics of humans. Elton recommended that ecology be a part of the training of biologists, as it was primarily this branch of science that would be 'more able to offer immediate practical help to mankind than all the others' in the present 'parlous state of civilisation', due to ecology's potential to see through the 'very complicated' nature of biological interdependencies to the 'definite economic laws' that governed life.[17] It was in this vision of 'ecology in the public service' that Clements and other ecologists would later mobilise as governments struggled to respond to the national crisis of the Dust Bowl during the 1930s, the greatest ecological disaster of human origin at that point in American history.[18]

Born to a pioneer farming family, Clements' early formative years and professional career were spent observing and documenting the historic transformation of the ecology of the vast prairies of the American Mid-West, as the Great Plains were unsettled by settlers and incorporated into the American economy. Best known for his attempt to establish general theories of ecological succession, Clements' organicist conception of the 'successional climax community' derived primarily from numerous detailed field studies of the Great Plains. Clements approached the recurring problem of the role of human communities in the economy of nature by drawing a sharp line between the Native Americans—whom he believed had relatively little capacity to influence the natural order—and the modern European colonists who began to decisively, and catastrophically alter the ecology of the Plains from the later nineteenth century.

Dead Zones: Social and Natural Disaster on the Grasslands

The ancestors of the Plains Indians who made their *oikos* upon the prairies and of the bison that likewise flourished there are thought by Western scientists to have arrived relatively recently in North America, crossing the Bering Strait from Eurasia between 17,000 and 10,000 years ago. Vast herds of bison roamed and grazed the rich grasslands of the Plains: the

[16] Elton, C. (1927). *Animal ecology*. New York: Macmillan, p. vii.

[17] Elton (1927, p. viii).

[18] Clements, F. (1935). Experimental ecology in the public service. *Ecology* (16), 342–343.

population is estimated to have numbered over sixty million in the mid-eighteenth century. As the frontier of the United States expanded westward, the government encouraged the slaughter of the buffalo, a form of ecological (i.e. economic) warfare waged upon the nations of the plains—including the Sioux, Cheyenne, Crow, Blackfoot, Comanche, Dakota, Cree, Ojibwe, Kiowa, Arapaho, and Pawnee peoples—yet another example of the role of politically induced starvation in constituting the new *nomos* of 'the free economy' and its allotments of private property.[19] By 1876, the US Army had effectively rid the Great Plains of its Indigenous people, enforcing a presidential order that they be concentrated and confined in reservations. As railway networks expanded through the mid-West, the American bison came under enormous pressure from thousands of 'sporting' shooters and commercial hunters. By the mid-1890s, the species was on the verge of extinction, reduced to a population of perhaps a thousand [Fig. 10.1].

Thus, the grasslands of the Great Plains lost the primary 'ecological engineers' and 'keystone species' (to use contemporary terminology) who had maintained the long-enduring complex of native grasslands and the soil structures upon which they depended. Through impressive weight and weight of numbers, the bison kept the plains open, turning soil over as they grazed, and increasing the bioavailability of nutrients in the soil through urine and manure.[20] Like the First Nations of Australia, Native Americans practiced their own sophisticated technologies of broadscale fire-based land management, burning the grasslands to generate fresh grass for grazing animals and keeping them open by limiting the recurrence of dense vegetation communities such as mesquite and forest. Among the several effects of intentional burning was the building up of soil fertility and its capacity to retain moisture. The nineteenth-century biochemist Justus von Liebig, a pioneer of modern soil science, noted the capacity of charcoal to readily absorb nutrients from animal excreta, binding carbon to phosphates and nitrogen to form what he held to be the most superior and sustainable of fertilisers.[21]

[19] Daschuk, J. (2013). *Clearing the plains: disease, politics of starvation, and the loss of Aboriginal life*. Regina, Saskatchewan: University of Regina Press.

[20] Knapp, A. et al. (1999). The keystone role of bison in North American tallgrass prairie. *BioScience*, 49(1), 39–50.

[21] Liebig, J. (1852). *Agricultural chemistry*. Philadelphia: TB Peterson; Wilson, K. (2014, Jul 25). Justus von Liebig and the birth of modern biochar. *Biochar Journal*, www.biochar-journal.org/en/ct/5.

Fig. 10.1 A pile of American bison skulls waiting to be ground into fertilizer, c. 1892 (Author Unknown: Burton Historical Collection, Detroit Public Library. Public Domain. https://en.wikipedia.org/wiki/File:Bison_skull_pile-restored.jpg)

Yet as the anthropologist Omer Stewart discovered in the mid-twentieth century, despite abundant historical observations of First Nations people burning their countries from sea to shining sea, neither the anthropologists nor the ecologists of his time—including Clements—were interested in developing an understanding these broadscale ecological engineering practices, informed as they were not by the one- to five-year-long studies of scientists, but by millennia of close empirical observations of native plant and animal communities, climate, and hydrology, and the intergenerational knowledge of the effects of human intervention in the fire regime.[22] Stewart concluded that racist and ethnocentric views and the

[22] Stewart, O. (1951). Burning and natural vegetation in the United States. *Geographical Review*, 41(2), 317–320.

prejudices against inter-disciplinary research of the academic division of labour prevented his contemporaries from considering the written records of American environmental history, and the detailed pyro-ecological knowledge retained by the Indigenous elders and communities that Stewart sought to understand and record—then alone among his peers as a humanist fire ecologist.[23] Academic ecologists subscribed to the common view that Indigenous people had neither the means nor the intention to modify their environments, which were held to be in a natural stasis or equilibrium prior to Western land appropriations. To the detriment of his career, Stewart's successful advocacy in court on behalf of Indigenous people in landmark Californian land rights cases conflicted with the Social Darwinist views of senior anthropologists called to testify on the other side, who portrayed Indian societies as transient wanderers belonging to superseded stages of human social development, and whose claims to land rights could not be justified as they had failed to make adequate economic use of the land.[24]

Following the suppression of Indian resistance and eradication of the buffalo, millions of wheat croppers were encouraged to migrate and establish farms on the prairies of the Great Plains. The period 1875–1886 was one of unusually high rainfall, and the grasslands grew green and abundant. Huge numbers of settlers migrated westward, taking up selections and fencing off sections of the prairies for intensive agriculture, and to a lesser extent in the dryer regions, grazing. During periods of relatively constant rainfall, politicians and land speculators encouraged migration to what was at best marginal farmland, under the optimistic belief that climate responded positively to agriculture. 'Rain follows the plough' was the common slogan of the estate agents and land-grant colleges, reflecting less the results of climatology than the exceptionalist confidence that American economic expansion was blessed by Divine Providence. Against all evidence and experience, the US Bureau of Soils proclaimed in 1909 that 'the soil is the one indestructible, immutable asset that the nation possesses. It is the one resource that cannot be exhausted: it can not be used up.'[25] Even in the short history of European occupation, the area had

[23] Stewart, O. ([1963] 2010). Barriers to understanding the influence of use of fire by aborigines on vegetation. *Fire Ecology*, 10(2). https://doi.org/10.4996/fireecology.1002001.

[24] Stewart, O. (2002). *Forgotten fires: Native Americans and the transient wilderness.* University of Oklahoma Press, pp. 17–22.

[25] Ponting, C. (1991). *A green history of the world.* London: Penguin, p. 260.

been known to be prone to extended drought. A severe drought during 1893–1896 led to a partial and temporary outmigration, as about half a million people abandoned the frontier regions and turned their wagons eastward. It was in these years that Clements began his early study as a botanist, as part of a project to discover which grazing and farming practices might be viable in the region. Between 1889 and 1930, the so-called 'virgin' prairie, with it fragile soils formed slowly over thousands of years by the complex interactions of native species and the law, lore and pyrotechnologies of the Plains Indians, was ploughed into wheat farms by small-holders, tenants and sharecroppers—many of whom were impoverished, and had learned their trade in the valleys and forests of the Appalachian hills—at an average rate of over a million acres per year, with no regard for techniques of soil conservation.

Destruction of the native grasslands was well underway by the time Clements began his field studies as a young botanist in the 1890s. In no small part, the vertiginous transformation of the prairies was accomplished by the expansion of transcontinental railways. Steam-locomotives rapidly transported US Army units to crush Indigenous resistance. They moved people and goods to and from the interior, and unified interstate commerce. In order to finance the construction of transcontinental railways, between 1850 and 1871 private rail companies were awarded major grants of land in junction cities and along the designated routes, up to twenty square miles per mile of track laid, which could be held or sold off to subsequent settlers in adjacent new townships and farms. This represented a privatization of almost 10% of the lands held in public trust as of 1850. Thus were formed the first truly giant financial and industrial corporations in American economic history, as railway companies became 'the largest landowners: rural, urban, sylvan, and mineral'. Indeed, Gaffney attributes the formation of the US school of neoclassical economics—which characteristically emptied the category of land of all content save individual allotments of property rights—to the investments of such landed interests in the careers of Richard Ely and J.B. Clark, whose research and public lecture tours vehemently opposed the taxation of land rents and the public regulation and ownership of utilities.[26]

[26] Gaffney. M. (1994). Neo-classical economics as a stratagem against Henry George. In M. Gaffney & F. Harrison (Eds.), *The corruption of economics* (pp. 29–164). London: Shepheard-Walwyn, p. 89.

Much of the railway corridors that first concretized the jurisdiction of the United States over the plains remained the private property of the railway companies—and thus were often the only places where native plant communities and seedbanks remained. Clements observed that where there was an unbroken line of seed and minimal disturbance, the prairie vegetation could over fairly short periods of time make what seemed to be complete recovery, arriving at a stable equilibrium or 'climax' state of maximum diversity and biomass after passing through various succeeding 'stages' of development.[27] Numerous field observations in the grasslands and prairies of the American West led Clements to formulate general principles of plant succession, applicable to widely different climactic environments and the different species that were adapted to them. Noting the functional interdependence of associated species in the process of succession, Clements designated the most important analytical unit of ecology as the 'complex organism', offering a detailed phylogeny and classification of climaxes in divergent plant communities.[28] In a state of maturity and health, the complex organism manifested as a 'climax community', as the ideal equilibrium state of nature to which living systems tended to return after a disturbance. The climax community manifested the highest levels of species diversity, biomass and population stability:

> The unit of vegetation, the climax formation, is an organic entity. As an organism, the formation arises, grows, matures, and dies. Its response to the habitat is shown in processes or functions and in the structures which are the record as well as the result of these functions. Furthermore, each climax formation is able to reproduce, repeating with essential fidelity the stages of development. The life history of a formation is a complex but definite process, comparable in its chief features with the life-history of an individual plant. The climax formation is the adult organism, the fully developed community, of which all initial and medial stages are but stages of development. Succession is the process of the reproduction of a formation, and this process can no more fail to terminate in the adult form of vegetation than it can in the case of the individual plant.[29]

[27] Kingsland (2005, p. 144).

[28] Clements, F. (1925). Phylogeny and classification of the climax. *Yearbook of the Carnegie Institution*, 24, 334–335.

[29] Clements, F. (1916). *Plant succession: an analysis of the development of vegetation*. Washington: Carnegie Institute, pp. 124–125.

For Clements, organicism was no mere metaphor, and he was criticised by many for insisting on the deterministic sequence of succession, which he thought of as the central fact of ecological study. Due to the organisational and self-equilibrating properties of natural plant associations, Clements argued that it was valid to describe them as a kind of higher order organism, or 'super-organism', which, like individual organisms, exhibited a fixed developmental sequence that could be predicted once detailed phytological studies of locally adapted plants had been completed. Within the limits dictated by climate and landscape, general principles of ecological succession of the complex organism, Clements thought, could be ascertained. Although these principles could only be derived from careful study in the field, without the possibility of a general theory and a coherent, accepted methodology, ecology would not be justified as a systematic science with a unique and defined research program Unless ecology could provide an objective, systematic account of how nature actually functioned it would remain nothing more than a collection of localised natural histories, each idiosyncratically couched in its authors' own notions of relevance.

For Clements the long-term stability or equilibrium of the mature climax community was ultimately determined by climactic patterns of temperature and rainfall. The long-term dynamic equilibrium of the complex community as it moved towards the climax was thus determined by the long-term equilibria of climate. An early member of the American Meteorology Society (est. 1919), in 1922 Clements contributed to discussions amongst economists and geographers seeking to identify periodicities in the cycling of drought, flood and other climatic phenomena. Although Clements was skeptical of such a premature and drastic reduction, some participants held that local climate changes could be correlated with the eleven-year sunspot cycle identified by Johann Wolf in 1852, and hoped to explain economic fluctuations and the behaviour of human populations in terms of such cycles.[30] This 'sunspot hypothesis' seems a veritable tradition amongst classical liberals seeking to naturalise the intensely political effects of land appropriations. The founding neoclassical economist W.S. Jevons proposed it to explain away the catastrophic famines of 1870s British-administered India.[31] Broadcast from the prairie state of

[30] Masutti, C. (2006). Frederic Clements, climatology, and conservation in the 1930s. *Historical Studies in the Physical and Biological Sciences*, 37(1), 27–48.

[31] As discussed in Chap. 5.

Illinois by the Atlas-affiliated Heartland Institute, it has been reinvented for the climate crisis era as the counter-science furphy that sunspot cycles determine the heat balance of the earth, and not the atmospheric gases emitted in vast quantities by the combustion of hydrocarbon fuels.[32]

In a 1935 article 'Ecology in the Public Service', Clements offered a program for the applications of ecological science, reflecting on an ecological disaster of unprecedented national importance, the Dust Bowl crisis. Unfolding from 1934 through 1939, this was the culmination of the degradation of the grasslands that Clements had spent much of his life studying. Triggered by a long drought and major heatwaves, an expanding 'dead-zone' of severely eroded farmland generated increasingly severe dust storms, as desiccated soil was wind-swept across the denuded country. In May 1934, the first of a long sequence of major dust storms picked up 350 million tons of topsoil and deposited it all over eastern United States. Twelve million tons fell on Chicago alone, and dust landed on ships 300 miles off the Atlantic coast. The destruction of vegetation through land clearing, overgrazing, and deep ploughing of soil was amplified in a self-reinforcing process by the dust storms, which stripped remaining topsoil and destroyed millions of acres of standing crops. The desolation of the ex-grasslands led to the partial collapse and depopulation of farming economies in west Kansas, south-east Colorado, north-west Oklahoma, north Texas, north-east New Mexico, and parts of Nebraska and the Dakotas. Impoverished environmental refugees, a quarter of whom suffered from respiratory illness, abandoned their broken farms in a mass exodus of 3.5 million people.[33]

The catastrophe opened up an institutional niche for the upstart science. By this time, Clements was inclined to think of ecology as a coherent scientific approach to the entire range of problems regarding life and the environment. The events of the Dust Bowl showed the future necessity of the role of the ecologist as an impartial source of knowledge on what forms of agriculture and settlement were permanently viable given the vagaries of regional climates and the as yet ill-understood ecological communities of the New World. The myopic individualism of farmers, lenders, and land speculators, each seeking to maximise profits in the boom

[32] See e.g. Contoski E. (2016, Dec 29). The sun, not CO2, determines our climate. The Heartland Institute. https://www.heartland.org/news-opinion/news/the-sun-not-co2-determines-our-climate-contoski. Accessed 23 April 2018.

[33] Ponting (1991, pp. 260–261).

periods accompanying high prices or high rainfall, led to the widespread adoption of farming practices which damaged the long-term reproductive potential of the 'complex organism'. The Dust Bowl disaster showed that without careful land-use planning, the cycles of boom and bust driven by the corrupting logics of politics and the chaos of short-term profit seeking would result in the 'ruin of the countryside', with a corresponding slide into moral degeneracy, such as the cattle-rustling and vigilante justice that he had witnessed amidst the less degraded landscapes of his youth.[34]

Clements was well positioned to exert influence on land use policies after the disaster of the Dust Bowl, urging administrators to discourage cropping in the drier areas of the plains, which would be more suitable for grazing. Ecology thus gained a vocation for assisting society to make the best economic use of land by demonstrating which modes of exploitation would be most in tune with the soil types, vegetation, and climate for that region. Worster suggests that Clements' notion of the climax community posits a stable external nature devoid of humans and suggests that this reflects a nostalgia for the 'virgin' expanses of prairie he encountered as a young man growing up on the uncrowded frontier of settlement. For Kingsland, his attempt to discover how natural systems developed in the absence of modern disturbances was motivated by a concern for the social consequences of rapid land degradation. She argues that Clements' hard-core insistence upon the organismic metaphor of the climax superorganism at the heart of his dynamic ecology—which he asserted as an objective fact rather than a heuristic concept—was crucial to the role he envisioned for ecology as an objective managerial science of civilised development. Clements' organismic determinism notably assumed two things: humans were not (or were no longer) part of nature; also, it was possible to detect which changes in the environment were 'natural' and which were not even after humans had arrived on the scene. Without knowledge of how nature functioned in the absence of human intervention, how could ecologists offer any advice on which kinds of development were more likely to be viable in the long term?

With the ability to make predictions came the possibility of control. Subject to the limitations of climate, which Clements assumed would forever remain beyond human influence, the ecologist armed with the knowledge of the process of succession would not only be able to accurately characterise natural processes but achieve commanding power over them.

[34] Kingsland (2005, pp. 149–151).

'In short', he argued, 'as an instrument for the control of the entire range of human uses of vegetation and the land, succession is wholly unrivalled'.[35] This implied all sorts of interventions. As Kingsland puts it,

> Once understood the natural process could be retarded, accelerated, telescoped, held in one stage indefinitely, or deflected along another course, perhaps even destroyed in order to allow the process to start again. It could be modified by the introduction of new species.[36]

Interestingly, this could more or less serve as a description for the knowledge praxis documented in recent histories of the sophisticated use of controlled burns by Indigenous peoples who, through a knowledge praxis dubbed 'firestick farming' in the Australian context, carefully arranged different plant associations and habitat mosaics of different age structures to maximise the abundance, predictability, and diversity of plants and animal communities.[37] Clements refused such possibilities, claiming that although 'fire has frequently been invoked to explain the presence of prairies, this is nowhere true of the climax grassland and it has much less significance than commonly supposed for the forest border and the postclimax of valleys or similar relicts'.[38] Omer Stewart's assembly of the ample written evidence of Native American pyro-ecological practices and their transformative effects on grassland and forest communities pointed to forms of ecological management and social organisation (e.g. to an *oikonomia*) far more complex than Western researchers were inclined to attribute to the Plains Indians. Completed in the mid-1950s but only published in 2002, Stewart's study contributes to the belated recognition that productive, long-abiding 'harmonies' with the biotic community have been achieved by the pyric technologies of non-Western ecologists intervening in the 'successional' development of the 'complex organism'.

Notwithstanding its ethnocentric historical blindspots, Clements' dynamic ecology of developmental holism implied an anticipatory practice

[35] Clements, F. (1935). Experimental ecology in the public service. *Ecology*, 16, 342–363.

[36] Kingsland (2005, p. 151).

[37] As discussed in Chap. 3. See: Gammage, B. (2011). *The biggest estate on Earth: how Aborigines made Australia*. Sydney: Allen & Unwin; Pascoe, B. (2014). *Dark emu. Black seeds: agriculture or accident?* Broome: Magabala; Steffenson, V. (2020). *Fire country: How Indigenous fire management could help save Australia*. Melbourne: Hardie Grant Travel.

[38] Clements, F., & Chaney, R. (1936). Environment and life in the Great Plains. Carnegie Institute: Supplementary publication, no. 24, p. 33. (Cited in Stewart, 1951, p. 351).

of land management, one perfectly in tune with the scientific optimism of the Progressive Era, manifest in the democratic social management of the New Deal. The super-organism fitted well with reformist models of society that saw society as a 'body politic' ideally directed by a 'brains trust' to secure healthy forms of social development.[39] Writing of the changes of policy inspired by Clements and like-minded professional land managers after the Dust Bowl, Bowler suggests that the possibility of regional management in the interests of the long-term health of the complex organism could only be conceived of given a government that was willing and able to over-ride individual gain for the public good. Linking the acceptance of ecology as a science of holistic management to political economy, Worster suggests that because the disaster of the Dust Bowl followed the financial collapse of Wall Street and coincided with the international crisis of the Great Depression, people were less inclined to accept the view that unregulated financial capitalism with its private monopolies and precarious commoners was an optimal, self-regulating 'system of natural liberty', and were predisposed to accept the ecologists' concern for maintaining the community of life in long-term stability through rational planning and intervention.[40] The difficulties of the Depression and the looming prospect of another world war inspired a mood of communalism and cooperation over competitive individualism, and widespread support for the use of scientific foresight and institutional planning to restore social stability and general welfare, as manifest in President Franklin D. Roosevelt's New Deal policies. These included funding for 200,000 employees of the Forest Service to plant trees as windbreaks, the establishment in 1935 of the Soil Conservation Service, and for a range of agencies to rationalise agriculture, grazing, and conservation.[41]

As Beeman notes in his study of the 'permanent agriculture' movement which gained significant cultural support and institutional footholds in the wake of the Dust Bowl crisis, 'a detailed account of the New Deal's attack on the soil crisis would require volumes'.[42] Suffice to say that ecology's entry into the political-institutional realm in the 1930s raised a now familiar dilemma with regard to its social function. Was it a 'pure' scientific

[39] Bowler, P. (1993). *Norton history of the environmental sciences.* W.W. Norton, pp. 520–521.

[40] Worster (1977, p. 232).

[41] Masutti (2006).

[42] Beeman (1995, p. 57).

profession that ought merely to accurately describe encounters with an objective nature in the field, carefully building theory and providing objective, value-free information to hand over without opinion to 'policy makers'? Or was ecology the basis for the transformative new 'land ethic' heralded by Aldo Leopold, an applied science with which to manage the entire agricultural and industrial interface with nature, providing insight into the most sustainable forms of value-creation and exploitation within the limits of the regenerative capacity of the complex organism?

In either case, most ecologists felt that their expert analyses of the consequences of destructive land use practices (such as those that led to the desertification of the Dust Bowl) were underutilised. The environmental and agricultural policies of the New Deal were primarily aimed at relief works and the stabilisation of agricultural economies. In the wake of the Dust Bowl crisis the management of the Great Plains was effectively transferred from marginal small-holders to the professional managers of larger farms and government agencies. Yet the vital importance of ecology's potential as a knowledge paradigm remained under-recognised. Through the crisis of the 1930s, the Chicago-trained ecologist Paul Sears criticised the mechanized and unplanned agriculture that emerged in the United States for undermining the complex organism and its equilibrium. Noting that the idea of balance was fundamental to biology, physics, and chemical theory, he warned there was 'no reason to think that human activities are exempt'. American farmers had ignored the evidence ecology provided, and 'failed to develop [their] artificial plant cultures in a way to simulate nature in holding and building soil'.[43] Few universities taught ecology as a subject. As late as 1964, Sears lamented the low status of ecology, which was generally perceived as of 'limited interest and utility'. This he attributed to an ideological milieu in which the 'current glib emphasis on economic growth' was presumed as 'the solution of all ills'.[44] Little has changed in this regard it would seem.

In summary, we can note that the constitutive metaphors that brought ecology into pantheon of sciences were organist and communitarian, a paradigm that would be superseded in a grander synthesis with the emergence of the ecosystem concept, which would subsume Clement's communitarian ecology within a dynamic systems theory grounded in chemistry and thermodynamics. In its engagement with

[43] Cited in Beeman (1995, p. 85).

[44] Sears, P. (1964). Ecology—a subversive subject. *BioScience* (14), 11–13.

the productivist ethic and historicist discourse of national development, through the Dust Bowl crisis, ecology itself became transfigured by its claim to provide a new way of perceiving the totality of human-environment interactions that could lead to a shift in social attitudes toward land, replacing exploitation with conservation as the overarching concern. Exposing the general blindness to the potentially apocalyptic consequences of environmental destruction, ecologists hoped to forge a new language of natural law and foster a transformative social ethic of interdependence with biotic community; paving the way for the environmental movements of the 1960s and anticipations of a coming 'age of ecology'.

CHAPTER 11

Energy, Ecology, and the Great World Engine

Imperial Systems

In this chapter we examine the route by which ecology arrived 'in the lobby of the energy hotel'.[1] A century after the neoclassical economists had claimed to develop a universal theory of value on par with energy physics in their *a priori* 'hedonic calculus' of the market as a self-equilibrating system, ecologists had consolidated their own discipline around a concept of 'the ecosystem' grounded in energy flows, if we take the widely read 1971 edition of E.P. Odum's *Fundamentals of Ecology* as a watershed. Of course the treatment of energy in systems ecology differed significantly from the botched mathematical appropriation of pre-entropic energy concepts by the marginal revolutionaries. Unlike the neoclassical economists, whose 'science' Thorstein Veblen mocked in grand literary style for abandoning the scientific requirement to treat empirical phenomena and relations of cause and effect, ecosystems ecologists were fully cognisant of the modern physics of energy and the second law, and attempted to incorporate it into the theoretical models driving empirical fieldwork.[2] In line with classical thermodynamics, the ecosystem concept promised

[1] To borrow a phrase from: White, M. (2004). In the lobby of the energy hotel: Jevon's formulation of the postclassical 'economic problem'. *History of Political Economy*, 36(2), 227–271.

[2] Veblen, T. (1909). The limitations of marginal utility. *Journal of Political Economy*, 17(9), 620–636.

J. Walker, *More Heat than Life: The Tangled Roots of Ecology, Energy, and Economics*,
https://doi.org/10.1007/978-981-15-3936-7_11

general modelling and the possibility of prediction on the basis of the precise quantification of biological life's transformations of matter and energy through anything identifiable as a 'system': a pond, an atoll or the biosphere as a whole, including its human populations and their machines. Ecology's rise to scientific prominence occurred at a time when the American paradigm of permanent economic growth had taken hold, a *nomos* which in practice was predicated on ever expanding fossil-energy combustion and globalised natural resource appropriation, a reality completely ignored in the anti-materialism of liberal economic theory.

Frederic Clements' 'successional climax community' began to be displaced as the central concept of professional ecology from 1935, when the Oxford botanist Arthur Tansley introduced the concept of the 'ecosystem'.[3] Clements 'superorganism' concept artificially limited the scope of ecology by confining it within an arbitrary boundary between the biotic community of living animals and plants and inorganic nature. A boundary drawn, for example, between prairie grasses and the soils in which they grow, cannot be empirically justified given the cycling of elements and energy through the interface of soil by microbial communities of primary producers and detritovores, in turn interacting with hydrological flows and the weathering of the surface minerals of the Earth's crust. As F.B. Sumner once put it, 'The organism and environment interpenetrate one another through and through, the distinction between them is only a matter of convenience.'[4]

The ecosystem concept allowed explorations far beyond the community dynamics of vegetation, auguring a grand synthesis with microbiology, biochemistry, pedology, geomorphology, climatology, and geophysics at the scale of the Earth as a Sun-orbiting planet, grounding ecology ultimately in the universal laws of thermodynamics. As Tansley wrote, 'Though the organisms may claim our primary interest, when we are trying to think fundamentally we cannot separate them from their special environment, with which they form *one physical system*.'[5] While the organicist approach was still pursued as late as 1949 in Alee and Park's textbook on animal ecology, by the mid-1950s the ecosystem had replaced the

[3]Tansley, A. (1935). The use and abuse of vegetational concepts and terms. *Ecology*, 16(3), 284–307.

[4]Cited in: Lotka, A. (1925). *Elements of physical biology*. Baltimore: William & Wilkins, p. 374n.

[5]Tansley (1935, p. 299).

super-organism as the central object and concept of ecological study.[6] Tansley famously coined the term in an article which specifically sought to address the 'use and abuse of vegetational concepts and terms' he detected in the cosmic organicism of Whitehead, in Clements' insistence on the successional climax, and the ambitious metaphysical holism of General Smuts. By the late 1960s, largely due to the educational and public careers of Howard and Eugene Odum, the ecosystem had become popularly recognised as the central concept not only of ecology as science, but as the political message and analytical paradigm of the modern environmental movement.

In his history of British ecology in the age of Empire, Peder Anker describes how Tansley's ecosystem concept arose from a 1930s confrontation between ecologists sympathetic to the organicist philosophical holism of General Jans Christian Smuts—a botanist and South African statesman active in the League of Nations and at the formation of the United Nations—and his critics amongst scientists of the anti-racist South African left, who were aligned with Tansley and the Oxford school of botany by academic training and patronage networks.[7] Yet despite the differences in perspective, both Smuts and Tansley worked within a British tradition of ecological science in the service of colonial administration, a milieu which had the problems of rational land, resource and population management in mind from the outset. 'Carrying capacity' for example, a key concept in population ecology, ties population numbers to the 'maximum sustainable yield' of a spatially defined bioregion. According to John Mackenzie, it was first developed by British researchers for the purposes of colonial political economy, a methodology to determine what part of native or peasant agricultural production was 'surplus' to the survival needs of its producers, in order to determine the maximum level of taxation that could be sustained without engendering rebellions or unwanted depopulations in occupied territories.[8]

Smuts' espoused a spiritual holism which arose from his practice of 'thinking like a mountain' in the sublime presence of nature, a philosophy

[6] Allee, W., Emerson, A., Park, O. & Schmidt, K. (1949). *Principles of animal ecology*. Philadelphia: W. B. Saunders, p. 728.

[7] Anker, P. (2001). *Imperial ecology: environmental order in the British Empire, 1895–1945*. Cambridge, MA: Harvard University Press.

[8] McKenzie, J. (1997). Empire and the ecological apocalypse: the historiography of the imperial environment. In T. Griffiths & L. Robin (Eds.), *Ecology and empire: environmental history of settler societies*. Melbourne University Press.

which informed his contradictory advocacy of peaceful harmony in international relations and authoritarian rule at home. Smut's holistic ecology of the complex organism offered a 'natural' justification of the racist policing and confinement of native African people to 'Bantustans' and 'homelands', where in a process of 'separate but equal development', their cultures could slowly evolve in accordance with their environment, 'protected' from the modernity they were not yet ready for in evolutionary terms. As Prime Minister, Smuts did not hesitate to deploy lethal police force to suppress rebellions against the South African ruling class and its *nomos* of the Earth; a political order like all settler commonwealths founded in the violence of a land appropriation. Whether it was trade unionists on strike, Jews seeking civil rights to observe the Sabbath and work only a six-day week, or Hottentots refusing to pay their dog tax, 'Smuts crushed it all by violent force', with mass shootings and aerial bombing raids.[9]

In his famous 1935 article, Tansley attacked as unscientific the superorganism concepts espoused by John Phillips, a protégé of Smuts and expositor of Clementsian ecology. Tansley confessed that 'Phillips' articles remind one irresistibly of the exposition of a creed—of a closed system of religious or philosophical dogma', one against which the urge to 'revolt becomes irrepressible'. Implicitly reclaiming the science of ecology from Smuts' speculative and reactionary metaphysics, Tansley expanded the remit of ecology far beyond theories of 'plant sociology' and community succession, arguing that

> [...] the more fundamental conception is [...] the whole system (in the sense of physics), including not only the organism complex, but also the whole complex of physical factors forming what we call the environment of the biome. Though the organisms claim our primary interest, when we are trying to think fundamentally we cannot separate them from their special environment, *with which they form one physical system* [my emphasis]. [...] Our natural human prejudices force us to consider organisms (in the sense of the biologist) as the most important parts of these systems, but certainly the inorganic 'factors' are also parts—there could be no systems without them, and there is constant interchange not only between the organisms but between the organic and the inorganic. These ecosystems, as we may call them, are of the most various kinds and sizes. They form one category of the universe as a whole down to the atom. [...] The fundamental concept is the

[9] Anker (2001, p. 51).

> ecosystem, which is a particular category among the physical systems that make up the universe. In an ecosystem the organisms and the inorganic factors alike are components which are in relatively stable dynamic equilibrium.[10]

The ecosystem concept was remarkable in the way that it opened up the scope and scale of ecology as a scientific methodology and a potentially applied science. In his concept of the ecosystem, Tansley foregrounded the dynamic interaction between the 'biocoenosis'—a term introduced by the German zoologist Karl Mobius similar to 'the community' in that it describes a group of living creatures—and the 'biotope', the inorganic material environments to which lifeforms adapt, and which are in turn affected by biological processes. Tansley's critique of Clementsian ecology noted that its dogmatically organismic model of the 'complex organism' bracketed living nature from non-living nature in a way that limited inquiry to established areas of biology—zoology, floristics, taxonomy, biogeography. Tansley proposed a research agenda could lead to the formalisation of general ecological principles, by means of an extension of its inquiry into climatology, geochemistry, physiology and thermodynamics. By including the inorganic sphere and blurring the distinction between living and non-living nature, the concept could be readily scaled, which eliminated the difficulties of defining the boundaries of different 'successional climax communities' under tags such as 'grassland' or 'forest', and of enumerating in detail the multiplicity of species that comprised them. All that was required was one or more species and its environment. An ecosystem could thus be defined according to the interests of the researcher. An ecosystem could be a puddle, a watershed, or the entire planetary surface of the Earth.

The development of the ecosystem concept was crucial to the development of ecology as an applied science of environmental management. By shifting the perspective from the natural 'communal' relations between organisms to the biogeochemical cycling of matter/energy through ecosystems, it opened the way to the analysis of landscapes that included significant human presences, and even of entirely anthropogenic ecosystems—such as the artificial life-support systems devised from the

[10] Tansley (1935, p. 299).

1960s to enable human spaceflight beyond the limits of 'spaceship Earth'.[11] As Kingsland puts it, ecosystems ecology

> sought to move beyond general conceptions of ecological processes by adding exact measurements, experiments, and tests of hypotheses. It welcomed application to new techniques in applied mathematics, looked to physics to provide basic principles for ecology, and pushed ecology into the computer age with enthusiasm. It was cautiously optimistic about the possibilities of environmental engineering. And it too sought a way to incorporate human activity into the analysis of ecological process, both on the local level and in the biosphere.[12]

In developing this biophysical approach, ecologists entered decisively into the policy arena, whether they liked it or not. Just as mathematical axiomatisation had for political economy, the institution of ecology as a recognisably 'physical' science conferred upon it the status of a candidate social physics. Simultaneously, dire revelations of environmental breakdown in the public sphere conferred a prophetic role upon the ecologist as an expert capable of anticipating the future consequences of present economic activity that threatened the order of life. In becoming public figures with increasing scientific authority, ecologists made a radical contribution, providing an analysis of the future of industrial society diametrically opposite to that assumed by economists, at a time when the neoclassical variant of the discipline was again achieving political importance in the context of the post-war preoccupation with economic growth and modernisation theory. Challenging the irrational and destructive millennialism of infinite economic growth, the systems ecology that would be developed in America could claim a superior legitimacy as a 'sane science' fully grounded in modern physics.

Living Energy: Lotka's World Engine

The impulse to rid natural history of its theological baggage and to integrate it with general theories of the physical universe was implicit in the career of the late-nineteenth-century German scientist Ernst Haeckel, who replaced the old phrase 'economy of nature' with the modern term

[11] Walker, J. & Granjou, C. (2016). MELiSSA the minimal biosphere: human life, waste and refuge in outer space. *Futures*, 92, 59–69.

[12] Kingsland (2005, pp. 178–179).

'ecology'. A marine biologist who avidly promoted Darwinism with his famous image of the evolutionary 'tree of life', Haeckel was also a populariser of the productivist doctrines of Helmholtzian energetics, which linked the conversion of solar energy by the biosphere to the dynamic forcefulness of modern industrial society in an unbroken chain of energy transformation and productivity:

> The sum of force, which is at work in infinite space and produces all phenomena, is unchangeable. When the locomotive rushes along the line, the potential energy of the steam is transformed into movement [...] The whole marvellous panorama of life that spreads over the surface of our globe is, in the last analysis, transformed sunlight. It is well known how the remarkable progress of the technical sciences has made it possible for us to convert the different physical forces from one form to another [...] Accurate measurement of the quantity of force which is used in this metamorphosis has shown that it is constant or unchanged. No particle of living energy is ever extinguished; no particle is ever created anew.[13]

Notable in Haeckel's bioeconomics of 'living energy' is a productivist model of nature as vast reservoir of energy awaiting conversion to useful 'work'. Energy conservation was claimed as a transcendental principal, and was 'productivist' insofar 'as it placed the metaphor of the machine at the centre of scientific explanation and the energy of the universe in the service of an order dedicated to the production of work.'[14] It is indeed ironic that the elaboration of the principle of the conservation of energy led to the ideology of productivism, which would be transformed in the American century to a quest for unlimited economic growth through unrestricted combustion of fossil-fuels and a general hostility to resource conservation.

One consequence of the import of the machine metaphor in replacement of the organismic and communitarian metaphors of the botanical, floristic, and zoological branches of ecology was that systems ecology was increasingly open to the import of machine metaphors—and other aspects of the dominant economic theory. What were the attractions of the machine metaphor for ecology, and what were its implications? What this energetic turn meant for the claims of ecologists in the sphere of political

[13] Haeckel, E. (1899). *Die Welträtsel*, Stuttgart, p. 87.

[14] Rabinbach, A. (1992). *The human motor: energy, fatigue, and the origins of modernity*. Berkeley: University of California Press, p. 56.

economy is a key consideration given the emergence and subsequent institutionalisation of the environment movement from the 1970s onward. Perhaps the most concise explanation of the rise of ecosystems ecology in post-war America has been provided by Golley, who says that

> In America…the ecosystem system concept appeared modern and up to date. In short it was a machine theory applied to nature.[15]

It seems that the attractions of this route for ecologists were similar to those experienced by the economists who had taken it a hundred years earlier. Both nineteenth-century economists and twentieth-century ecologists saw the way out of the inherently political subject matter of the discipline by appealing to the rationality and universality of energy physics. They thus hoped to provide objective foundations for the management of human interactions with their environments, bypassing politics and cultural specificities by accessing a pure nature external to social relations, in turn allowing a scientific analysis of the forms of social organisation most in accordance with the laws of nature.

If Tansley provided the term with which ecology was to be synthesized with energetics, the American actuary, demographer and population biologist Alfred Lotka (1880–1949) helped prepared the ground for its reception in the American academy. Lotka's (1925) opus *Elements of Physical Biology* was among the first inter-disciplinary works in English to rigorously approach the relation between biological evolution, thermodynamics, and technoindustrial expansion.[16] Lotka corresponded with Vernadsky in the 1920s, as both men developed their signature monographs.[17] And although neither cited the other, both Lotka's biophysics and Vernadsky's biosphere concept filtered into the intellectual milieu from which American systems ecology would emerge. Vernadsky was a brilliant synthesizer of the natural sciences of his day, under-recognised in part due to cultural and political distance between Russian and Anglo-American research communities. Lotka's influence on the development of global systems ecology was facilitated by membership in the latter, but his approach was more

[15] Cited in: Ingerson, A. (2002). A critical users' guide to 'ecosystem' and related concepts in ecology. Arnold Arboretum, Harvard University: Institute for Cultural Landscape Studies, p. 11.

[16] Lotka (1925).

[17] Bobulescu, R. (2015). From Lotka's biophysics to Georgescu-Roegen's bioeconomics. *Ecological Economics*, 120, 194–202.

eclectic and interdisciplinary. In proposing a novel science called 'biophysics', Lotka drew upon sources from across the natural and the social sciences, in an idiosyncratic approach informed by his employment as a statistician and demographer for the Metropolitan Life Insurance Co. Lotka's *Elements* was an original, ambitious and influential book, but also 'too eclectic to become part of the canon of any field.'[18] Lotka nevertheless contributed, if indirectly, to the development of systems ecology, and by another route to the ecological economics inaugurated by Nicholas Georgescu-Roegen—whose classic critique of orthodox economics *The Entropy Law and the Economic Process* (1971) often cited Lotka, but never any professional ecologists.

The biophysical perspective developed by Lotka can be traced to an 1886 lecture given by the Austrian physicist Ludwig Boltzmann to the Imperial Academy of Science in Vienna.[19] Twenty years after it had been formalised by Clausius, Boltzmann offered a fresh interpretation of the second law of thermodynamics, proposing to unify it with Darwin's evolutionary biology. Boltzmann argued that the 'struggle for existence' of biological life was essentially a struggle for 'available energy' ultimately deriving from the sun.[20] In so doing, Jackson argues, Boltzmann created a 'metaphorical bridge' between Darwinian evolution and thermodynamics. Lotka pursued Boltzmann's link between Darwinian selection and energy appropriation in his *Elements of Physical Biology* (1925)—which offered a unified theory of geochemistry, population growth, ecological dynamics, industrial development, and civilizational consciousness in the context of a rigorous, if idiosyncratic, discussion of the evolutionary development of the bioenergetic economy of nature. His work is of particular interests from the perspective of our current analysis of the shared genealogies of orthodox economics and ecology, as it is with Lotka that liberal economics intersects directly with ecology on the ground of energy physics. Neither discipline would be left unchanged by the encounter. Lotka was amongst the first, in a 1914 article, to apply mathematical models from economics to ecological questions, attempting to mathematise ecology in a way that could in turn yield an objective standard of economic value derived from

[18] Kingsland (1994, p. 232).

[19] Jackson, T. (2003). Sustainability and the 'struggle for existence': the critical role of metaphor in society's metabolism. *Environmental Values*, 12(3), 289–316.

[20] Boltzmann, B. ([1886] 1974). The second law of thermodynamics. In B. McGuiness (Ed.), *Theoretical physics and philosophical problems*. Dordrecht: Reidel, p. 19.

the principle of evolution.[21] In conceiving of this ambitious project to ground evolution in thermodynamics (and value theory in evolution) Lotka was inspired directly by liberal economic theory. He had been impressed by Herbert Spencer and his grandiose efforts to formulate a general law of evolution that could unify all historical processes. Jevon's *Theory of Political Economy*, which advanced the neoclassical price theory and proposed a universal theory of economic mechanics, provided a direct model for Lotka's mathematical analysis of population oscillations around equilibrium values. Vilfredo Pareto, who strongly asserted the identity of economic and physical 'law', provided the mathematical approach to competitive equilibrium distributions with the concept of marginal ophelimity or 'optimality'.[22] Lotka was also influenced by his teacher, the physical chemist Friedrich Wilhelm Ostwald, a founder of the German 'social energetics' movement and a keen writer on questions of industrial organisation. Ostwald saw the entire universe as a cosmic fund of labour power governed by the laws of energy. The whole purposeful activity of civilisation was converting available energy into useful work, redeeming energy flows that would otherwise be dissipated by making them socially useful.[23] Progress could be measured by the increasing efficiency of energy conversion and the elimination of waste. Lotka was among the first to pursue this vision into general biology. As Kingsland reports

> Physical chemistry emphasized thermodynamic principles and mathematical analyses, so Lotka imagined that physical biology must treat the organic world as a giant energy transformer. [...] Lotka believed it would be profitable to analyse the evolution of the entire world system as a whole. He compared the world to a giant engine or, using an image familiar to thermodynamics, to a giant mill wheel.[24]

Reflecting on the link between evolution and energy appropriation, Lotka conceptualised his ecology in terms of the 'selfish effort of each organism and species to divert to itself as much as possible of the stream of available

[21] Lotka, A. (1914). An objective standard of value derived from the principle of evolution. *Journal of the Washington Academy of Science*, 4, 409–418.

[22] Kingsland, S. (1994). Economics and evolution: Alfred Lotka's economy of nature. In P. Mirowski (Ed.), *Natural images in economic thought: markets read in tooth and claw* (pp. 231–246). Cambridge, MA: Cambridge University Press.

[23] Rabinbach (1992, p. 182).

[24] Kingsland (1994, p. 158).

energy'.[25] This competition for available energy could provide a rigorous point of analysis for the mathematical treatment of oscillations in population numbers, and Lotka's work with Vico Volterra in developing a system of equations to model predator-prey population dynamics is more frequently cited by ecologists than his more general work on biophysics. Relating evolution to energy transformation (in Spencer's sense of the macro-evolution of the social organism, as distinct from Darwin's account of the competitive 'micro-economics' of speciation), Lotka made a heroic attempt to derive a general law of evolution from a single extremum principle. Through a series of theoretical moves, he freshly interpreted evolution as a way of increasing energy flow and efficiency. Thermodynamic principles, he believed, supported his view that natural selection would not only increase the total mass of biological systems, but also the energy flow through the system, an explanation which linked the directionality of evolution to the irreversibility of the second law of thermodynamics.

> [...] the operation of natural selection tends to increase the biomass and embodied free energy in the biosphere. It also tends to increase the rate of circulation of both matter and energy flux through the system.[26]

Lotka's work anticipated the study of food chains, the cycling of water and nutrients between consumers and producers, and the mathematics of energy transfer through ecosystems. Physicists, however, were not interested in his work, and did not pursue his calls for a new science of physical biology. And whilst his contributions were recognised by senior ecologists such as Charles Adams, who invited Lotka to join the Ecological Society of America in 1925, it is difficult to ascertain the extent to which Lotka was read by the early ecosystems theorists, as his work was not widely cited by ecologists until it was republished in 1956 as *Elements of Mathematical Biology*, a book widely consulted in the 1960s and 1970s 'age of ecology'.[27]

[25] Cited in Jackson (2003, p. 6).

[26] Cited in: Ayres, R. (1994). *Information, entropy and progress: A new evolutionary paradigm*. New York: American Institute of Physics, p. 117. See: Lotka, A. (1922). Contribution to the energetics of evolution. *Proceedings of the National Academy of Science*, 8, 147–151; Lotka, A. (1922). Natural selection as a physical priniciple. *Proceedings of the National Academy of Science*, 8, 151–154.

[27] Golley, F. (1993). *A history of the ecosystem concept in ecology*. New Haven: Yale University Press, p. 58.

Ecosystems Energised

A key figure in the development of ecosystem theory was George Evelyn Hutchinson, who fostered post-war ecology in America through the careers of his many students, which included H.T. Odum, Robert Macarthur, and Donna Haraway. His influence extended to many areas of systems theory and environmental science, and not least because Hutchinson was among the first to recognise the importance ofVernadsky's writings, which he had been introduced to in English translations provided by Vernadsky's son, a faculty colleague at Yale. Attempting to transcend the polarised debate between the supporters of Tansley and of Clements over whether or not communities could be legitimately identified as higher-level superorganisms, in a 1940 review of *Bio-Ecology* (a textbook by Frederic Clements and Victor Shelford), Hutchinson argued that 'general principles are largely classificatory. […] if the community is an organism it should be possible to study its metabolism.'[28]

Hutchinson encouraged this approach in the promising research of a young postdoctoral student by the name of Raymond Lindemann, who tragically died before the 1942 publication of his important study of 'The Trophic-dynamic Aspect of Ecology.'[29] The essential achievement of this paper was its presentation of a rigorous analysis of energy flows through the food cycle of the relatively well-bounded ecosystem of Cedar Bog Lake, meticulously documented over a five-year long study which included analysis of core samples of lake sediments. Bringing Tansleys's critique to fruition, Lindeman cited Vernadsky's 'biogeochemical' approach in arguing that

> […] the discrimination between living organisms as parts of the 'biotic community' and dead organisms and inorganic nutritives as part of the 'environment' seems arbitrary and unnatural. The difficulty in drawing clear-cut lines between the living community and the non-living environment is illustrated by the difficulty in determining the status of slowly dying pondweed covered with periphytes, some of which are also continually dying. […] [M]uch of the non-living nascent ooze is rapidly reincorporated through "dissolved nutrients" back into the living "biotic community." This constant organic–inorganic cycle of nutritive substance is so completely integrated that to consider such a unit as a lake primarily as a biotic community

[28] Hutchinson, G. E. (1940). Review: Bio-Ecology. *Ecology*, 21, 267–268.
[29] Lindeman, R. (1942). The trophic-dynamic aspect of ecology. *Ecology*, 23(4), 399–417.

> appears to force a "biological" emphasis upon a more basic functional organization.[30]

With this paper a new scientific paradigm of systems ecology emerged, one that explicitly deployed energy as an economic concept. In reviewing this paper, Hutchinson himself deployed economic terms, explaining that Lindeman had demonstrated 'that the most profitable method of analysis lay in the reduction of all the interrelated biological events to energetic terms.' 'Here for the first time' Hutchinson noted, 'we have the interrelated dynamics of a biocoenosis presented in a form that is amenable to a productive abstract analysis.'[31]

Lindemann's paper grouped all the different species into 'trophic levels', characterised by their function within nature's energetic economy as primary producers, primary and secondary consumers, and decomposers (or recyclers). Primary producers are 'autotrophs' producing their own 'food' through photosynthesis; the rest are 'heterotrophs', animals and bacteria which feed on other organic tissues. A key insight in Lindemann's study was that at each level of the trophic pyramid, much of the energy embodied in the previous level was lost as heat to the atmosphere. Only a fraction of the solar energy arriving in an area was converted into plant matter, only a part of the energy incorporated into plant biomass was converted into the bodies of herbivores when eaten, and carnivores at the top of the food chain were similarly limited in number and biomass by predator-prey dynamics and the efficiency of energy transfers below them in the pyramid. By quantifying these losses, the ecologist could measure the 'productivity' of each trophic level in terms of the energy embodied as biomass (minus the energy expended in respiration, digestion or hunting) and thus the 'efficiency' of transfers of energy between trophic levels.[32]

The ecologist could then calculate Gross Primary Productivity for her chosen ecosystem—the rate at which it accumulated biomass—and then subtracting the high proportion of energy dissipated as waste heat in the metabolic process, arrive at a figure for Net Primary Productivity. With Lindemann we see the 'economy of nature' metaphor fulfilled for the modern period, with the natural world analysed through the medium of

[30] Lindeman (1942, p. 399).

[31] Worster, D. (1977). *Nature's economy: the roots of ecology*. San Francisco: Sierra Club Books, p. 306.

[32] Worster (1977, pp. 306–311).

energy conversion efficiencies and the statistical analysis of productivity. Here again we see the use of a fertile metaphor in opening up new avenues of scientific inquiry. The 'metabolic' approach to the physical analysis of ecosystem processes was precursory to the more recent emergence of such vital disciplines as ecological economics and industrial ecology, which followed Lotka in identifying the biological basis of economic activity, and which through the method of material flows analysis provided a substantive approach to the analysis of national economy in terms of the appropriation and dissipation of measurable material flows of energy, minerals and biomass products traded across borders.[33]

Lindemann's paper developed a methodological approach to the energetic analysis of ecosystems which was taken up by Hutchinson's student H.T. Odum and his brother Eugene in *Fundamentals of Ecology* (1953). It was thus enormously influential in propelling ecology to scientific prominence in the 1960s—a story which we take up in more detail in the following chapter. Lindeman's research arguably catalysed one of the biggest projects ever organised in biology: the International Biological Program (IBP) which ran between 1964 and 1974, which contributed to the emergence of contemporary Earth systems science.[34] The IBP was established following the successes of the International Geophysical Year (1957–1958) in establishing transnational networks of scientific collaboration and the collection of global scale observations for data-driven research. [35] The IGY led to the discovery of the Earth's 'Van Allen belt' of radiation, ground breaking studies in Antarctica, and the launch of artificial satellites such as *Sputnik*. The continuity between the two programs was assured through the oceanographer Roger Revelle—an able science organiser who had in 1956 called upon the IGY to systematise research into the anthropogenic carbonisation of the atmosphere and oceans. Revelle lent his support to the young geochemist Charles Keeling, securing long-term funding for his routine recording of atmospheric concentrations of carbon dioxide. By

[33] See e.g.: Magalhãesa, N., Fressoz, J-B., Jarrige, F., Le Roux, T., Levillain, G., Lyautey, M., Bonneuil, C. (2019). The physical economy of France (1830–2015): the history of a parasite? *Ecological Economics*, 157, 291–300.

[34] Coleman, D. (2010). *Big ecology: the emergence of ecosystem science*. Berkeley: University of California Press.

[35] Aronova, E., Baker, K., & Oreskes, N. (2010). Big science and big data in biology: from the International Geophysical Year through the International Biological Program to the Long Term Ecological Research (LTER) Network, 1957–present. *Historical Studies in the Natural Sciences*, 40(2), 183–224.

1960, Keeling had painstakingly assembled scientific proof that CO_2 concentrations were rising, drawing international attention to anthropogenic global warming.[36] Revelle was the lead author of the first public warning of the potentially catastrophic risks of planetary heating, a 1965 report to President Lyndon Johnson which was communicated in the same year to the senior oil executives of the American Petroleum Institute.[37]

Revelle introduced many of the IGY ideals in the planning of the American contribution to the IBP. Set up after years of negotiation between the presidents of the International Council for Scientific Unions (ICSU) and the International Union of Biological Sciences (IUBS) over what might be an appropriate topic for biologists to study, Revelle's argument that 'the time for ecology has come' won the day, and the IBP was initiated in 1964 around the unifying theme of 'The Biological Basis of Productivity and Human Welfare', chosen by the IUBS leader Conrad Waddington. Waddington later explained this choice of focus:

> [We] felt that the only possible line would be to formulate a programme around something which was indubitably of major social and economic importance for mankind as a whole [...] The most attractive field [for the IBP], I thought, was something to do with the way in which solar energy is processed by the biological world into the formation of complex molecules which man can use, as food or otherwise.[38]

The IBP was a major 'big push' investment in systems ecology, organizing thousands of biologists, their universities and national academies to measure the 'productivity' of the photosynthetic conversion of solar energy into biomass across an exemplary sampling of different ecosystems, from arctic tundra, to deep ocean floors, to peasant agrosystems and modern industrial monocropping. True to its origins in the Great Plains, among the major American investments in ecology's transformation by the IBP into a data-driven science of Earth systems analysis was the

[36] Weart, S. (2010). The idea of anthropogenic global climate change in the 20th century. *Wiley Interdisciplinary Reviews: Climate Change*, 1(1), 67–81.

[37] As noted in the Introduction.

[38] Waddington, C. (1975). The origin. In E. Worthington (Ed.), *The evolution of IBP* (pp. 4–11). Cambridge: Cambridge University Press (Cited in Aronova et al., 2010, p. 199).

Grassland Biome modelling project, run by the ambitious and polarising science manager George Van Dyne.[39]

The IBP's Big Science project to measure the Gross World Product of the biosphere gave a huge boost to systems ecology and the wider Earth systems sciences, establishing standardised protocols, for example, for the use of satellites to systematically map and measure biological phenomena from space. However, the IBP was criticized by many scientists. Although laudable as a goal, many felt the focus on human welfare, human populations, and their interaction with biological 'productivity' smacked of the social sciences, and did not readily translate into a strictly scientific research program. At least not one capable of enabling ecology, inherently a field science addressing complex open systems with enormous numbers of variables and no clear temporal, spatial, or physical boundaries, to catch up to the impressive breakthroughs of laboratory-based biology, such as the 1953 characterisation of DNA by Watson and Crick and their unfairly unacknowledged collaborator Rosalind Franklin. Molecular biologists and community ecologists, sceptical of the systems approach itself, were critical of the IBP's hierarchical and centralised forms of management and organisation, which required intensive cross-coordination of national and international bodies. Others attributed its inconclusive results to an overconfidence in the application of computerised mathematical models to problem definitions (e.g. measuring 'system productivity') too vague to justify the considerable research funding it soaked up.

The Currency and Climate Crises of Ecological Energetics

Reflecting on the transformation of ecology effected by the analysis of energy as the fundamental 'currency' of ecosystems in his classic *Economy of Nature* (1977), Worster offered the following critical assessment:

> [...] to a great extent, ecology today has become 'bioeconomics': a cognate, or perhaps even subordinate, division of economics. [...] The metaphors used [in a 1967 paper analysing the 'typical river' in terms of biological capital, energy and raw material supply, assembly lines, efficiency, productivity, output, finished products] are more than casual or incidental; they express

[39] Golley (1993, pp. 109–140); Kwa, C. (1993). Modelling the grassland. *Historical Studies in the Physical and Biological Sciences*, 24(1), 125–155.

> the dominant tendency in the scientific ecology of our time. In their theoretical models, ecologists have transformed nature into a reflection of the modern corporate industrial system.[40]

Yet Worster's history did not deal in depth with the history of economic concepts, and importantly, not at all with the pretensions of orthodox economists in appropriating the legitimacy of physics through the selective appropriation and half-baked application of physics metaphors. The critical potential of systems ecology to provide a civilizational counter-narrative to that of conventional growth economics is clearly diminished by the naturalisation of the corporate-industrial system that is the consequence of such a metaphorical repertoire, derived from a selective reading of the social sciences that privileges economic theory as 'scientific'. This recurring tendency of ecology to appear as a 'subordinate' subdivision of economics was arguably informed by the ambiguous legacy of Lotka's 1920s attempt to unify biology, energy physics, and economics.

For Lotka, the phenomenon later termed 'economic growth' was not a social activity that occurred independently of and against the backdrop of a static and unchanging nature, but the world historic evolution of humanity as a biophysical force of nature, a social organism equipped with increasingly powerful 'extra-somatic prostheses': the thermo-industrial technologies that, for Lotka, had evolved naturally from the struggle to maximise the consumption of energy:

> Man and machines today form one working unit, one industrial system. The body politic has its organs of sight and hearing, its motive energies, its moving members, in close copy of the primitive body of man, of which it is a magnificent and intensified version.[41]

Lotka's account of the thermodynamic harnessing of the worlds' energies to an ever more powerful social machine was progressive and optimistic. As Kingsland puts it, 'Lotka's notion of the body politic was intended to show that the evolution of the social organism through technological expansion was part of a natural process that contributed to the individual's unity with nature.'[42] Modern civilisations' appropriation of the maximum available energy was merely an extension of the laws of nature that operated

[40] Worster (1977, p. 292).
[41] Cited in Kingsland (1994, p. 239).
[42] Kingsland (1994, p. 240).

on each species and drove the evolutionary process. Despite that fact that the competitive-individualist lineage of classical liberal political economy can be traced in Lotka's influences through Pareto, Jevons, Spencer and Darwin back to Malthus and Senior, Lotka imagined that in rising above selfish individualism and becoming 'collaborators with nature', people could embrace modernity and contribute to an orderly and rational expansion of the social organism. Kingsland argues that Lotka was advocating a kind of secular millennialism, the response of the educated middle-classes to the anxieties of the interwar period. Though some blamed the nationalist and class-based antagonisms of modernity on the social-technical organisation of the industrial system, for Lotka more science and technology, not less, would guarantee progress. His vision of the new industrial society was one of transcendent efficiency and social harmony, as the body politic was steered ever onward by an elite intelligentsia that would actively fulfil 'the great World Purpose' by harmonising social institutions with the laws of nature.[43]

Writing in the aftermath of the Great War, keenly attuned to a sense that human societies were on the verge of massive technological transformations, Kingsland argues that Lotka's work was intended to show the unity of 'man' and nature, by intimately tying human social evolution—to be fostered into the future by a 'natural aristocracy' of the scientifically literate—to the inexorable operation of the vast world engine. In Lotkas's view, the species most successful at diverting available energy would 'tend to grow in extent (numbers) and this growth will further increase the flux of energy through the system'. He admitted the possibility that sparing use of the energy flow and a more careful husbandry of resources might 'work to the advantage of a species talented in that direction', but continued to insist that the general tendency in such systems is to appropriate the '*maximum possible share of the available energy resources* [my italics]'.[44] Such was his confidence in this 'maximum power principle' that he proposed it as a candidate fourth law of thermodynamics.

While the substantive elements of Lotka's evolutionary biophysics and the absurd abstractions of neoclassical economics are worlds apart, Lotka's conflation of the neoclassical image of market exchange as 'utility maximisation' under equilibrium constraints with energy maximisation as 'the

[43] Lotka (1925, p. 428).
[44] Cited in Jackson (2003, p. 6).

struggle for existence' of biological life writ large demonstrates the shared provenance of both paradigms. The irony of such resonances is that while Lotka's techno-scientific optimism prefigured the millennialism of the post-WWII American ideology of infinite growth, it proceeded (unlike the radical subjectivism of Austrian and neoclassical economics) from a biophysical analysis of human social organisation grounded firmly in the scientific materialism of his day. G.E. Hutchinson described Lotka's *Elements* as a 'book [...] written to provide for biology, or at least parts of biology, a basis comparable to that given by theoretical physics to experimental physics. Lotka probably knew a smaller proportion of the relevant biology of his time than a theoretical physicist usually knows of experimental physics [which] gives parts of his book a curiously naïve character. In spite of this limitation it is a great work, one of the foundation stones of contemporary ecology.'[45] This was an approach which would influence the coterminous development of American ecosystems ecology, (especially the work of H.T. Odum), and the critical bioeconomics of Nicholas Georgescu-Roegen. Taken together, these provided a rigorous basis for the environmental movement's critique of the fossil-fuelled infinite growth paradigm, and its priesthood amongst academic economists.

Written almost a century ago, certain passages of Lotka's *Physical Biology* appear strikingly modern to the contemporary reader, in particular his discussion of the global carbon dioxide cycle. This he begins by recalling the Promethean endeavour of human fire mastery, going on to note that 'Arrhenius builds his conception of the future of industrial development of our race under the expectation that the atmosphere is gaining in carbon dioxide, under the present regime of evaporating our coal mines, as it were, into the air'. Presciently, Lotka observed that:

> Economically we are living on our capital; biologically we are changing the complexion of our share in the carbon cycle by throwing into the atmosphere, from coal fires and metallurgical furnaces, about ten times as much carbon dioxide as the natural biological process of breathing [e.g. respiration]. How large a single item this represents will be realised when attention is drawn to the fact that these human agencies alone would, in the course of about five hundred years, double the amount of carbon dioxide in the entire atmosphere, if no compensating influences came into play.[46]

[45] Hutchinson, G. (1978). *Introduction to population ecology*. New Haven: Yale University Press, p. 2.

[46] Lotka (1925, p. 225).

Underestimating the exponential curve of future industrial fire, Lotka's discussion of the carbon cycle failed to mention Arrhenius' application of the earlier work of Fourier and Tyndall establishing the heat absorbing qualities of atmospheric gases, nor his alarming 1896 estimation that a doubling of atmospheric CO_2 could raise the average temperature of the Earth's surface by 5–6°C.[47] Arrhenius had estimated that CO_2 doubling was a distant prospect, which might take place five thousand years in the future. With more up-to-date data, Lotka shortened this extrapolation to five centuries hence, but screened out from his opus consideration of the catastrophic consequences of such a radical thermal shift in the conditions of life on Earth. For Lotka, it still seemed a remote possibility, perhaps to be realised in some distant future age—and not, as it appears to us, in a few short decades. Thus his early observations on the anthropogenic forcing of the carbon cycle did not trouble his productivist confidence that thermo-industrial expansion was a natural and rational 'self-organising' process, one essentially in unity with 'the great trend of all Nature.'[48]

H.T. Odum once summarised Lotka's 'maximum power' thesis as '[t]he idea that over time a network that draws more resources and uses them better toward maintaining that network will tend to replace designs that have fewer resources with which to work.'[49] Translated into geopolitical terms and applied with the benefit of hindsight to the neoliberal era, Odum's words serve as an explanation consistent with Lotka's thesis of the success of oil billionaires and powerful hydrocarbon lobbies in repressing citizens' attempts to respond to the deepening catastrophe of runaway climate change by legislating for a progressive reduction of hydrocarbon combustion and its replacement with renewable energy technologies that do not further destabilise the carbon cycle. From the outset a political project of the private wealth-holders in control of the hydrocarbon-based industries at the core of an ever-expanding thermoindustrial economy, neoliberalism appears as an intentional form of network organisation

[47] Weart (2010, p. 68). The IPCC estimates a slightly lower range of 3–5°C, but as current research suggests, may have systematically underestimated the amplification of the runaway heating trend by the release of methane to the atmosphere from agriculture, thawing permafrost and marine clathrates, and petroleum gas and coal-bed methane extraction. See: Bendell, J. (2018, July 27) Deep adaptation: a map for navigating climate tragedy. IFLAS Occasional Paper 2.

[48] Lotka (1925, p. 433).

[49] Odum, H. (1995). Self-organization and maximum empower. In C. Hall (Ed.), *Maximum power: the ideas and applications of H.T. Odum* (pp. 311–330). Niwot, Colorado: University of Colorado Press, p. 311.

seeking 'maximum power'. Defended in the name of the freedom of the 'spontaneous' global economic order by an 'elite intelligentsia' (the Mont Pèlerin Society), this project has been pursued through a transnational political machinery (the Atlas Network) which has fostered public ignorance of the abundantly confirmed findings of the Earth sciences, and which has proved highly effective in capturing state power to immunise destructive corporate activity from democratic control. Yet there are of course limits to Lotka's apolitical analysis of 'maximum power' networks. If a form of social organisation mobilised to neutralise all opposition to the destruction of the Earth's biosphere as a viable life-support system serves some 'Great World Purpose' beyond the temporary enrichment of a narrow and callous elite, it is difficult to imagine what that might be.

CHAPTER 12

Ecologist as Cyborg: The Military Origins of the Subversive Science

In the early twentieth century, the sciences of ecology and economics were properly and safely demarcated around ontologically distinct research objects. Whilst acknowledging internal disciplinary differences, we may generalise and say that ecologists dealt with a biological community tending towards equilibrium and devoid of humans, and economists with the naturally self-equilibrating phenomena of exchange in a market economy with no material interactions with 'the environment'. Practitioners of these arcane arts had infrequent cause for contact, let alone contestation. This tidy division began to unravel in the post-WWII period, as dawning consciousness of ecological risk on a biospheric scale emerged, triggered by widespread fears of nuclear annihilation made possible by the vast energies unleashed by atomic technologies. Thus was inaugurated what Ulrich Beck describes as *riskgesselschaft* ('risk society'), a social order shot through with the unprecedented spectre of incalculable and uninsurable ecological hazards. Since then, the pervasiveness of global ecological risk has undermined the authority of the nation-state as the guarantor of the safety and well-being of its citizens, and with it the very idea of an external natural world independent of human activity. One consequence of this, according to Beck, is the pitting of science against science and the erosion of the epistemic and moral authority of scientists by the scientific creation of unforeseen hazards.[1]

[1] Beck, U. (1992). *The risk society: towards a new modernity*, trans. M. Ritter. London: SAGE.

J. Walker, *More Heat than Life: The Tangled Roots of Ecology, Energy, and Economics*,
https://doi.org/10.1007/978-981-15-3936-7_12

Despite the popular association of 'ecology' with counter-cultural radicalism, back-to-the-land Arcadianism, a critical view of Promethean technoscience, and a quest for the re-enchantment of wild nature, the reality is that the process by which systems ecology claimed the status of a 'hard' science was anything but subversive. The scientific and technological milieu which engendered the first truly global crisis of technologically generated ecological risk—the US-Soviet nuclear arms race—was inseparable from the development and institutionalisation of the ecosystem concept at the core of the new integrative science of global systems ecology, which fused the biological and geophysical sciences with thermodynamics by foregrounding the primacy of solar energy conversion in the economy of nature. As his classic 1935 article cites only Anglophone sources, it appears that Tansley was unaware of Vernadsky's holistic and integrative account of the biosphere (1926) when he proposed the ecosystem as a term with which to approach the differentiated panorama of life—from 'vegetation-clad weathering profiles and hillslopes [...] to Earth's whole terrestrial surface'—as 'one physical system'.[2]

The formative history of systems ecology—from which the radical environmentalist critique of the American ideology of infinite economic growth claimed its validity and drew its paradigmatic metaphors—was characterised by individual and institutional accommodations to military funding, in the context of a formidable unleashing of the vast energies of the atom. Ecosystems ecology emerged from the intellectual environment of the American Cold War military-academic complex, a particular expression of a new general systems theory of self-adjusting systems and self-regulating mechanisms dubbed 'cybernetics'.

Like cybernetics, networked digital computation, game theory, and behavioural microeconomics, American systems ecology was ultimately an artefact of the Manhattan Project and the Cold War expansion and reorganisation of Big Science in the United States around the linked geopolitical imperatives of nuclear security and securing access to energy for economic growth. Thus were the ideological fault lines established which define our present crisis, in which the political organisations of the neoliberal era strive to secure the 'spontaneous self-organisation' of fossil-fuelled economic growth from counter-movements attempting to protect civilisation from self-annihilation in an inferno of carbon-induced planetary

[2] Richter, D. & Billings, S. (2015). 'One physical system': Tansley's ecosystem as Earth's critical zone. *New Phytologist*, 206(3), 900–912.

heating, a consequence of mass combustion announced by Earth system scientists in the mid-twentieth century.

Cybernetics as Social Physics

We have shown so far how intertwined the development of ecology was with prevailing economic paradigms and questions of national development and economic growth. Ecosystems ecology sits in an ambiguous position with regard to this historical theme. While its designation as the 'subversive science' is coterminous with the arrival of an energetic approach to ecosystem theory, it is not often recognised that ecosystems ecology would not have gained its initial funding or paradigmatic methods without the centralisation of scientific research accomplished by the imperatives of national security, in a research environment dominated by military funding.[3] In its full energetic determination, the ecosystem as the foremost analytical unit of the order of living nature was derived from the wartime study of control and information in human-machine interfaces. From its chthonic WWII origins in the closed intellectual workshops of the US total war effort and the subsequent Cold War national security state, cybernetics produced an ambitious research agenda that, while unsuccessful in providing a unifying metascience capable of overcoming disciplinary methodological differences, was nonetheless highly influential in the heterogeneous cyborg sciences spawned by the approach: operations research, communication and computer engineering, artificial intelligence, robotics. Cybernetics was applied as a form of systems analysis and control theory in numerous spheres: anthropology, economics, the behavioural social sciences, and also to strategic questions of industrial and military organisations. In the post-war period, both economics and ecology underwent profound transformation as a result of exposure to the post-war development of systems theory.

Nailing down exactly what cybernetics is is a difficult question. Perhaps the simplest way to think of it is as its post-war practitioners did: as a cutting-edge metascience with the potential to bring a whole gamut of technical, scientific, and social problems within the ambit of formal analysis by reframing them as problems of input and output in dynamic systems regulated by positive or negative feedback. Numerous projects were

[3] Kingsland, S. (2005). A subversive science? *The evolution of American ecology, 1890–2000* (pp. 179–205). Baltimore: John Hopkins University Press.

gathered together under the concentrated reorganisation of research for the 1941–1945 war effort under senior science organisers such as Vannevar Bush and Warren Weaver. Under the general ambit of 'operations research', which involved the development and application of mathematical techniques to the prerogatives of the large-scale organisation of the total war effort, social scientists, economists, and physicists worked together in a centralised, well-funded environment that accorded scientists with a new importance and respect. Numerous advances occurred due to the cross-fertilisation of interdisciplinary work required to solve the many logistical problems of the total war economy: that is the need for the rapid development of new military prototypes, of a means for selection among different weapons platforms for mass production and deployment, the importance of intercepting and decoding encrypted enemy signals, and the need for highly sophisticated strategic deception once the enemies' communications had been penetrated.

Along with key figures such as Allan Turing, John von Neumann, and Claude Shannon, the MIT mathematician Norbert Wiener looms large in the origin stories of cybernetics. His 1948 book *Cybernetics: or Control and Communication in the Animal and the Machine* represents the first explication of the new systems theories under that term.[4] Wiener argued for the utility of the approach in the following terms: 'Cybernetics extends and widens the circle of processes which can be controlled—this is its special property and merit.'[5] During WWII, Wiener worked on gun-aiming servo-mechanisms for anti-aircraft artillery. Out of this research came a co-authored 1943 paper titled 'Behaviour, Purpose, and Teleology'.[6] The paper defined 'purposeful behaviour' in both machines and living organisms in terms of information feedback. Bryant explains the fertile metaphor posited in this founding paper of the new metascience:

> Certain machines were like organisms in that they received inputs and produced behaviour or output in turn. When a portion of that output circled around to become input again, the machine acquired a means for responding to the effects of its own behaviour. Its behavior became purposeful when this circuit of information feedback guided the machine toward a goal. A

[4] Wiener. N. (1948). *Cybernetics: or control and communication in the animal and the machine*. Cambridge: MIT Press.

[5] Wiener (1948).

[6] Rosenblueth, A., Wiener, N. Bigelow, J. (1943). Behaviour, purpose and teleology. *Philosophy of Science*, 10(1), 18–24.

> torpedo with a target-seeking mechanism exhibited purposeful behaviour, for example, whereas a bomb that simply fell to earth from an airplane did not. Structurally, organisms and self-directed machines were far different, but in terms of behaviour they could be classified together, as an order of things that were purposeful and predictive. 'The broad classes of behaviour', the authors argued, '*are the same in machines and in living organisms.*' Such behaviour, whether exhibited by an organism or a machine, depended upon the circular feedback of information. [my emphasis][7]

The historical literature on cybernetics often divides its avatars into two camps. The first, exemplified by von Neumann, were consummate insiders within what Eisenhower termed the 'military-industrial complex', with access to academic prestige, high-level government committees, and substantial military funding. Embedded in the military mindset of RAND think tanks and Air Force strategic command, these men were the social and atomic engineers of the Cold War, intensely engaged in a hair-trigger nuclear confrontation with a formidable and inscrutable enemy, in a bunker atmosphere 'slick with dread and heavy with doom'.[8] The intense demands of the wartime work environment ended for many scientists with the demobilisations of 1945, and those that returned to civilian research applied the technological optimism of the military engineer to the problems of peacetime. It is among this second camp that we find Norbert Wiener, Gregory Bateson, G.E. Hutchinson, and H.T. Odum; figures optimistic about the expansion of knowledge, technology, and the application of rational systems of governance in the humane service of civilisation. As time wore on some of these voices became increasingly critical of the irrationalities arising from the absorption of American scientists and research funding in the development of weapons of mass destruction. This pattern of Cold War science was evident as early as 1946 to Norbert Wiener, who attempted to have his own wartime papers on missile control removed from libraries in protest at the 'tragic insolence of the military mind'.[9]

A full genealogy of cybernetic concerns—involving the study of self-regulating systems and interactions between human labourers and

[7] Bryant, B. (2000). Nature and culture in the age of cybernetic systems. Paper delivered to the American Studies Association Annual Conference, Detroit, p. 1.

[8] Mirowski (2002, p. 56).

[9] Wiener, N. ([1946] 1983). A scientist rebels: letter to Mr. George Forsythe. *Science Technology, and Human Values*, 8(3), 36–38.

industrial machines, a fascination with the formal and analogical similarities between machines and organisms to the point of outright conflation, with attempted construction of automatons to the building of computing machines—could be taken back at least as far as Charles Babbage's *The Economy of Machines and Manufactures* (1832) and his collaboration with Ada Lovelace in a failed attempt to construct the Analytical Engine—a steam-powered computing machine composed of brass cogs with which to solve complex equations.

More pertinently, this history of fascination with 'thinking' machines is intimately connected to discussions of the second law of thermodynamics amongst physicists occasioned by the figure of 'Maxwell's demon': a thought experiment proposed by James Clerk Maxwell, which posed the question of the relationship of entropy to knowledge, or 'information'.[10] Maxwell imagined an impossibly disembodied entity that could resist the entropic dissipation of energy in an ideal gas chamber by instantaneously sorting gas molecules of higher temperatures from those of lower temperatures, opening a nano-scale valve to emit cool particles from a chamber whilst preventing the hot ones from escaping. Whilst this thought experiment is strictly impossible in actual thermal systems, Maxwell's demon came to symbolise the potential of 'information' (as orderliness) to temporarily counter, and perhaps transcend, the inexorable forces of entropic disorder: thus 'information' was reconceived of as 'negative entropy' or 'negentropy'. The concept of negentropy deployed by the physicist Ervin Schrodinger in his *What is Life?* (1944) would contribute to the discovery of molecular DNA as the biophysical concentration of morphological order in the compressed format of the genetic programme. The Bell Labs electrical engineer Claude Shannon, by contrast, borrowed the concept to define information in terms of the 'signal to noise ratio' of a message in transmission, in a casual informatic analogy which precluded the material implications of thermodynamics. This cybernetic conceit was crucial to the fashion for speaking of a post-industrial 'information society' or 'knowledge economy' since the World Wide Web was constructed in the 1990s, as if the evolutionary trajectory of modern industrialism was now one of informatic dematerialisation and no longer of exponential growth in material scale, resource demands, and waste outflows. This is a

[10] Mirowski, P. (2002). Some cyborg genealogies, or how the Demon got its bots. In *Machine dreams: economics becomes a cyborg science* (pp. 26–93). Cambridge, MA: Cambridge University Press.

vision which finds its zenith in the Silicon Valley eschatology of the coming event-horizon of 'the Singularity'—a concept first advanced by von Neumann—after which select individuals will achieve immortality by uploading consciousness to a networked post-human artificial intelligence destined to expand throughout the Solar System, and even beyond. The origins of this particular manifestation of the cybernetic paradigm—as an ascent-trajectory of escape from the dependence of 'meat machines' on the biosphere—can be seen in the origins of the term 'cyborg'. The term was coined by Manfred Clynes and Nathan Kline to describe to a US Air Force conference the possibilities of experimental augmentation of laboratory animals that might allow them to engage in feedback stabilisation and control of an artificial metabolic environment. Published in a 1960 edition of *Astronautics*, these ideas were crucial to the development for the space programme of the artificial life-support systems which would allow astronauts to survive beyond the Earth in the abiotic environment of outer space. As Clynes would later recall:

> I thought it would be good to have a new concept, a concept of persons *who can free themselves from the constraints of the environment to the extent they wished*. And I coined this word cyborg. [my emphasis][11]

The original vision of the 'post-industrial' was conceived around the hope that computing machines would lead to the realisation of a social physics for a new era of infinite energy and infinite information. In his 1973 'venture in social forecasting', the conservative sociologist Daniel Bell argued that the seeds of the post-industrial society were sprouting around 1950:

> The period since the end of World War II has produced a new consciousness about time and social change. […] To begin with, the transformation of matter into energy by the creation of the atom bomb in 1945 made the world dramatically aware of the power of science. […] In 1946, the first digital computer, the ENIAC, was completed […], and it was soon followed by the MANIAC, the JOHNNIAC, and within a decade ten thousand more. […] In 1947, Norbert Wiener published his *Cybernetics*, which spelled out the principles of self-regulating mechanisms and self-adjusting systems. If the atom bomb proved the power of pure physics, the combination of the computer and cybernetics has opened the way to a new *'social physics'*—a set of techniques, through social and communications theory, to

[11] Cited in: Gray, C. (1995). *The cyborg handbook*. New York: Routledge, p. 29.

> construct a *tableau entière* for the arrangement of decisions and choices. [emphasis in original] [12]

The social physics projects of the post-war years prioritised the question of engineering economic growth and national development. As we have seen with the 1950s Paley Commission, this meant marketing the growth ideology of endless consumption at home and a geopolitical strategy abroad that would ensure the 'free movement' of natural resources from the global South into the US industrial system. The post-industrial thesis of the 1970s foretold an exciting future of technological transformation which would yield an abundance of leisure time, liberating the mass of worker-consumers to become self-directed and artful 'prosumers'. But there is also something darkly prophetic here, as Bell notes the potential of social control through networked ICT—something which rings true in our present era of sophisticated public opinion management and ubiquitous digital surveillance.

Elsewhere in the book, Bell took dismissive note of the nascent environmental movement:

> [...] ecological reformers like Rachel Carson and Barry Commoner, [...] invoke the institutional charisma of science in making moral or political judgments. What we have here is the resumption of the prophetic claims of science in setting forth truth as against self-interest. [...] Paradoxically, the vision of Utopia—a fully automated production economy with an endless capacity to turn out goods—was suddenly replaced by the spectre of Doomsday. In place of the early-sixties theme of endless plenty, the picture by the end of the decade was one of a fragile planet of limited resources whose finite stocks were rapidly being depleted, and whose wastes from soaring industrial production were polluting the air and waters. Now the only way of saving the world was zero growth. What was striking in this change is the shift in attention from machinery to resources, from man's mastery of nature to his dependence on its bounty, from Harrod-Domar-Solow growth economics to Malthusian-Ricardian scarcity economics. And the principle of diminishing returns, rather than increasing returns to scale, becomes the analytical motif.[13]

[12] Bell, D. ([1973] 1976). *The coming of post-industrial society.* New York: Basic Books, p. 404.

[13] Bell (1976, p. 464).

For someone so optimistic about the social physics project, it is notable that Bell's defence of the growth orthodoxy against the biophysical critique of ecologists was grounded in subjectivity, not in physicality. For Bell, the economic process was driven by the need of 'men' for status, and as status by definition requires inequality, 'growth' is ultimately a psychological arms race of conspicuous consumption, which due to the insatiability of desire can never be completed.[14] This Hegelian subjectivism mirrors the underlying creationist ontology of liberal economics, which denies any necessary role for the appropriation of natural resources in the production process: ontological greed and 'technology' produces an endlessly expanding stream of commodities *ex nihilo*.

Among the most important technical developments of the post-war period were nuclear fission and the digital computer. These were not at all separate projects: the computer was initially but a subordinate daemon, devised to release a more awesome djinn from its bottle, the thermonuclear bomb. One of the key objectives driving the design of early punch-card computing machines was the need to perform the thousands of complex calculations that went into the design of atomic weapons. In 1945 von Neumann (who not incidentally redirected the neoclassical research programme into game theory in a book published with the Austrian economist Oskar Morgenstern the previous year)[15] arranged for the use of the ENIAC to begin the simulation. One million IBM cards were produced, each carrying one of the initial values for each point in the computational mesh, and 'the computations to be performed required the punching of intermediate output cards which were then resubmitted as input'.[16] Well before the Manhattan Project was in sight of completion, Edward Teller was already demanding research funds for the development of the hydrogen bomb, and the immediate priority was the improvement of computation. As Mackenzie reports, '[t]he complex calculations of hydrogen-bomb simulation exceeded the capabilities of the punched card machine operation' used for the design of the first atom bomb.[17] From the outset, computer modelling technologies were also engaged with the

[14] Bell (1976, p. 464).

[15] von Neumann, J. & Morgenstern, O. (1944). *Theory of games and economic behaviour.* Princeton University Press.

[16] Stern, N. (1981). *From ENIAC to UNIVAC: an appraisal of the Eckert-Mauchly computers.* Digital Press, pp. 62–63.

[17] MacKenzie, D. (1996). *Knowing machines: essays on technical change.* Cambridge, MA: MIT Press, pp. 112–113.

quantification of technological hazards at the biospheric scale. Prior to the detonation of the first A-bomb at Alamogordo, a question arose regarding the rate at which deuterium would decay in an atomic reaction. There was a fear that if this chain reaction gained too much momentum it might set the whole of the Earth's atmosphere on fire, a risk which the Los Alamos IBM machine calculated to a probability deemed by Teller as acceptably remote for the test to proceed.

Economics would be profoundly transformed by the intellectual environment of the early Cold War, as science funding ballooned and the organisation of research was restructured from the top-down by the national security state to meet the formidable technological challenges of the nuclear confrontation with the USSR. In his *Machine Dreams* (2002), Mirowski recounts the historical transformation of neoclassical economics through the intellectual contact of economists with physicists mobilised for the war effort in venues such as the RAND Corporation.[18] The prototypical 'think-tank', RAND was a hybrid institution that emerged at the interface of the Air Force and its avionics contractors: part military strategy research unit, part university, part private corporation. With no clear commitment to any specific branch of the social or physical sciences, RAND was contracted to advise on the rationalisation of scientific research into strategic thermonuclear command structures and man-machine interfaces, organised around the prerogatives of 'C3I'—communication, command, control, and information.[19] These intellectuals were continually exposed to the pressure of extreme consciousness of catastrophic risk to the future of the globe, which was to be met by a similarly extreme confidence in the power of mathematical rationality:

> [RAND] was supposed to be a haven for steely eyed technocrats with ice in their veins, pursuing the mandate to think through a real apocalyptic conflict, one fought with unimaginably devastating nuclear weapons. These layers of looming apocalypse upon apocalypse [...] had the effect of escalating the drive for an abstract rationality well beyond that expounded in any previous social theory. It sought reassurances that could not rest satisfied with a mechanism for giving people whatever it was they thought they wanted, as with some pie-in-the-sky Pareto optimality; neither would it accept that the consequences of present actions could be extrapolated into the future using some mechanical inductive procedure. In an inherently unstable situation

[18] Mirowski (2002).

[19] Mirowski (2002, p. 158).

> fraught with extremes, where meanings were elusive and one slip might mean the catastrophic loss of everything that ever mattered for humanity, what was required was ironclad standard of rationality imposed upon the threatening chaos by means of some external governor. The establishment of control was the essential precept in a nuclear world: passive resignation would never suffice, laissez-faire was out of the question.[20]

Incorporated into the military reorganisation and funding of scientific research, American social science underwent profound reconfiguration around cyborg themes. If the 1870s saw the original interest in social physics, during the Cold War American neoclassical economics assumed its current mathematical format and scientific pretensions. Economists became less interested in theories of collective action, welfare, and institutions and more fascinated by formality, abstraction, statistical inference, and algorithmic representations of individualised 'rationality'. Mirowski refers to this moment as the second stabilisation of the neoclassical conceptual orthodoxy: like the first occurring through an oblique historical connection with physics, mediated by the most up-to-date machine metaphors, and mobilised against the threat of 'socialism'.[21]

Neoclassical theory's description of competitive general equilibrium underwent some major rethinking during the early Cold War, as a direct effect of the interdisciplinary efforts of influential American thinkers employed in highly covert military research. The theoretical outcomes were the hyper-rational, behaviourist models of information processing produced within decision theory and game theory, which sought to codify and mathematise the logical structure of decision-making under conditions of extreme risk and uncertainty. Developing the ideas proposed by von Neumann, the mathematical economist John Nash reconstituted the 'foundations' of microeconomic 'rationality' in game theory. The Nash 'equilibrium' updated the ageing Walrasian equilibrium as the basis of general equilibrium theory. Unlike earlier formulations of neoclassical economics, where the assumptions of perfect rationality, perfect information, and frictionless markets lead to a global equilibrium with a serendipitous welfare function—a Smithian 'natural harmony of interests' between a multiplicity of formally equal market agents, none of which can monopolise control over the system as a whole—game theory internalised a

[20] Mirowski (2002, p. 313).
[21] Mirowski (2002, p. 157).

decidedly paranoid 'hermeneutics of suspicion' into economics. The logic of the famous 'prisoner's dilemma' unfolds from the ontological assumption of information asymmetries, played out as the power relations of dominance and submission in the fearful context of the interrogation cell. However atypical or decidedly anti-social such a scenario might seem, it has been claimed by economists aligned with Vernon Smith (MPS member and professor at the Koch-funded George Mason University economics department) that 'game theory, with the Nash equilibrium as its centerpiece', has become 'the most prominent unifying theory of social science'.[22]

For Mirowski, this reinvention of American economics arose from the need for an ideologically acceptable control theory within the decidedly apocalyptic Cold War research programme. Game theory models of rational choice brought into the same sphere of analysis the choices faced by Cold War nuclear strategists with those made by the atomised economic actor of neoclassical economics in the struggle to maximise 'utility', although now at the direct expense of other competitive units. Nash's economic agents are engaged in competition with inscrutable and deceptive opponents, a competition carried to the point of total annihilation of the (socialist) enemy.

Exceedingly remote from any concept of the 'balance of nature' or 'optimal welfare', this notion of 'equilibrium' rather reproduces the thermonuclear strategy formulated in a 1958 paper by RAND strategist Albert Wohlstetter, who criticised the idea that an 'automatic balance' or technological stalemate between the nuclear powers would ensure peace, arguing instead for a militant policy of actively 'maintaining the delicate balance of terror' amidst a situation of 'extreme instability'.[23] Formalised by US Secretary of State Robert Macnamara in the early 1960s as the policy doctrine of MAD (mutually assured destruction), this led not to a 'balance of forces' or an 'optimal' level of expenditure on nuclear security but to the runaway feedback of the ever-expanding arsenals of the nuclear arms race.

[22] Holt, C. & Roth, A. (2004). The Nash equilibrium: a perspective. *Proceedings of the National Academy of Science*, 101(12), 3999–4002.

[23] Wohlstetter A. (1959, Jan). The delicate balance of terror. *Foreign Affairs.*

Ecology as Social Physics

A watershed moment in the characterisation of ecology as the science of self-organising and self-correcting mechanisms in the idiom of first-order cybernetics was the delivery of a paper by G.E. Hutchinson to the New York Academy of Sciences, entitled 'Circular Causal Systems in Ecology'.[24] In this 1948 paper, Hutchinson outlined the biogeochemical approach of global systems ecology, which interfaced the self-organising physiology of living ecosystems with the global circulation of geochemical elements.

Whilst his primary intention was to posit the existence of an array of self-regulating mechanisms at the scale of the biosphere, Hutchinson's paper was to have deeply prophetic significance for our own time: he developed a carbon budget for the biosphere, and attempted to balance it. A diagram of the carbon cycle presented in this article notably incorporated the impacts of thermoindustrial civilisation upon the Earth's primary 'circular causal system' (Fig. 12.1).

Amongst the earliest attempts to illustrate the global carbon cycle in a single image, it included a quantitative estimate of the anthropogenic transfer of 'buried organic carbon' to the atmosphere as carbon dioxide, 'returned by combustion of fuel'. As we have seen, by the mid-twentieth century it was clear to many that atmospheric and oceanic concentrations of carbon dioxide were increasing. While he included this fact in his diagram, this was inconsistent with the text that accompanied it. Citing Callendar's ground-breaking work which emphasised industrial fire as the prime cause of an estimated 10% increase in atmospheric CO_2,[25] Hutchinson ventured that this was an overestimation and that the increase would better be attributed to land clearing and soil erosion by the 'technological cultures' of the North, which progressively reduced the biosphere's ability to reabsorb atmospheric carbon through photosynthesis.

Of course as we know now, this was never an either/or proposition. Focused on explicating the concept of circular causal systems for global ecology, Hutchinson chose not to discuss Callendar's parallel concern

[24] Hutchinson, G. E (1948). Circular causal systems in ecology. *Annual of the New York Academy of Sciences* (50), 221–246.

[25] Callendar, G. (1940). Variations of carbon dioxide in various air currents. *Quarterly Journal of Research in Meteorology* (66), 395–400.

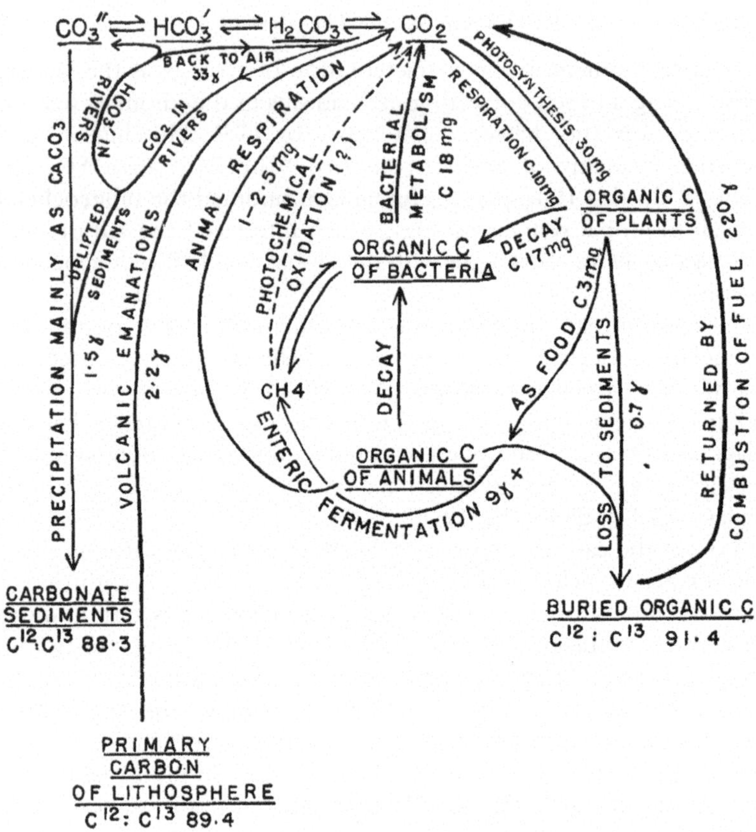

Fig. 12.1 Hutchinson's diagram of the global carbon cycle. (Source: Hutchinson, G. E. (1948). Circular causal systems in ecology. *Annual of the New York Academy of Sciences*, 50, 221–246. Reprinted with permission of John Wiley & Sons)

with the observed global accumulation of heat, [26] and thus to consider the role of the natural carbon cycle in auto-stabilising the heat balance of the Earth.

Whilst circumspect about the possibility of the fossil fuel combustion throwing the climate out of 'balance', Hutchinson's 1948 paper at least

[26] Callendar, G. (1938). The artificial production of carbon dioxide and its influence on climate. *Quarterly Journal of Research in Meteorology* (64), 223–240.

implied that the self-regulatory mechanisms that the history of life had evolved to stabilise the great biogeochemical cycles of the Earth had been altered, potentially pushing the global ecosystem out of balance. Not only was this one of the earliest attempts to apply a systematic analysis of global geochemical cycles of carbon, nitrogen, and methane within the framework of the ecosystem concept, it was also perhaps one moment when global ecology became a social physics, via an early quantification of the physical impact of the 'world industrial organism' on the biosphere's grandiose self-stabilising cycles.

The paper raised the profile of ecology by showing it to be an exciting discipline capable of providing important insights into the problems of basic science and also of addressing the relationship between industrial society and the biosphere as a whole. It was also among the first papers in ecology to incorporate the new metascience of cybernetics. Hutchinson was a key participant in the highly influential Macy conferences (1946–1953). An interdisciplinary open-house gathering of academic luminaries attended these conferences, whose purpose was to discuss 'Circular Causal and Feedback Mechanisms in Biological and Social Systems', a term later condensed to 'cybernetics'.[27]

Exemplary of post-war optimism, these academics advanced cybernetic theory in the hope for a unified science that could promote orderly, peaceful economic development in the anxious first decades of the nuclear arms race. Machines able to perform rapid computation offered to radically extend the epistemological reach of interdisciplinary research and to revolutionise business, government, and human-environment relations. In a keynote speech, one of the convenors of the Macy conferences, the social scientist Lawrence Frank, declared: '[W]e are engaged, today, in one of the major transitions or upheavals in the history of ideas. When the social sciences accept these newer conceptions [...] and learn to think in terms of circular causal processes, they will probably make amazing advances.'[28] In a 1948 paper that applied 'physical thinking' to social problems, Gregory Bateson expressed hope that the new systems theory might unify the biological and human sciences by allowing the success of physics flow

[27] von Foerster, H. (Ed.) (1954). *Cybernetics: circular causal and feedback mechanisms in biological and social systems.* Proceedings of the Macy Conferences, Josiah Macy Jr. Foundation.

[28] Frank, L. (1948). Foreword. *Annual of the New York Academy of Sciences*, 50, 192–195.

into other fields.[29] In a paper entitled 'Social Theory and Social Engineering', Hutchinson expressed similar sentiments.[30]

As Taylor notes, ecology reprocessed by cybernetics meant the replacement of older organismic metaphors with those derived from heat engines and computers:

> Although vitalism was a defeated force in biology, it was nevertheless a radical step to unify the study of living and non-living systems. The new theorists of feedback systems conceived of nature as a machine, and the same time acknowledged the purposive and regulatory character of the nature-machine. A theory of 'teleological mechanisms' could not only abolish vitalism but also the old cause and effect determinism. Furthermore the same terms could be applied to all systems, whatever their components; living and non-living could be intermeshed, eliminating the biological relations from physical factors. [...] It would not be long [...] before purely physical theories, such as those of thermodynamics—or, even more abstractly, of information theory—would be taken up as organising principles for ecology.[31]

Just as cybernetics vigorously broke down the nature-society dichotomy with its metaphorical conflation of machines and organisms, so too did it blur the lines between the social and natural sciences. Exemplifying a shift in American sociology from a qualitative communitarian approach to the quantitative functionalism enabled by high-speed computer analysis of data, Warren McCulloch was of the confessedly utopian view that a cybernetic social science would allow 'Man' to 'learn to construct for the whole world a society with sufficient inverse feedback to prevent another and perhaps last holocaust'.[32] To this purpose, the powerful data storage and analytic capabilities of the digital computer suggested that long-range data sets could allow scientists to determine which mechanisms of negative feedback could account for the stability and purposive behaviour of social groups. This would allow an all-encompassing system of governmental feedbacks to be constructed. The cybernetic systems approach to

[29] Bateson, G. (1946). Physical thinking and social problems. *Science*, 103, 717–718.

[30] Hutchinson, G. E. (1946). Social theory and social engineering. *Science*, 104(2694), 166–167.

[31] Taylor, P. (1988).Technocratic optimism: H.T. Odum and the partial transformation of ecological metaphor after World War II. *Journal of the History of Biology*, 21(2), 221–246.

[32] McCulloch, W. (1948). A recapitulation of theory with a forecast of several extensions. *Annual of the New York Academy of Science*, 50(4), 259–277. See p. 264.

understanding nature and machinery was easily translatable to a systems theory for control-engineering society.

And so the systems ecologist emerged as a candidate social physicist, capable of surpassing the economist as the engineer of society by providing the knowledge with which to manage the integrated totality of interactions between the economy and the Earth. The ecosystem, as Bryant writes,

> [...] was a comprehensive model of nature that integrated humans into the environment and accounted for the ecological effects of technologies, a model that could be employed to evaluate the relationships among people, nature and machines.[33]

Bryant argues that the ecosystem concept enabled the distinction between two levels of technological interaction between nature and society: the physical and the informational.

> In the first case, critics of modern technologies used the ecosystem concept to make comprehensible, and to legitimate alarm over, the environmental threats posed by the Bomb, chemical pesticides and other technologies characterised by the physical force they exerted on the world. [...] At the same time, the ecosystem concept served as a model of nature compatible with, even necessary to, the cybernetic technologies of the information age. These technologies were characterized by their ability to monitor their own internal and external states and adapt in response to changing conditions. They exhibited self-regulation through information feedback, a capability held to be fundamental to all complex self-organizing systems, natural or artificial. The ecosystem concept naturalized the features of cybernetic system-ness characteristic of post-war information technologies and attributed them to nature on a macro scale.[34]

The Atomic Promise: Infinite Energy and Geoengineering

Up until the mid-1960s, when images of the 'mad scientist' such as Stanley Kubrick's military sociopath *Dr. Strangelove* (1965) began to penetrate the popular lexicon, physicists commanded an unqualified public respect

[33] Bryant (2000, p. 1).
[34] Bryant (2000, p. 2).

and prestige over other scientists.[35] In the wake of the Manhattan Project, nuclear engineers were proof of the incredible power of scientific knowledge. Although this effort was doomed to failure, much government propaganda was developed in order to redirect public anxiety about atomic war towards the hope that their efforts would soon harness 'the peaceful atom' to the great benefit of all. In the 1950s, the Atomic Energy Commission was tasked with both the domestication of atomic technologies and propaganda efforts to shape a favourable climate of pro-nuclear public opinion. With a mandate to plan civilian atomic infrastructures for electricity production, the AEC also produced a plethora of 'educational' materials on the subject of 'Our Friend the Atom', even going so far as to promote the viewing of an atmospheric H-bomb detonation at the Nevada testing range as an exciting event. Local government assisted a roaring tourist trade, publishing road maps directing motorists to the best vantage points for the awful spectacle of the flash and the mushroom cloud rising.

Ignoring the problem of radiological pollution, proponents of the atomic millennium proclaimed that energy would become so abundant that it would be 'too cheap to meter', which would massively increase productivity. In a series of articles, David Dietz, the science editor of a widely read syndicate of American newspapers, explored the utopian possibilities of atomic energy in the coming era. Americans would run their Cadillacs for a year on 'a pellet of energy the size of a vitamin pill'. Unlimited energy would allow Americans to control the climate: 'Summer resorts would be able to guarantee the weather', and 'artificial suns will make it as easy to grow corn and potatoes indoors as on the farm'. Nuclear-powered mechanical hearts were proposed by some medical visionaries. Boundless free and clean energy was promised by future techniques of nuclear fusion, where 'self-regenerating' breeder reactors would produce no pollution apart from hot water.

Also on the drawing board of policy-makers were the uses of specifically designed nuclear devices for large-scale environmental engineering in the interests of economic growth—for the excavation of harbours, dams, canals, or surface mining. Feasibility studies were drawn up for a 'Panatomic Canal' that would join the Atlantic and the Pacific at sea level, obviating the need for locks. Instead of cutting through the mountain ranges that

[35] Debate continues as to whether Strangelove was modelled on John von Neumann, Herman Kahn, or Edward Teller, with the smart money on Teller. Goodchild, P. (2005). *Edward Teller: the real Dr. Strangelove*. Harvard University Press.

traversed the Isthmus of Panama with ordinary equipment, nuclear row charges would cut huge incisions through the bedrock to 200 feet below sea level in less than a minute.[36] This project was mothballed after the Test Ban Treaty of 1962, which banned atmospheric bomb tests. Significant opposition to the Panatomic Canal was also given for ecological reasons, as the effects of exposing the separately evolved biota of the two oceans to one another were entirely unpredictable. One wonders what local residents thought of the plan.

This kind of geoengineering was actually carried out by Communist development agencies. In 1965 the Soviets detonated an atomic charge under a riverbed in Kazakhstan. The resulting crater formed a reservoir, which was intended to bring irrigated agriculture to the arid surrounding plains.[37] In addition to the diversion of rivers and the effects of radioactive fallout from the Kazakh test site, the mass use of chemical fertilisers, pesticides, and heavy agricultural machinery has since destroyed much of the region's fragile soil, insect pollinators, and aquatic life, leading to near total ecological and hydrological collapse of the Aral Sea. Once a huge freshwater lake bustling with fishing boats and merchant haulage, the waters have receded so far that one might still witness ship skeletons slowly rusting atop sand-dunes far from the shrinking sea.

The systems theories that accompanied the development of the digital computer led some to an incredible optimism regarding the ability to produce models of complex and stochastic phenomena that were more or less isomorphic with the natural (or social) phenomena in question. It was hoped that computer models simulating open environmental or social systems could lead to the development of instruments of control from the predictions of the model. Cybernetics was key to all of this by virtue of its ambition to provide a universal key to all the other sciences. Chunglin Kwa has described the period from the mid-1950s to 1973 as the 'heyday of macromodelling', when the computer modelling of complex systems—even though the computing power and mathematical complexity invested in these models were incredibly simple by today's standards—was thought to confer almost unlimited potential for control over the phenomena

[36] Nye, D. (1994). *American technological sublime.* Cambridge, MA: MIT Press, pp. 234–236.

[37] Adushkin, V. & Reith, W. (2001). The containment of Soviet underground nuclear explosions. Open File Report: US Geological survey.

being modelled.[38] Take for example von Neumann's project for controlling the weather. In 1956 he declared he had 'no doubt one could intervene on any desired scale, and ultimately achieve rather fantastic effects'.[39] As Gleick reports:

> [von Neumann] recognised that a complicated dynamical system could have points of instability—critical points where a small push can have large consequences, as with a ball balanced on top of a hill. With the computer up and running, [he] imagined that scientists could calculate the equations of fluid motion for the next few days. Then a central committee of meteorologists would send up airplanes to lay down smokescreens or seed clouds to push the weather into the desired mode.[40]

In a more spectacular example, Kwa reports that von Neumann once proposed that the famine-inducing drought conditions of the Sahel could be fixed by the judicious detonation of atomic bombs off the coast of Africa. Storms would be summoned to bring crops to the deserts, or to overwhelm the communist enemy.

If to some nuclear energy and computers seemed to offer the power to control the climate, by the mid-1950s it was becoming apparent in other scientific circles that conventional hydrocarbon fuels would progressively undermine the natural stability of the carbon cycle, rendering the climate ever more prone to chaos. In 1965, US President Lyndon Johnson was presented with the earliest public study on the accumulation of carbon dioxide in the atmosphere and warned of its potentially devastating effects.[41] The study was discussed in a keynote address to senior oil executives at the 1965 conference of the American Petroleum Institute, who had been warned of the potentially catastrophic effects of the greenhouse effect at least as early as 1959, by no less than the H-bomb physicist Teller.[42] In keeping with the optimism of the day, the President's Science Advisory Committee opined that global warming could be fixed very

[38] Kwa, C. (1994). Modelling technologies of control. *Science as Culture*, 4(3), 363–391.

[39] Kwa, C. (2001). The rise and fall of weather modification. In C. Miller & P. Edwards (Eds.) *Changing the atmosphere: expert knowledge and environmental governance* (pp. 135–165). Cambridge: MIT Press.

[40] Gleick, J. (1987). *Chaos*. London: Abacus, p. 18.

[41] Revelle, R. et al. (1965). Atmospheric carbon dioxide. In President's Scientific Advisory Committee, Restoring the quality of our environment. Washington: White House.

[42] Franta, B. (2018). Early oil industry knowledge of CO_2 and global warming. *Nature Climate Change*, 8(12), 1024–1025.

cheaply, by geoengineering. Littering vast areas of the world's oceans with shoals of floating Styrofoam particles would make a kind of globally distributed mirror, lightening the colour of the sea, raising the Earth's albedo, and offsetting the thermal effects of radiative forcing.

Although numerous (non-nuclear) experiments in rainmaking and hurricane steering and suppression had been attempted for civilian purposes, the revelation from the leaked Pentagon Tapes that the skies of Vietnam were being weaponised in efforts to destroy the rice crop provoked worldwide outrage, leading to the UN Environmental Modification Convention (1978) which forbade weather warfare and ended official support for geoengineering. This was not because people believed that weather engineering did not work but that it might work too well, causing catastrophic flooding or the permanent loss of rainfall by messing up seasonal rainfall patterns.

If digital computers fostered a machine dream of atmospheric discipline, they also enabled the meteorological studies of Edward Lorenz (1963), whose foundational work on deterministic chaos ultimately established the impossibility of accurate, long-range local weather forecasting.[43] The science of non-linear dynamics exposed the problems of verifying the effects of weather modification experiments: as all points in a weather system are highly sensitive, no causal model can demonstrate (with what quantum physics and tort law call 'counterfactual definiteness') what local weather events would have occurred in the absence of a local intervention. (Contra the denialists, this does not invalidate the predictive capacity of long-range climate models dealing with global physical parameters such as solar radiation, atmospheric chemistry, and the heat balance of the Earth's atmosphere and oceans.) Even though the power of computation was extremely limited by today's standards, the belief that weather engineering was 'working' exemplifies a faith in the isomorphism between symbolic representation and the phenomena being represented, whereby the model was thought to confer almost unlimited potential for control.

If the sheer radiant force of nuclear technology briefly widened the horizon of millennial expectation, yielding machine dreams of infinite growth in energy availability coupled with complete technical power and control, the frightening reality of its hair-trigger deployment for global warfare and the emerging account of the longevity of harmful radiation

[43] Lorenz, E. (1963). Deterministic non-periodic flow. *Journal of the Atmospheric Sciences*, 20, 130–141.

after accidents like Three Mile Island and Chernobyl meant that the 'peaceful atom' has rarely been fully trusted or welcomed into civilian life. The visceral awe and spiritual inspiration at human achievement witnesses reported at the contemplation of the mushroom cloud too quickly collapsed into terror.[44] The emergence of the 'nuclear sublime' was accompanied by a sense of the permanent technological violation of nature through military hubris. The MAD bipolar logic of 'mutual deterrence', with its destructive potlatch of increasing 'throw-weight' producing arsenals capable of destroying this and other worlds many times over, meant that for the first time in history the continuing stability of the conditions of Earthly life as a whole were subject to an executive decision. Automated communications and control gave the commander-in-chief only fifteen minutes to decide whether early warning system reports of an incoming missile attack were credible enough to irreversibly put the entirety of the nuclear war machine into simultaneous action. The biopolitical reach of the sovereign decision had come to encompass the totality of the 'critical zone' of planetary life.

Against the official energetic optimism of the 'peaceful atom', ecologists were crucial in countering the common view that atomic weapons were simply extra-large sticks of dynamite, by publicising the utterly catastrophic effects of radiation and drastic climate change in the nuclear winter that would follow an atomic war. Not only would vast numbers of humans perish in a full-scale nuclear exchange but the biosphere itself could be threatened with collapse and mass extinction. A hypothetical situation 'in which all organisms were exposed to a million rads of radiation could result in the extinction of every species but the hardiest of bacteria, wiping out more than a billion years of evolution and returning the Earth to the "green-slime" ecosphere of the very distant past'.[45] Even relatively low doses of radiation, such as strontium-90 and caesium-137 finely dispersed throughout the Earth's stratosphere by the multi-megaton airbursts of the late 1950s and early 1960s, were shown to threaten human beings directly. As studies of Innuit and Lapp nomads far removed from test sites showed, radionuclides tended to bioaccumulate in body tissues,

[44] Nye (1994, p. 235).

[45] Arthur, W. (1990). *The green machine: ecology and the balance of nature*. Oxford: Basil Blackwell, p. 225.

increasing their concentration as they moved up the food chain.[46] The daily possibility of nuclear annihilation penetrated the psychic terrain of people everywhere.

The consequences of the Cold War turning hot, communicated to publics by ecologists and popularised by the nascent environmental movement, effectively arrested the naïve technological millennialism of unending progress. As humanity became godlike in its masterful capacity to unleash the Promethean energies of physical nature, living nature, which had since the Old Testament laid claim to a certain ferocious inexhaustibility in Western mythology, was revealed as utterly fragile and contingent. In his diary, Harry Truman noted the end of the innocent myth of perpetual human progress through the war on nature. 'The human animal and his emotions change not much from age to age. He must change now or he faces absolute and complete destruction and maybe the insect age or an atmosphereless planet may succeed him.' Josef Stalin was more direct: 'Atomic weapons [...] can hardly be used without spelling the end of the world.'[47]

Thus it fell to the ecologist to take up a public role as the 'sane scientist', speaking truth to power in the public interest and warning of the destruction threatened by pyrotechnical hubris. If ecologists first rebelled against military discourses of omnicompetence, it was to economic millennialism that they were to give their most profound and astutely unheeded challenge. Out of the terror of an anticipated nuclear war arose a more insidious set of warnings regarding the prospects of 'peaceful' economic expansion. Through the depletion of forests, fisheries, soils, biodiversity, minerals, and energy resources at one end of the economy and the accumulation of pollution at the other, the very success of the industrial apparatus threatened the stability and regenerative capacity of the biosphere, and with it the survival of human civilisation.

Taking H.T. Odum as archetypal and indicative of this new public vocation of ecology, the following chapter outlines his contributions to the ecosystem concept between the early 1950s and the early 1970s. We consider his subsequent 'subversion': the enunciation of an ecological critique of prevailing economic ideology. Of particular interest here is Odum's 1971 attempt to construct an energy theory of value from systems ecology

[46] Odum, H. T. (1971). Radiation ecology. In E. P. Odum, *Fundamentals of ecology* (pp. 451–467). W.B. Saunders.

[47] Gaddis, J. L. (2005). *The Cold War*. London: Allen Lane.

with which to ward off the catastrophic risk of ecological meltdown. Odum offered nothing less than a new social physics for the Age of Ecology, in an ethical vision of global cybernetic control over the interactions between large-scale technological systems and a biosphere conceived of as an essential partner to 'Man' in ecological and energetic terms. We then finish by comparing Odum's eco-energetics with the devastating critique of neoclassical economics given by Nicholas Georgescu-Roegen, who in the same year devoted a classic text to the ignorance of economists regarding the consequences of the entropy law for thermoindustrial society. As I will argue, 1971 represents a moment when the long journeys of the alienated twin sciences of 'economy of nature' came full circle, arriving at the same crossroads, although without recognising each other as blood relations from an intricate genealogy of Western thought.

CHAPTER 13

Power and Entropy: The Limits of Ecological Economics

Power and the World System: H.T. Odum's Economy of Energy

A science of commerce and government, political economy has since its inception been a topic of intense public debate. This is not the case with ecology. As we have seen, ecology decisively entered the public sphere only in the late 1960s, at a time when the concept of the ecosystem became central to its scientific legitimacy and social role. The ecosystems ecology developed by the brothers Eugene and Howard T. Odum separately and together from the mid-1950s redefined the discipline at a time when ecology was gathering attention and importance as a 'crisis discipline'. It was from this point that ecological science came of age in its new social role as the analyst of humanity's home. Eugene is primarily remembered as an educator, but it was his brother Howard Thomas Odum who brought ecology boldly into the analysis of the industrial causes of the global environmental crisis. Despite numerous collaborations in fieldwork and publication, the brothers tended to favour different metaphors. As Kingsland says, 'Eugene thought of the ecosystem in organic terms as though it were an organism in a state of homeostasis, Tom deviated from this organic

J. Walker, *More Heat than Life: The Tangled Roots of Ecology, Energy, and Economics*,
https://doi.org/10.1007/978-981-15-3936-7_13

analogy and increasingly thought of the ecosystem as a machine governed by feedback mechanisms.'[1]

On their way to becoming prominent ecologists, the Odum brothers were influenced by the British born scientist G.E. Hutchinson, appointed to faculty at Yale in 1928. Eugene studied physiological ecology for his doctorate, and learned from Hutchinson via Howard, whose doctoral thesis on the global cycle of naturally occurring strontium was supervised by Hutchinson.[2] The topic was timely: the dispersal of radionuclides in the environment had become a vital concern due to the unnatural quantities of strontium-90 circulating the globe, a result of atmospheric nuclear weapons tests conducted between 1945 and the Test Ban Treaty of 1962. At a time when the civilian atomic energy industry was expanding in the United States, the high importance placed upon the domestication of nuclear power in the upper echelons of national security and planning institutions opened a series of professional opportunities for ecologists.[3]

Ecosystems ecology enjoyed the continued support of research funding from the Office of Naval Research and the Atomic Energy Commission (AEC), as researchers pitched ecology to military funders as an indispensable framework for the emerging literature on 'health physics': the effects of radiological pollution on human physiology. This research was crucial to the question of safety and thus to the advanced utilisation of the technology. As humans could absorb contamination indirectly from its diffusion in the environment, ecology could assist in determining the maximum 'safe' load of radiation for whole ecosystems. 'Radiation ecology' played an important role in the elaboration of systems ecology, as ecologists were hired to study atomic test sites, the discharge of effluent from plutonium reactors, and the process of plant succession in the hundreds of abandoned farms compulsorily resumed as buffer zones around experimental nuclear facilities, and to analyse food chains by tracking the diffusion of radionuclide tracers inserted into various points of the ecosystem.[4] Eugene had worked with the AEC-directed Oak Ridge National Laboratory, under the ambit of 'health physics', a major research centre for systems ecology, and with Howard on the Eniawatak Atoll H-bomb test sites. Here they

[1] Kingsland, S. (2005). *The evolution of American ecology, 1890–2000*. Baltimore: John Hopkins University Press, p. 195.

[2] Kingsland (2005, p. 189).

[3] Bocking, S. (1997). Ecologists, ecosystems and the atom: environmental research at the Oak Ridge National Laboratory. *Journal of the History of Biology*, 28, 1–47.

[4] Kingsland (2005, p. 192).

employed a biogeochemical method to analyse the processes of ecosystem succession on the islands and reefs as the sites began to recover from the blast. These studies also contributed to the knowledge of how certain forms of radiation became concentrated at the top of the food chain. Slowly fading radioactive isotopes allowed the tracing and mapping of energy and nutrient flows through the ecosystem as it approached a new climax stage over the long term—providing that 'excessive' radiation did not retard 'normal' growth altogether. This metabolic analysis conducted through the mass-balance accounting of material flows became a key method of systems ecology. Another AEC project undertaken in the rainforests of Guatemala involved the meticulous classification of biota in a set acreage, the measurement of respiration, the removal of everything living from the site, followed by physico-chemical quantification of the now-dead 'biomass' through the measurement of nutrients, and finally of the embodied energy released by its combustion. A barrel of radioactive material giving off 10,000 curies was then placed in the centre of the fenced off area and left for three months. Radioecologists were then brought in to analyse the effect of the surge of radiant energy on the early stages of forest succession.[5]

From this period, H.T. Odum took the ecosystem in a more mechanistic direction, in 1960 introducing symbols from his undergraduate training in electrical engineering to depict ecosystem functions and socio-material organisation with his unique energy circuit diagrams. A crucial point of inspiration for the thermodynamic ecology of Odum was Lotka's *Elements of Physical Biology*, which attempted to develop a rigorous 'economy of nature' by joining thermodynamics to evolution. Where Darwin had provided the key evolutionary principle of natural selection, Odum argued that Lotka had raised natural selection to the status of a general energy law, as Neufeld reports:

> Odum [...] attempted to incorporate evolutionary thinking into his research, looking for patterns that may have arisen by natural selection, but at the ecosystem level. According to Lotka, evolution was not just species changing through time, it was an *overall accumulation and distribution of energy within a system*. Natural selection tended to maximize the flow of energy and matter through a system, a concept little dealt with by traditional evolutionary

[5] Odum, H. T. (1971). *Environment, power and society*. New York: Wiley Interscience, pp. 294–295; Odum, H. T. (1971). Radiation ecology. In E. P. Odum (Ed.), *Fundamentals of ecology* (pp. 451–467). W.B. Saunders.

> ecologists, perhaps because most of them work at the level of the individual and population, and not the ecosystem. Odum took Lotka's idea and called it the '*maximum power principle*', and it became the cornerstone of the bulk of his research. [emphasis original][6]

Thus Odum provided a means for conceptualising 'economic growth' as the 'overall accumulation and distribution of energy within a system' according to the dynamism of the 'maximum power principle'.

In *Machine Dreams* (2002), Mirowski argues that the hardcore instrumentalisation of rationality in American neoclassical economics was a controlling response to the apocalyptic tensions and looming chaos of the US-Soviet policy of mutually assured destruction (MAD), a result of economists' contact with the decision sciences of the Cold War military-academic complex. Odum's 'machine theory' of ecology, derived from similar intellectual and funding environments and actual field studies of the bomb's effects, followed a comparable pattern. Yet the affective dimensions of the nuclear threat were not the only source of his drive to subject the world to an 'iron-clad standard of rationality': Odum saw another apocalypse emerging from the ordinary activities of economic development. His development of an energetic language for ecology was an attempt to situate ecology between economics and physics in such a way as to provide a conceptual language and a set of engineering tools for the management and stabilisation of the interface between complex technological and ecological systems. For H.T. Odum, the linking of control theory and thermodynamics promised to lead to a 'consilient' decision theory worthy of this civilisational task.

Cyborg metascience had opposite effects on the different 'economies of nature'. As a result of its reprocessing by cybernetics, American economic science further entrenched itself as part of the order of the 'free' society to be defended at all costs. By contrast, systems ecology became increasingly 'subversive'. By 1970, the Odum brothers had forged direct connections between systems ecology and the environmental movement's critique of the American oil-based growth paradigm, analysing the metabolism of industrial society as a global system of industrial technomass competing for energy and space with the global ecosystem. 'Cybernetics', wrote

[6] Neufeld, H. (2001). The history of the ecosystem concept: Part 3, post-WWII ecology. http://www.appstate.edu/~neufeldhs/Ecosystems/OriginsPart3.htm. Accessed 23 Feb 2019.

E.P. Odum in *The Crisis of Survival*, a popular collection of environmentalist essays, 'provides a convenient way of viewing the environmental problem as a whole'. Noting that economies and ecosystems, like machines and organisms, are functioning systems of dependent parts, he argued that such systems:

> [...] can exist in two general states: (1) a 'transient' state in which the whole is growing or otherwise changing in time; and (2) a 'steady state' in which the system is maintained in equilibrium, a system of inputs and outputs. [...] A growth, or youthful stage is under the influence of what is called 'positive feedback' in that each increase accelerates another increase, often in geometric progression (a doubling followed by a doubling, etc.) For the individual or the population, as for a new business, growth and positive feedback are necessary for survival. [...] However, growth does not and cannot continue unrestricted because 'negative feedback control' also comes into play, either due to some limitation imposed by the external environment or due to the action of an internal 'governor' that brings about an orderly slowdown and establishes a 'set point' at which growth stops. As a living system becomes larger and more complex more of the energy that it transforms must be 'fed back' to maintain and control the intricate structure; quality maintenance replaces mere quantitative growth as the strategy of survival in the mature system.[7]

H.T. Odum was emerging as a critical voice pointing to the lack of a scientific perspective amongst economists:

> Ecologists are familiar with both growth states and steady state, and observe both in natural systems in their work routinely, but economists were all trained in their subject during rapid growth and most don't even know there is such a thing as steady state. Most economic advisors have never seen a steady state even though most of man's million year history was close to steady state. Only the last two centuries have seen a burst of temporary growth because of temporary use of special energy supplies that accumulated over long periods of geologic time.[8]

In his fascinating and idiosyncratic *Environment, Power and Society* (1971), Odum takes up the problem of energy with regard to what Alfred

[7] Odum, E. P. (1970). Introduction: the attitude revolution. In F. Scott et al. (Eds.), *The crisis of survival*. Glenview, IL: Scott, Foresman & Co., p. 13.

[8] Odum, H. T. (1973). Energy, ecology, and economics. *Ambio*, 2(6), 220–227. See p. 222.

Marshall referred to as 'the world industrial organism'. In this book, Odum analyses the fossil-fuel-based 'system of Man' and its interaction with the global ecosystem revealed by the macroscopic 'maximum power principle'. Both the biosphere and the 'system of man' can be measured and represented with 'pathways of power': conduits of energy flow. After describing the biosphere in solar-energetic terms, Odum accounts for socio-energetic order in agriculture, Roman imperialism, international relations in the age of fossil fuels, democracy, religion, and the forces of history and the soul, all liberally illustrated with his energy circuit diagrams (See e.g. Figs. 13.1 and 13.2). In a classic piece of scientism, he conflates 'power' as 'energy available to do work' with a sociological meaning of power, claiming that 'economic, political and social power flows are just as measurable as those of the simple physical and chemical systems'.[9] Analysis of the contemporary 'system of man and nature' shows that the unbalancing of energetic flows by human activity has manifested

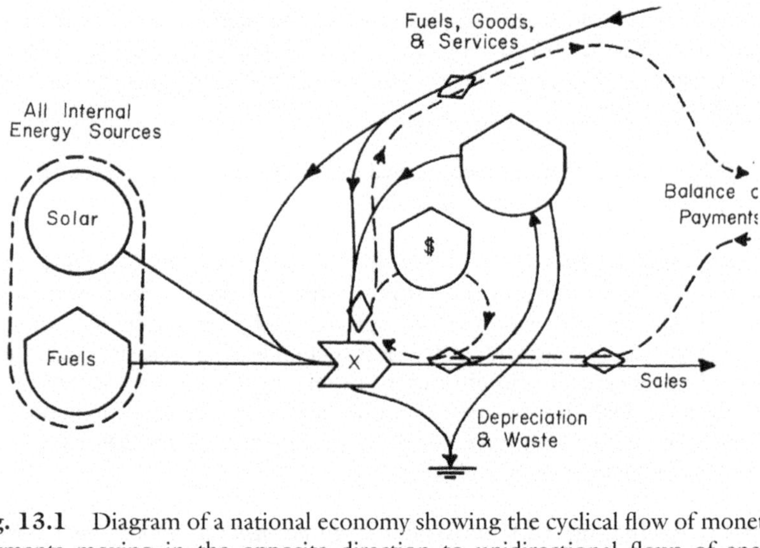

Fig. 13.1 Diagram of a national economy showing the cyclical flow of monetary payments moving in the opposite direction to unidirectional flows of energy. (Source: Odum, H.T. (1973). Energy, ecology, and economics. *Ambio*, 2(6), 220–227. Reprinted with permission of the Royal Swedish Academy of Sciences)

[9] Odum (1971, p. 54).

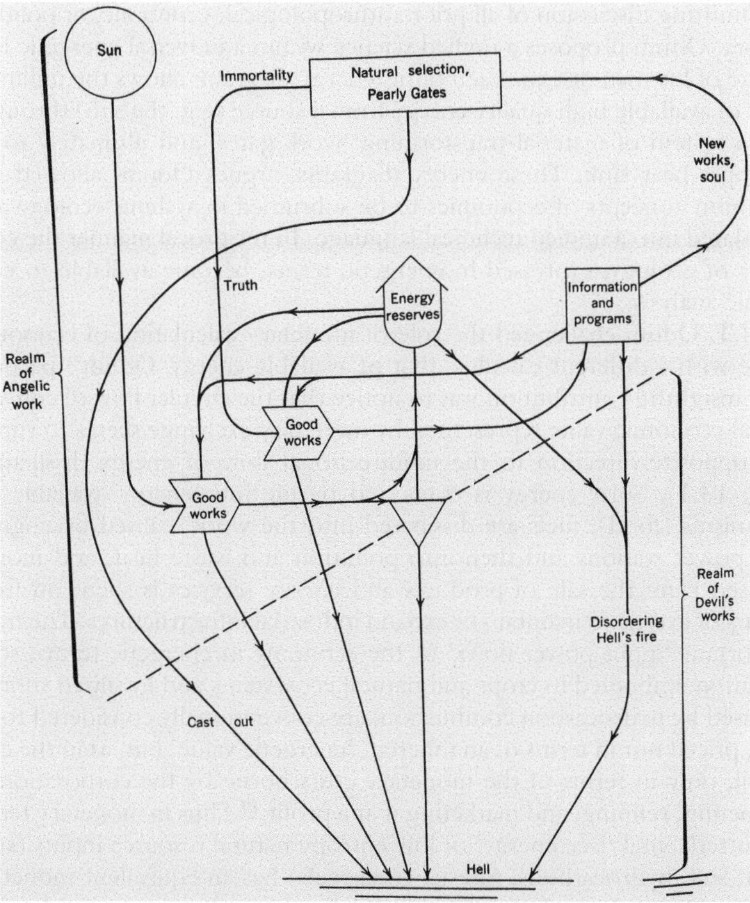

Fig. 13.2 Odum's theology of energy. Diagram showing 'angelic operations of order, evolution and selection of information above and the evil processes of disorder, dissipation and heat death below'. (Source: Odum, H. T. (1971). *Environment, power and society.* New York: Wiley Interscience, p. 252. Copyright © 1971 John Wiley & Sons Inc. Reprinted with permission of Wiley)

from the point of view of the biosphere as 'cancer at many levels'. For Odum, this planetary dilemma requires not merely that ecology should come to exert some influence on political and economic practice but should entirely subsume it.

Omitting discussion of all prior anthropological, economic, or political theory, Odum proposes a unified science within a universal energetic language of his own design. Each of his energy diagrams shows the unilinear flow of available high quality energy from a source (e.g. the sun) throughout a system of material-transforming 'work gates' and ultimately to an entropic heat sink. These energy diagrams, argues Odum, allowed the common concepts of economics to be subsumed in systems ecology and translated into a unified technical language. In reciprocal manner the concepts of ecology, expressed in energetic terms, become available to economic analysis.

H.T. Odum challenged the role of monetary calculation of economic value with a different calculus, that of available energy. Odum's particularly insightful contribution was to notice that the circular flow of conventional economic value represented by monetary exchange seems to run in the opposite direction to the unidirectional flow of energy dissipation (Fig. 13.1). Solar energy is converted by life into energy available for organisms (food), fuels are dissipated into the work realised by engines and power stations and then into pollution and waste heat, and money realised from the sale of products and energy services is spent on fresh inputs of exergy to maintain or expand industrial infrastructures. The most important 'input power flows' to the economy in energetic terms, solar radiation embodied in crops and natural ecosystems and fossilised sunrays released by hydrocarbon combustion, are conventionally considered to be free, priced not in terms of an inherent 'energetic value' but, as in the case of oil, only in terms of the monetary costs borne by the corporation in extracting, refining, and marketing it at a profit.[10] Thus in monetary terms the utterly vital 'free energy' of low entropy natural resource inputs (sunlight, soil, hydrocarbons, forests, coral reefs) has an equivalent monetary value as the useless exhaust flows and damaging pollution accumulated at the entropic end of the economic process.

As Georgescu-Roegen observed in *The Entropy Law and the Economic Process* (1971), published in the same year as Odums's treatise, industrial growth is 'completely tributary' to the transformation of solar and fossil energy: 'the issue of returns boils down to that of returns to mining and agriculture. [...] it is the pace at which low entropy is pumped from the environment into the economic process that limits the pace of this

[10] Odum (1971, p. 182).

process.'[11] Odum's energy diagram of a national economy (Fig. 13.1) gives graphic form to this insight, illustrating the irreversible unidirectional flow of solar and mineral energy through the 'work gates' of agricultural and industrial systems, entering for a time the circular flow of commodity exchange and exiting the circuit of value as waste to the sink of entropic dissipation.

From Odum's macroscopic point of view, money was a communicative disorder, sending erroneous value signals to 'Man' which accelerated the cancer of the biosphere. Odum devoted many years to producing an actionable method of value accounting derived from thermodynamics. As the economic system transformed and 'embodied' the energy dissipated from fuels, ores, forests, and farms into human populations, infrastructure, and capital goods, in a self-reinforcing cycle of positive feedback, it increased the flows of energy required to maintain the structure, in turn further expanding its capacity to access and dissipate more energy. Transferring Lindeman's concept of the 'trophic-levels' of a food pyramid to the analysis of thermoindustrial civilisation, he proposed the concept of 'emergy' (embodied energy) to theorise the energetic costs of maintaining structure and function (or 'orderliness', negentropy, or low entropy) in social and natural systems.

Odum's energy theory of value was developed to counter the environmentally destructive tendencies of classical economic value calculation, by bringing ecosystems and economies into a unified framework of physical analysis. Moving from 'natural law' to the 'positive law' of fiscal and monetary policy, Odum argued that analysis in terms of computable dollar/energy ratios could inform the urgent task of re-engineering of social institutions away from growth into the maturity of the steady state. This would involve ensuring that nature is 'paid' for her services, in terms of investments of human effort to restore and conserve the integrity of ecosystem functions. While his attempts to synthesise a unified metric of energy and dollar values would come to be classified with those of other cranks pursuing an energy theory of value, Odum nevertheless anticipates the recent literature on 'ecosystem services', which seeks to establish 'biodiversity banks' and financial markets trading in the 'ecosystem services' (e.g.

[11] Georgescu-Roegen, N. (1971). *The entropy law and the economic process.* Massachusetts: Harvard University Press, p. 196.

pollination, water purification, hydrological cycling, and carbon storage) rendered to the economy by undestroyed ecosystems.[12]

Unaware of the history of value theory recounted in this book, Odum replaces a pseudo-energetic theory of value, the neoclassical mechanics of market equilibrium under which 'value is conserved', with an explicit theory of energy value updated by the second law, in which it might be said that 'value tends toward dissipation'. As we have seen, the liberal economists of the 1870s reduced social (re)production to the problem of exchange, representing market prices as a non-entropic realm of 'subjective utility' maximisation under equilibrium constraints, conflating pre-thermodyamic 'energy' with the 'utility' of money in an attempt to evade the moral questions of mass industrial poverty and colonial land appropriation with a pure science of the laws of the 'price mechanism'. H.T. Odum's energy theory accomplishes the reverse, aiming to re-engineer the information signals given by the 'price-system' so that it reflects not subjective values but the 'real value' of the underlying bio-energetic flows of agriculture and industry, reorienting value theory to the 'real' expenditures and losses which increase the order and 'technomass' of the economy while exporting chaotic disorder to natural ecosystems and reducing the biomass of the biosphere.

Despite Odum's dissent with the orthodoxy of his time, his 'subversion' was tempered by a technocratic optimism regarding the ability of science to cut through ignorance and provide objective knowledge and solutions for the restructure of society needed to restore and sustain a long-term global equilibrium. Perhaps due to his commitment to the equilibrium concepts of systems ecology, his energetic analysis of planetary metabolism downplayed the catastrophic risks attending the fiery destabilisation of the carbon cycle implicit in the research of Callendar, Keeling, and Revelle. Odum noted such concerns, but like his teacher Hutchinson, he apparently believed that the biosphere was capable of re-absorbing anthropogenic carbon flows from combustion and re-establishing a new planetary equilibrium:

> Presumably the rising concentrations of carbon dioxide [...] released by consumption have also accelerated excess [photosynthetic] production in

[12] Walker, J. (2016). Bringing liquidity to life: markets for ecosystem services and the new political economy of extinction. In K. Kohli & M. Menon (Eds.), *Business interests and the environmental crisis* (pp. 5–37). SAGE: New Delhi.

> enough places to bring about a new overall balance in the biosphere at the higher level of carbon dioxide concentration.[13]

Whilst he did not cite his sources, Odum noted the 'much discussed possibility' that should the then current atmospheric CO_2 concentrations of 340 ppm rise to 2000 ppm 'the greenhouse effect would raise the earth's temperature, presumably melting the icecaps and flooding the coastal plains of the world'.[14] At the time of writing, we are at 415 ppm, and far less innocent. Annual global emissions are still increasing, and new records for global temperatures are set year on year. If the melting of polar icecaps and the consequent rising sea levels have long been the focus of anticipatory fears of global heating, a gothic list of less-anticipated catastrophic symptoms are now proliferating, such as the melting of the methane-rich Arctic 'permafrosts', the collapse of kelp and coral reef ecosystems in the wake of marine heatwaves, the intensification of cyclones and droughts, the disappearance of insect populations, the loss of once fire-resistant rainforests and swamplands to catastrophic wild fires, and the ominous deoxygenation, acidification, and heating of the oceans.

Having revealed the fundamental energetic dimensions of the biosphere crisis, Odum moves to the redesign of the anthroposphere. An effective global political organisation was needed, he argued, to redesign the interface between agricultural, industrial, and conservation zones on a global scale and to ease the inevitable transition to a steady state, which would come either through an orderly reduction of resource demands or chaotically through the inexorable logic of depletion. If the United Nations could be re-engineered in a manner consistent with energetic principles, Odum argued, world peace might be more than a dim hope. However, his vision of an energy-based system of international law looks less like a democratic forum for the preservation of the global commons than a form of energetic apartheid which would reinforce the stratified authority and privileges of industrial societies with the most voracious appetites for combustion. Votes in UN forums, Odum says, ought to reflect 'the power budget of a country', which is:

> [...] readily computed and could be determined annually. Instead of yearly votes in proportion to population they could be assigned in proportion to

[13] Odum (1971, p. 18).
[14] Odum (1971, pp. 278–279).

> the energy budget of the previous year. Thus the voting power would always correspond to real ability to influence the world.[15]

In what appears to be a veiled reference to the calls of the post-colonial G77 nations for a UN-mandated New International Economic Order based on the principle of permanent national sovereignty over natural resources, Odum warns that 'for the rich countries to give up their present power control to a system of untried capabilities may be very risky for the whole world's progress'.[16] Yet at other points in the book he warns that oil corporations are 'generally draining the greater body of the world towards its collapse' and must be brought to heel by 'cutting off the fuel supply' in order to restore public control.[17] As time went by, Odum would increasingly critique the structural inequity of international economic relations, which exported the natural resources of poorer nations to other lands for the consumption of affluent others.

Since the 1970s, the neoliberal recipe of privatisation and the rollback of social and environmental law has convincingly maintained the stratified correlation of political and economic power with highly unequal energy consumption, in a process Hornborg has analysed as the 'thermodynamics of imperialism'.[18] Manifest in the structure of international trade and sovereign debts, Hornborg argues that it is the wide differentials in the market value of human labour (as opposed to the uniformity of oil prices traded on world commodity exchanges) that facilitates the continuous asymmetric transfer of natural resources and fuels from South to North. This situation has effectively been normalised since the Brundtland Report (1987) and its compromise of 'liberal environmentalism', in which business lobbies traded acknowledgement of environmental crisis for agreement on the centrality of 'sustainable growth' as the solution. As we approach the present, the modest norms of liberal environmentalism established in UN treaties have been progressively undermined by the victories of corporate 'investors' in displacing the international public laws of the United Nations with a private system of supranational economic law,

[15] Odum (1971, p. 209).

[16] Odum (1971, p. 203).

[17] Odum (1971, p. 54).

[18] Hornborg, A. (2001). *The power of the machine: global inequalities of economy, technology and environment*. California: Altamira Press.

limiting the scope of national governments to pursue ambitious environmental policies.[19]

Although it seems impossibly remote from our own time, in the early 1970s ecologists believed that most people would soon understand the truths revealed by ecology and accept the necessity of a revolutionary transformation of social life.

> General realisation that the supply depot and the living space functions of one's environment are interrelated, mutually restricted and not unlimited in capacity has amounted to a historic 'attitude revolution' that is a promising sign that man may be ready to apply the principles of ecological control on a large scale.[20]

This did not mean merely the protection of remnant ecosystems, the application of pollution controls to industry, and a shift from fossil-fuelled growth to the renewable energies of the steady state, but the redesign of the entire interface of social and ecological systems. Odum expressed great confidence in the future of molecular biotechnology and whole-ecosystem engineering, envisioning comprehensive tools with which to design a new, regenerative economy of nature in which long-term human well-being was linked to an ethic of biological co-existence:

> The millions of species of plants, animals and micro-organisms are the functional units of the existing network of nature, but the exciting possibilities for great future progress lie in manipulating natural systems into entirely new designs for the good of man and nature. The inventory of the species of the earth is really an immense bin of parts for the ecological engineer.[21]

In addition to technoscience, the conscious redesign of world systems for the age of limits would require a transformation of the ethical foundations of social organisation. Accordingly, Odum proposes a political theology of energy value, replacing the biblical Ten Commandments with another ten better suited to the survival of the biosphere.

[19] Tienhaara, K. (2017). Regulatory chill in a warming world: the threat to climate policy posed by Investor-State Dispute Settlement. *Transnational Environmental Law*, 7(2), 229–250.

[20] Odum (1970, p. 13).

[21] Odum (1971, p. 259).

> 1. Thou shalt not waste potential energy.
>
> ...
>
> 6. Thou shalt judge value by the energies spent, the energies stored, and the energy flow which is possible, turning not to the incomplete measure of money.
>
> ...
>
> 7. Thou shalt not take from man or nature without returning service of equal value. [22]

Beyond the limits of Moses, Odum's prophetic biophysical ethics remoralises 'the laws of nature' to enable transition to the *oikonomia* of the coming Age of Ecology. In another of his fascinating energy circuit diagrams, the intimate relations of energy to information, of photosynthesis and fire to good and evil, and of economics and ecology to millennium and apocalypse are compressed into a single image through a unified theory of value (Fig. 13.2).

Insofar as we depart from the 'angelic works' of an enlightened solar-based society sustaining a harmonious relationship of reciprocity with the biosphere, the Devil's works will prevail, and we will be cast out of the Garden of Eden into the endless torment of 'disordering Hell's fire'.

Obscure and pedantic at the time, it must be admitted that this image now has a prophetic air, an anticipation of a world on the brink of thermal catastrophe. A century ago, Irving Berlin had a hit song with 'The Devil has bought up all the coal'. In 1975, the Venezuelan oil minister and OPEC founder Juan Pablo Perez Alfonzo cursed oil as 'the Devil's excrement' for its ruinous effects on peaceful, democratic development.[23] As is suggested by the contemporary agents of fossil capital doing the 'Devil's work' of deceiving the public, obstructing transitions to solar energy, and exhuming ever more fuel to pour on a world on fire, Odum's theology of heat and life discloses some essential moral truth of our age. In this he was anticipated by the poet William Blake, who at the dawn of the thermoindustrial revolution declared rebellion against the 'dark satanic mills', praying for salvation from 'single vision, and Newton's sleep', and warning that:

[22] Odum (1971, p. 244).

[23] Coronil, F. (1997). *The magic of the state: nature, money and modernity in Venezuela*. Chicago: University of Chicago Press.

> The ancient tradition that the world will be consumed in fire at the end of six thousand years is true, as I have heard from Hell.[24]

Odum and Georgescu-Roegen as Epistemological Twins

Separated at the dawn of the modern era, economics and ecology were momentarily re-united by the crisis of the 1970s in the collision of machine metaphors, an ontological crash between incommensurable intellectual cultures of energy. In keeping with my argument that the history of economics can be revealed more transparently if seen through the mirror of ecology, in what follows, I offer a comparison of Odum's systems ecology with the internal critique of economic orthodoxy mounted by the economist Georgescu-Roegen. Odum and Georgescu-Roegen make a fascinating pair, non-identical epistemological twins speaking to us from either side of the dialectical interplay between the twin sciences of *oikonomia*.

H.T. Odum and Georgescu-Roegen hold different positions in the canon of ecological economics. An eclectic ecologist, Odum made almost no impression on mainstream economic thought. His anti-growth activism and all-encompassing political ecology would become something of an embarrassment to those professional ecologists who insist that the social role of the scientist must be restricted to the production of value-free knowledge to be handed over to 'policymakers' without comment on 'what is to be done'. Take, for example, the strictures of Christian Lévêque, for whom ecology must never be *oikonomia*:

> A danger for ecology would be to appear as the science that will provide solutions. [...] If one can regret that a society destroys biodiversity, and feel that it is unfair, we cannot claim conversely that the protection of biodiversity deals with social justice among humans. [....] Scientific ecology and ecosystem ecology in particular *are not sciences of management* but sciences of knowledge. This distinction must be preserved. [...] It is not for scientists to propose [...] the choices needed to reduce the greenhouse effect [...]. It is the policy makers who make decisions on the basis of the information available to them. [my emphasis][25]

[24] Blake, W. (2008). *The complete poetry and prose of William Blake*. Berkeley: University of California Press, p. 39.

[25] Lévêque, C. (2003). *Ecology: from ecosystem to biosphere*. New Hampshire: Science Publishers, p. 9–10.

This refusal of the social responsibilities of expert knowledge for a safe position of radical innocence exposes a naïve theory of politics which is increasingly untenable. What are we to do when the knowledgeable hold their tongues and the 'policy makers' espouse an anti-science creed in the service of an unjust and irrational *oikonomia*, committing future generations to live the consequences of a general policy of unnecessary destruction maintained for the temporary enrichment of the few?

As a respected mathematical economist and one-time protégé of the Austrian economist Joseph Schumpeter, Georgescu-Roegen has had a mixed reception. His public resignation from the American Economics Association over the scientific value of mouse torture, and his learned exposé of the pseudo-scientific pretensions of 'equilibrium-path' growth economics in *The Entropy Law and the Economic Process* (1971) have meant that economists praise his early work on consumer theory but astutely ignore his mature work on agrarian development and bioeconomics. Senior economists have routinely failed to engage seriously with Georgescu-Roegen's timely and scientific contributions to economic theory—a far more deserving candidate for a Nobel prize than the roster of Mont Pèlerin Society economists ennobled as 'laureates'—and historians of economic thought have focused on his alleged personality defects to explain his outsider status and exclusion from the canon of the Great Economists.

Dealing directly with the orthodox theory of 'production' and the consequences of its ignorance of the second law, Georgescu-Roegen demonstrated that the transformation of solar energy is the ultimate limit and basis of industrial 'production'. Since thermoindustrial 'technology' is dissipative, and not generative as the biological metaphor of 'growth' suggests, 'the value of the standard form of the [neoclassical] production function as a blueprint of reality is nil'.[26] Brilliantly critiquing orthodox economics from within, Georgescu-Roegen established a philosophically sophisticated bioeconomics of fundamental limits that secures his place as a progenitor of contemporary ecological economics.

Drawing upon authorities across the natural and social sciences, Georgescu-Roegen arrives at an anti-mechanistic conclusion: 'Apt though we are to lose sight of the fact, the primary objective of economic activity is the self-preservation of the human species.'[27] This resolves to a demo-

[26] Georgescu-Roegen (1971, p. 244).
[27] Georgescu-Roegen (1971, p. 277).

cratic and humanist plea for the irreducible and untranslatable value of cultural and social experience, which unlike 'fitness', 'utility maximisation', or 'maximum power' is for Georgescu-Roegen the highest goal of the economic process. Unlike Odum, he resisted the technocratic temptations of the social physics project, refusing to contemplate a calculable theory of value:

> Were we to set out the balance-sheet of value [...] we would arrive at the absurd conclusion that the value of the low entropy flow in which the maintenance of life itself depends is equal to the value of the flow of waste, that is, to zero. The apparent paradox vanishes when we acknowledge that the true 'product' of economic processes is not a material *flow*, but a psychic flux – the enjoyment of life by every member of the population.[28]

It is either worrying or a tribute to the sophistication of this volume that even though *The Entropy Law* is considered a founding document of ecological economics, the index of Georgescu-Roegen's opus contains not a single reference to a professional ecologist or to the Earth system sciences of the day. One can only wonder how much more powerful his critique would have been had he been inspired less by Lotka's project to formalise the evolutionary laws of biophysics and had rather incorporated the biosphere-scale empirical holism of the 'Vernadskyan revolution' into the scientific framework of his opus.

The bibliography of Odum's *Environment, Power and Society* is similarly bereft of references to the social sciences, with the exception of the economist Wassily Leontif, then engaged in a comparable mass-balance analysis of intersectoral inputs and outputs. These striking omissions in coterminous works treating the same foundational problems plainly indicate the deep academic estrangement of the twin sciences across the natural and social divide, despite the perpetual trading of key metaphors, concepts, and terms. It is also reflective of different temperaments, with Georgescu-Roegen displaying the critical pessimism of the humanist intellectual and Odum the techno-optimism of Nature's cyborg, decoding and recoding a monistic universe of feedback mechanisms with an arcane circuitry of work, energy, and information. Despite mutual disinterest in the details of each other's fields, both made the straightforward claim that economic theory and practice should respect biophysical reality and the

[28] Georgescu-Roegen (1971, p. 284).

proper parameter for the analysis of production, growth, and distribution is not the abstract models of statistical mechanics but the second law, the domain of natural law in which 'life matters' and Earth's history happens. Both called for a transvaluation of values such that decisions are made in recognition that the Earth's rich 'endowment' of 'natural capital' is humankind's true source of wealth, a gift which must be conserved to the greatest extent possible.[29]

Some tentative explanations of the differing fates of our epistemological twins are worth considering. Georgescu-Roegen is one of the most-cited authors of the journal *Ecological Economics*, an intellectual enterprise yet to awaken the slumber of the average economist. In contrast, Odum's attempt to unify the 'system of man and nature' through ecological energetics is rarely referred to either in economics or in ecology. Neither of these prescient 1971 works are mentioned in the standard economics or ecology textbooks.

Georgescu-Roegen was a relentless critic of American economists' enthusiasm for Arrow and Debreu's 'proof' of general equilibrium (which he ridiculed for its initial assumption that all economic actors have a guaranteed lifetime income) and Solow's standard growth model (for obvious reasons). Both of these 'models' were awarded Bank of Sweden 'Nobel Prizes', while his own profoundly scientific work on the biological and thermodynamic dimensions of the economic process went ignored by all and sundry. A perennial candidate for the Prize, Georgescu-Roegen became embittered and uncooperative in his later career, believing himself to be the victim of a reflexive institutional lockout. The academic interpretation of his work was delayed until a later generation of organisers and networkers including Herman Daly and Robert Costanza took up the task.[30]

Odum, for his part, endured a common fate of the technocratic social physicist: failing to gain popular support for a grand programme of socio-ecological transformation couched in the terms of a physical theory unfamiliar to the layperson. As Chunglin Kwa argues, explicit programmes of social and ecological engineering would be abandoned in the 'decade of crisis':

[29] Gowdy, J. & Messner, S. (1998). The evolution of Georgescu-Roegen's bioeconomics. *Review of Social Economy*, 56(2), 136–156. See p. 152.

[30] Ropke, I. (2004). The early modern history of ecological economics. *Ecological Economics*, (50), 293–314.

> Modelling projects which aimed at large-scale control were abandoned in meteorology, economics and ecology at about the same time [1973]. The search for models that would convey power converging on omnipotence stopped. Concepts that had informed the search for global control, such as equilibrium and stability, lost much of their intuitive plausibility. [...] The relinquishment of the projects mentioned set in motion a redefinition of the idea of control, involving both the scale and the technologies of power.[31]

Despite this, Odum continued to work on the thousands of calculations needed to apply his emergy synthesis to concrete planning situations. Whilst he influenced an international coterie of environmentalists and energy policy analysts, by the 1980s evolutionary biologists and ecologists had largely turned away from the analysis of group dynamics as systems of feedback and coordination and instead shifted theoretical inquiry onto the competitive individualism of neo-Darwinian evolutionism, completing the occupation of ecology by the Walrasian social physics.[32] For our narrative, the fatal problem of Odum's heroic unification of nature and society through a general 'emergy synthesis' has been hinted at by Manssen and McGlade.

> [...] the various forms and uses of energy bound up in essential ecosystem processes present a formidable obstacle to obtaining an operational definition of a general, aggregated available-work concept, a prerequisite for the systems approach of Odum and others. [...] The prototypical derivations of the maximum power principle, and its interpretation, are contradicted on many scales both by empirical data and models, thereby invalidating the maximum power principle as a general principle of ecological evolution. *The conclusions point to the fundamental problem of trying to describe ecosystems in a framework which has a one-dimensional currency.* [my emphasis][33]

Georgescu-Roegen, for his part, maintained a fierce critique of mechanistic philosophy, and particularly of the mathematisation of value theory on the model of statistical mechanics. After studying economics at the Sorbonne in the 1930s, he concluded that 'economic phenomena cannot

[31] Kwa, C. (1993). Modelling technologies of control. *Science as Culture*, 4(3), 363–391. See p. 364.

[32] Taylor (1988, p. 244).

[33] Manssen, B. & McGlade, J. (1993). Ecology, thermodynamics and H.T. Odum's conjectures. *Oecologia*, 93(4), 582–596.

be described by a mathematical system, a faith I have never renounced'.[34] Georgescu-Roegen shared the view of his mentor Schumpeter that the most important vectors of the economy were qualitative and not quantitative: for him this meant that theory must be open, dialectical, and self-reflexive in its relationship to the institutional and biophysical effects of policies derived from abstract theories, which exist in a 'dialectical penumbra' of materiality and concept. That is, he believed that social and historical context was indispensable and that economics must remain open to other disciplines to retain legitimacy. Despite hammering home the inadequacies of a social science ignorant of the entropy law a century after it had revolutionised the natural sciences, Georgescu-Roegen avoids a reduction of economic value to measurable ratios of available energy, or, in his terms, 'low entropy'.

Low entropy, he insisted, is a *necessary* but not *sufficient* condition of economic value. A poisonous mushroom has the same degree of low entropy as a fungus renowned as a delicacy, but one has negative and the other positive utility for the human enjoyment of life. Similarly, an object that obviously possesses economic value (i.e. the sun or hydrocarbon minerals left unburnt in the subsoil) may remain forever beyond the circulation of private property in market exchange and thus never attract a price. Thus, despite his insistence that the second law reigns supreme over biological and industrial life, Georgescu-Roegen refused to contemplate a physical measure of economic value, as is clear in a passage that reads as a direct rebuttal of H.T. Odum's approach:

> It would be utterly wrong to equate the economic process with a vast thermodynamic system and, as a result, to claim that it can be described by an equally vast number of equations [...] which allow no discrimination between the economic value of an edible mushroom and a poisonous one. [...] All this, however, does not affect the thesis [...] that the basic nature of the economic process is entropic and that the Entropy Law reigns supreme over this process and over its evolution.[35]

Crucially, in their 1971 works, it seems that neither Odum nor Georgescu-Roegen were as well informed of the planetary accumulation of heat arising from the fiery destabilisation of the Earth's carbon cycle as

[34] Cited in Gowdy and Messner (1998, p. 138).
[35] Georgescu-Roegen (1971, p. 282).

were the oil executives of the American Petroleum Institute, who had by that time received ample briefing on the warnings of Teller, Revelle, Keeling, and others.[36] Rather, the immediate concern was with the inexorable depletion of natural resources and accumulation of pollution: 'For the earth as a whole, there is no disposal process of waste.'[37] By 1975, however, Georgescu-Roegen had become aware that the 'continuous accumulation of carbon dioxide in the atmosphere has a greenhouse effect which should aggravate the heating of the globe'. Yet this insight, which should be at the core of any account of the prospects of the 'growth' of a fossil-fired world economy, was not developed beyond a footnote in a classic paper devoted to communicating the far more abstract concept of entropy.[38]

In his 1959 address to the API, Edward Teller had canvassed an array of future energy R&D projects, including fuel cells, which 'could be run on solar energy'. Teller opined that the 'conversion of solar energy directly into electricity would be a most wonderful invention'.[39] Yet the photovoltaic effect, first described in 1839 by the French physicist Alexandre-Edmonde Becquerel, had at the time of Teller's speech already been utilised in the first viable solar cells developed at Bell Labs in the mid-1950s and installed on Vanguard 1, the first solar-powered space satellite launched in 1958.[40] Looking to the future, Georgescu-Roegen likewise stressed the importance of non-combustive, solar-based energy systems:

> [...] waterfalls – an energy produced indirectly by the sun's radiation – are increasingly being used to provide a source of free energy in the form of electric power. This trend is sure to become more accentuated. If the scattered efforts to use directly the solar radiation as a source of power succeed in making the idea operational, perhaps we shall not be as astounded as when we learned, in a macabre way, of man's harnessing of the atom. [...] such a success would represent a far greater, because more lasting, benefit to mankind. [...] it is bound to come under the pressure of necessity. [...] At

[36] Franta, B. (2018). Early oil industry knowledge of CO_2 and global warming. *Nature Climate Change*, 8(12), 1024–1025.

[37] Georgescu-Roegen (1971, p. 305).

[38] Georgescu-Roegen, N. (1975). Energy and economic myths. *Southern Economic Journal*, 41(3), 374–381. See p. 374.

[39] Teller, E. (1960). Energy patterns of the future. In *Energy and man: a symposium* (pp. 45–72). New York: Appleton-Century Crofts, p. 71.

[40] Rappaport, P. (1959). The photovoltaic effect and its utilization. *Solar Energy*, 3(4), 8–18.

> least, the industrial energy we derive or may derive from solar radiation does not produce by itself noxious wastes. Automobiles driven by batteries charged by the sun's energy are cheaper both in terms of scarce low entropy and healthy conditions – a reason why I believe they must, sooner or later, come about.[41]

Fifty years later, we are now in a position to realise the mature technological potential of distributed large-scale renewable energy systems composed of wind turbines, solar photovoltaics, batteries, and renewable hydrogen fuel production to replace fossil combustion as the primary source of industrial energy. Harnessing the universally available free energy of solar radiation and wind, such systems require a vanishingly smaller mining footprint to construct and operate, and produce no waste flows of particles, toxins, or carbon gases, at least throughout their working life. Thanks to the innovations of engineers and early policies (especially in Germany) supporting their widespread application, renewables are now the least-cost option for electricity generation at scale—with new builds of utility scale wind and solar yielding clean energy priced at less than the marginal operating costs of already-existing coal, gas, and nuclear power plants—and are now eminently capable of retrenching the politically entrenched and heavily subsidised fossil fuel sector as the energy source for perhaps the majority of industrial applications.[42] Again fossil capitalists were ahead of the academics in the 1970s. Working for Exxon, the chemist Stanley Whittington developed the original lithium-ion battery in 1973. In 1979, Exxon prototyped the first low-emissions electric-hybrid car.[43] This at a time when their terrifyingly accurate internal scientific forecasting of our present climate crisis was yet to be publicly refuted in the permanent, saturation level corporate communications campaign advancing the 'the Exxon position', which since 1988 has falsely proclaimed uncertainty of the science of global heating and the unbearable 'costs' to 'the economy' of legislation mandating energy efficiency, carbon taxation,

[41] Georgescu-Roegen (1971, pp. 305–306).

[42] Lazard (2019, Nov). Lazard's levelized cost of energy analysis, version 13.0. https://www.lazard.com/media/451086/lazards-levelized-cost-of-energy-version-130-vf.pdf. Accessed 2 Dec 2019.

[43] Banerjee, N. (2016, Oct 5). For Exxon, hybrid car technology was another road not taken. *Inside Climate News*.

decarbonisation, and renewable energy investment.[44] We can only imagine what might have been had Exxon chosen to devote its vast resources to the rapid development of zero-carbon energy technologies. Instead, Exxon executives have for three decades taken a leading role in poisoning the public sphere with a constant flood of 'business propaganda', collaborating with the Atlas Network and the wider fossil fuel sector to capture states or otherwise disable the capacities of democratic governments to respond to the long-anticipated thermal crisis of the Earth, doing immense harm to all life in the process.

Writing at the dawn of the neoliberal counter-revolution against social and environmental democracy, Odum insisted that the control of fossil fuel supplies:

> […] must be placed in the public sector, for the public good […] there must be enough international agreement to prevent runaway efforts of one country bent on competitive exponential growth which seems temporarily good for them but cancerous for all in the long run. This is a fearsome task with most of the fuel supplies in the hands of small partly developed countries and oil companies both of which are under short range planning only. We have a chance if first we can explain to the citizens of the world the underlying energy causes of our situation. Then the energy reforms of international law can follow […] This will require the incorporation of the energy ethic in all religious programs […] The rising levels of education and one-world television communication may make all this possible.[45]

Coupled with the long-term disinformation campaign pursued by fossil fuel interests, the neoliberal creed of privatisation and deregulation and the undermining of the parliamentary sovereignty and the international laws of the United Nations through a global constitution of private investor rights have effectively prevented Odum's transformative vision of a planetary energy democracy from being realised. If 'the citizens of the world' can still be hoped to 'have a chance', Odum's insights remain sound. The 'fearsome task' remains before us, and is infinitely more urgent. Oil, coal, and gas companies must somehow be brought under public control in the public interest. Long-term human survival requires

[44] Franta, B. (2018, Sept 19). Shell and Exxon's secret 1980s climate change warnings, *The Guardian*; Supran, G. & Oreskes, N. (2017). Assessing ExxonMobil's climate communications (1977–2014). *Environmental Research Letters* (12), 1–18.

[45] Odum (1971, p. 306).

nothing less than a global constitution for climate justice, one capable of permanently leaving hydrocarbon minerals where they perform their most economically valuable 'thermal services' to human society—in the ground, as geological internments of carbon safely removed over long eons from the atmosphere and oceans.

A Subordinate Science: Ecology in the Mirror of Economic Value

The Age of Ecology foretold in the late 1960s failed to eventuate. Undoubtedly, this was to a large extent the goal of formidable interventions made in defence of an entrenched 'pollutocracy'. Yet we might also consider the failure of ecologists to imagine the value of their science in terms truly independent of the cultural dominance of economic thought. Picking up a standard textbook on ecology, we turn to a chapter which introduces the ecosystems analysis of 'productivity', entitled 'The Flux of Energy through Communities'. Here we learn the definitive criteria which distinguish the human species from all others. This is not (as I would suggest) the social harnessing of vast energy flows through a plethora of 'extra-somatic' pyrotechnologies but rather the range of 'resources' that are constantly 'substituted' one for another in exchange: 'gold for housing, food for work, even cattle for wives!'[46] These Smithian forms of 'primitive barter' do little justice to our present condition, in which it is possible to exchange financial derivatives hedging against the political risk of oilfield investments for a leafblower, a tank of fuel, and a pair of plastic earmuffs. Commensurating the incommensurable, dollar-denominated sovereign debts can be traded for pledges to not destroy rainforests in the South, a promise then sold on as a 'carbon credit' which secures additional licenses to emit for industrial corporations in the North. Unwilling to open the Pandora's box of what could be meant by 'substitution', Begon et al. simply note with evident bemusement that:

> [...] in effect, dollars and pounds have come to be surrogates for resources.[47]

[46] Begon, M., Harper, J. & Townsend, C. (1995). *Ecology*. Cambridge, MA: Blackwell, p. 648.

[47] Begon et al. (1995, p. 648).

Casting a backward glance at systems ecology from the other side of the impossible metaphor <utility = energy = value = price>, the biologist Richard Levins has observed that:

> The notion of energy as the fundamental thing to look at as the universal medium of exchange is clearly brought into biology by analogy with economic exchange. [...] There was a hope [...] that we could ignore all the complexity of interacting species, the heterogeneity of populations, the complexities of competition and symbiosis, of mutation and predation, and reduce everything to a single medium of ecological exchange, which was designated as 'energy'.[48]

Yet in rejecting Odum's energetic ecology for theories which displace the quest for knowledge (*logos*) of the laws (*nomoi*) of the estate (*oikos*) to another reductionist register (e.g. to the functionalism of neo-Darwinian genetics or of complexity theory), ecologists have too often followed the economists in neglecting the social, ecological, and geopolitical importance of the interaction of carbon fixation and fire. Photosynthesis and the combustion of ex-biomass were integral phenomena in the non-reductionist, bio-thermodynamic paradigms of Vernadsky and Georgescu-Roegen, which although early and incomplete renditions of civilisation's prospects within the 'critical zone' of life on 'pyro-Gaia'[49] have nonetheless proved to be essentially congruent with the contemporary Earth system sciences and their vast documentation of our thermo-ecological predicament. The basic insight of thermodynamics, that 'temperature effects everything, [...] setting boundaries on what an organism can and cannot do',[50] now calls into question of the future viability of millions of species long adapted to a thermal envelope fast becoming a relic of geological history and the immediate life prospects of millions of people exposed to the slow violence of global heating.

In conclusion, perhaps nothing signifies the relegation of ecology to the status of a subordinate science in the neoliberal era—a 'subdivision of economics'—than the name of the scientific organisation tasked with the

[48] Cited in: Callebaut, W. (1992). *Taking the naturalistic turn*. Chicago: University of Chicago Press, p. 262.

[49] Clark, N. (2017). Pyro-Gaia: planetary fire as force and signification *Ctrl-Z: New Media Philosophy* (7).

[50] Clarke, A. (2017). *Principles of thermal ecology: temperature, energy and life*. Oxford: Oxford University Press, p. 385.

vital work of collating and reporting international research on the extinction of biodiversity and the erosion of ecosystems to 'policy makers': the Intergovernmental Science-Policy Platform on Biodiversity and Ecosystem *Services*.[51] The claim that economics should be downgraded in status to a subset of general ecology was explicit in the biocentric critiques of Odum and Georgescu-Roegen. This is an account of the proper epistemic hierarchy of the twin sciences we find completely reversed in the subsequent rise of the 'ecosystem services' concept, through which ecologists have accommodated themselves to the institutional dominance of a wholly anthropocentric and subjectivist economics. Birds and bees, insects and trees appear in this discourse as substitutable (and expendable) components of 'natural capital' infrastructures, merely one among many several classes of capital yielding 'services' to the market economy and its sovereign investors. The primary metaphor <the ecosystem is an economy> remains, but the incongruities that once alerted us to the presence of the metaphor have been eroded by its 'currency'. Familiarity and overuse has drained the metaphor of its figurative sense, such that it now appears to be a feature of the world.

The following chapter attempts an explanation for the extraordinary resilience of the economists' creed to the ecological critique, from the 1970s into the present. With tragedy and irony, it turns out that this concluding chapter in our 'universal history' can be attributed to the career of a particularly seductive and powerful metaphor—one which has displaced the constitutive machine metaphors of classical systems ecology and neoclassical economics. This metaphor, expounded most influentially by the arch-neoliberal Friedrich Hayek, can be concisely formulated: <the economy is a complex adaptive ecosystem>.

[51] The IPBES was established by the United Nations in 2012.

CHAPTER 14

Genealogies of Resilience: From Conservation to Disaster Adaptation

Developed within systems ecology in the 1970s, 'resilience' as a metascience of complex adaptive systems and as an operational strategy of risk management has flourished, progressively asserting itself as a dominant discourse in natural resource management. The concept of resilience has rapidly infiltrated vast areas of the social sciences, becoming a regular, if under-theorised, term of art in discussions of international finance and economic policy, corporate risk analysis, the psychology of trauma, development policy, urban planning, public health, and national security. Since the 1990s, global financial institutions such as the International Monetary Fund, the World Bank, and the Bank for International Settlements have increasingly incorporated strategies of 'resilience' into their logistics of crisis management, financial regulation, and development economics.[1]

This chapter is a rework of an article co-authored with Melinda Cooper. I am grateful for her generous permission to revise it for this book. All the changes are mine alone. Walker, J. & Cooper, M. (2011). Genealogies of resilience: from systems ecology to the political economy of crisis adaptation. *Security Dialogue*, 42(2), 143–160.

[1] Nsouli, S. et al (1995, Dec 1). Resilience and growth through sustained adjustment: the Moroccan experience. IMF Occasional Paper No. 117; World Bank (2006, Apr 27). Social resilience and state fragility in Haiti: a country social analysis. Caribbean Country Management Unit, Latin America and the Caribbean Region Report No. 36069–HT; Bank

J. Walker, *More Heat than Life: The Tangled Roots of Ecology, Energy, and Economics*,
https://doi.org/10.1007/978-981-15-3936-7_14

With the post-9/11 revolution in 'homeland security', resilience became a byword among agencies charged with co-ordinating security responses to climate change, critical infrastructure protection, natural disasters, pandemics, and terrorism, reorienting these once distinct policy arenas towards a horizon of critical future events which (we are told) we cannot predict or prevent but merely adapt to by 'building resilience'.[2] Abstract and malleable enough to encompass the worlds of high finance, defence, urban infrastructure, and the global environment within a single analytic, the concept of resilience has become pervasive in the language of global governance.

This chapter traces the genealogy of 'resilience' from its first formulation in ecosystems science to its recent proliferation across disciplines and policy arenas loosely concerned with the logistics of crisis management, investigating its premises in and generalisation from complex systems theory.

The contemporary usage of the term 'resilience' originated in the work of the ecologist C.S. Holling and retains definitive links to this field. This should direct our attention to the exemplary function of 'ecological risk' within contemporary imaginaries of security, which can increasingly be taken as the foundational context of security in our era of thermal crisis. The success of this ecological concept in colonising multiple arenas of governance can be attributed to its intuitive ideological fit with a neoliberal philosophy of complex adaptive systems, the under-acknowledged legacy of Friedrich Hayek. Where once it was commonly assumed that neoliberalism was of a piece with the neoclassical method of the Chicago school, this crudely positivist current of neoliberalism has increasingly been overtaken in intellectual influence by the mature Austrian philosophy of Hayek. Further, it would appear that the science of complex adaptive systems has become a discursive reference point for the full spectrum of contemporary risk interventions. Whereas energy physics provided a foundational role in modernist theories of economic and ecological organisation, and the homeostatic systems of first-order cybernetics dominated the economic and military sciences of the Cold War, complexity science now

for International Settlements (2008). Basel Committee on banking supervision announces steps to strengthen the resilience of the banking system. Geneva: BIS.

[2] World Bank (2008). *Climate resilient cities: a primer on reducing vulnerabilities to climate change impacts and strengthening disaster risk management in East Asian cities.* Washington DC: World Bank; UK Cabinet Office (2007). Operations in the UK: the defence contribution to resilience. Joint Doctrine Publication, 2e.

serves as a source of naturalising metaphors for contemporary practices of security, functioning to neutralise critical inquiry into the disastrous consequences of neoliberal policy in the arenas of financial regulation, urban planning and crisis response, development, environmental management, and climate change.

This conceptual genealogy of 'resilience' first considers Holling's innovations in ecology, and then Hayek's in economics broadly defined. Inspired by very different concerns, Holling and Hayek made profoundly influential contributions to their respective fields, and these have ended up coalescing in uncannily convergent positions. In their writings of the early 1970s, Holling and Hayek were simultaneously preoccupied by questions of the epistemic limits to prediction and assertions of ecological limits to growth. In common is their rejection of metaphors from classical thermodynamics, their early adoption of the lexicon of 'complex adaptive systems', their pessimism about the management of complex systems according to predictive models, and the rejection of the *Limits to Growth* report as an example of all that was wrong with the image of their respective sciences in the public sphere. Importantly, in their late careers both figures sought to universalise the significance of their projects well beyond the natural/social science boundary. The two perspectives, originally informed by antagonistic concerns, have ended up merging in the contemporary discourse of crisis response through resilience. At stake in this tacit union is a governmental philosophy of nature and society so all-encompassing and resilient to critique that the effects of political interventions (and non-interventions) made in its name, no matter how catastrophic, seem as inescapable as the weather.

This chapter argues for the critical importance of the proximity between the emergent discourse of 'resilience' and contemporary neoliberal practices of political management, which is demonstrated through an analysis of the rise of resilience in the specific cases of international finance, critical infrastructure protection, and 'sustainable' development. Returning to the arts of ecological management from which 'resilience' arose, I conclude with a reflection on the evolution of complex ecosystems theory, from critique to collaboration.

C. S. Holling's Innovations in Ecology: Origins of Resilience Science

The work of the ecologist Crawford 'Buzz' Holling represents a crucial shift in the annals of systems ecology. Holling did some of the most important work in the early 1970s to modernise the classical ecosystems model of ecological dynamics in terms of the new 'complexity science': away from mechanistic assertions of equilibrium typical of post-war cybernetics towards the contemporary 'complexity science' view of ecosystems. In the 1990s, Holling went on to found the consortium of environmental scientists called the Resilience Alliance. More recently, these initiatives have been brought together within the Stockholm Resilience Centre, a high-profile think-tank which promotes the uses of resilience theory in international environment and development projects. What follows is a brief outline of Holling's innovations in applied ecology in the 1970s and his subsequent efforts since the mid-1990s to incorporate 'social systems' and 'economic systems' into a general complexity science of 'socio-ecological resilience'.

Abel and Stepp have described the interface of complexity science with ecology:

> […] what actually constitutes complex systems science is not yet settled. Although there are many threads, we and others […] see an integrated, evolutionary science of complex systems emerging from the synergy between new computational paradigms (chaos theory, agent-based modelling, and self-organization), dramatic breakthroughs in the venerated field of non-equilibrium thermodynamics, empirical research into large, complicated systems such as weather, earth systems, and ecosystems, and innovation in evolutionary theory. […] As an emerging field, some researchers claim their part as the whole, but we prefer to see the connections and the possibilities of an open, multi-disciplinary, evolutionary, and integrative systems science.[3]

Whilst a sharp distinction between classical systems ecology and its post-1970s complexity turn effaces much that is continuous in the discipline, the influence of the new complexity science on Holling's school

[3] Abel, T. & Stepp, J. (2003). A new ecosystems ecology for anthropology. *Ecology and Society*, 7(3).

of ecosystem management is profound.[4] The key image of science that propelled the formalisation of economics (in the 1870s) and ecology (in the 1950s) was of smooth and continuous returns to equilibrium after shock, an image derived from different vintages of classical mechanics, thermodynamics, and systems theory. Holling's widely cited paper 'Resilience and Stability of Ecological Systems' (1973) destabilised the notion of 'equilibrium' as the core of the ecosystem concept and the normal terminus of ecosystem trajectory, signifying a shift amongst ecologists away from the notion that there exists a 'balance of nature' to which life will return eventually if left to self-repair.[5] Having worked for years in the field as a resource manager and conservation ecologist, Holling began his classic 1973 paper on resilience by noting that:

> [...] traditions of analysis in theoretical and empirical ecology have been largely inherited from developments in classical physics and its applied variants [...] there has been a tendency to emphasize the quantitative rather than the qualitative, for it is important in this tradition to know not just that a quantity is larger than another quantity, but precisely how much larger. [...] But this orientation may simply reflect an analytic approach developed in one because it was useful and then transferred to another where it may not be.[6]

Holling went on to distinguish between an existing notion that he calls 'engineering resilience' and his alternative, a properly 'ecological' resilience. Engineering resilience, associated with the mathematical models of the systems ecology of the day, is an abstract variable, simply the time (t) it takes a system to return to a stable maximum (or equilibrium position) after a disturbance. The return is simply assumed, and the equilibrium state is taken as equivalent to long-term persistence.[7] What Holling sought

[4] de Laplante, K. (2005). Is ecosystem management a postmodern science? In K. Cuddington & B. Beisner (Eds.), *Ecological paradigms lost: routes of theory change* (pp. 398–416). Amsterdam: Elsevier.

[5] Holling, C. S. (1973). Resilience and stability of ecological systems. *Annual Review of Ecology and Systematics* (4), 1–23.

[6] Holling (1973, p. 1).

[7] Odum, E. P. (1969). The strategy of ecosystem development. *Science*, 164 (3877), 262–270; Lewontin, R. (1969). The meaning of stability. In *Diversity and stability of ecological systems* (pp. 13–24). Brookhaven, NY: Brookhaven Symposia in Biology, vol. 22; May. R. (1973). *Complexity and stability in model ecosystems*. Princeton, NJ: Princeton University Press.

rather to define is a complex notion of resilience which might account for the capacity of an ecosystem to remain cohesive even while undergoing extreme perturbations. If stability refers to the familiar notion of a return to equilibrium, 'ecological' resilience designates the complex biotic interactions that determine 'the persistence of relationships within a system': thus resilience is 'a measure of the ability of these systems to absorb changes of state variables, driving variables, and parameters, and still persist'.[8]

Holling pointed to the dangers of the management theory of 'maximum sustained yield' (MSY), long dominant in industrial forestry and fisheries, with its claims to enumerate a fixed quantity of 'surplus' cod or spruce that can be harvested year in year out, without undermining the ability of the ecosystem to regenerate its own reproductivity. Holling's argument here (mirroring Hyman Minsky's post-Keynesian account of financial crises) is that the long-term anticipation of stability may be inherently destabilising. The expectation of a constant yield of (re)productivity may fragilise a natural resource to such an extent that it undermines the complex factors supporting the resilience of the system as a whole: for example, by lowering the average age of an overharvested fish population to a point below sexual maturity. 'The very approach [...] that assures a stable maximum sustained yield of a renewable resource might so change these deterministic conditions that the resilience is lost or reduced so that a chance and rare event that previously could be absorbed can trigger a sudden dramatic change and loss of structural integrity of the system.'[9] Holling's perspective on resource management reflects the emerging critical voices which, in the early 1970s, insisted that intensive methods in agriculture and resource extraction would at some point transgress inherent limits to sustainability, resulting in mass extinctions and intolerable over-pollution. For Holling, the equilibrium approach was dangerous in its abstraction: glossing over the unknowably complex interdependencies of specific landscapes pressed into the conditions of maximised yield, it accelerated the process of fragilisation, leading to the potentially irreversible loss of biodiversity and ecosystem function. The urgent focus for the conservation manager in a significantly humanised world should not be the equilibrium of a pristine ecosystem but the resilience of biotic communities exposed to severe economic pressures.

[8] Holling (1973, p. 17).
[9] Holling (1973, p. 21).

In contrast, Holling's perspective seeks to open up a management approach capable of sustaining productivity even under conditions of extreme instability. Its ability to adapt to and deflect from particular limits derives from the fact that it has abandoned long-term expectations:

> A management approach based on resilience [...] would emphasize the need to keep options open, [...] and the need to emphasize heterogeneity. Flowing from this would be not the presumption of sufficient knowledge, but the recognition of our ignorance: not the assumption that future events are expected, but that they will be unexpected. The resilience framework can accommodate this shift in perspective, for it does not require a precise capacity to predict the future, but only a qualitative capacity to devise systems that can absorb and accommodate future events in whatever unexpected form they may take.[10]

The above passage, taken from the conclusion of Holling's 1973 article, is significant because it so clearly anticipates the guiding ideas of contemporary complex systems theory and its practical applications in crisis response. Under the sign of resilience, this is an approach to risk management which foregrounds the limits to predictive knowledge and insists on the prevalence of the unexpected, seeking to 'absorb and accommodate future events in whatever unexpected form they may take'.

Holling's later contributions to the practices of adaptive ecosystem management earned him a wide professional following. Following consensus-building work with senior neoclassical economists, Holling and fellow ecologists formed the Resilience Alliance and conceived of an ambition to expand the insights of the resilience perspective well beyond ecology.[11] Emblematic in the name change of the house journal from *Conservation Ecology* to *Ecology & Society*, the Alliance was no longer concerned with resilience as a property of ecosystems but of the coevolution of societies and ecosystems as a single system. This new research into 'social-ecological resilience' aspires to set the ground rules for a general systems theory capable of integrating society, the economy, and the biosphere. This totality is dubbed the 'Panarchy':

[10] Holling (1973, p. 21).

[11] Arrow, K., Bolin, B., Costanza, R., Dasgupta, P., Folke, C., Holling, C.S., Jansson, B., Levin, S., Mäler, K., Perrings, C. & Pimentel, D. (1995). Economic growth, carrying capacity, and the environment. *Ecological Economics*, 15(2), 91–95.

> the structure in which systems, including those of nature (e.g., forests) and of humans (e.g., capitalism), as well as combined human-natural systems (e.g., institutions that govern natural resource use such as the Forest Service), are interlinked in continual adaptive cycles of growth, accumulation, restructuring, and renewal.[12]

There is a significant difference in scope and tone between this later definition of socio-ecological resilience and Holling's earlier work. Holling is no longer arguing that some ecosystems may not recover from extreme fluctuations, nor that ecosystems under the stress conditions of maximum sustained yield might be better preserved with a more 'resilient' management approach, but rather that *all* social-ecological system dynamics can be approached as non-linear iterations of an 'adaptive cycle', in which four distinct phases can be identified.

Where classical ecology focussed only on the phases of rapid successional growth (*r*) followed by the conservation phase of a stable climax equilibrium (K), the Resilience Alliance argue that these phases are inevitably followed by collapse (Ω) and then a spontaneous reorganisation that leads to a new growth phase (α) [see Fig. 14.1].

What unites these diverse systems and allows Holling to propose a common theorisation of their dynamics is the proposition that each can be defined by a concept of 'capital'. This capital, be it financial, organisational, or biophysical, is 'the inherent potential of a system that is available for change, since that potential determines the range of future options possible'.[13] In short, Holling seeks to independently theorise an abstract dynamics of capital accumulation, one not predicated on the progressive temporality of classical political economy but rather on the inherent crisis tendencies of complex adaptive systems. In this respect Holling's later work becomes much more closely aligned with Hayek's mature theory of spontaneous market order and complex social evolution. Although Holling never cited Hayek, and Hayek routinely ignored the environmental sciences, I argue that it is Hayek's influential philosophy of free market dynamics that has made the contemporary policy arena so receptive to the overtures of the Resilience Alliance. If the Mont Pèlerin Society (MPS) and the Resilience Alliance have anything in common, it is the attempt to

[12] Gunderson, L. & Holling, C.S. (Eds.) (2002). *Panarchy: understanding transformations in human and natural systems.* Washington, DC: Island Press.

[13] Holling, C. S. (2001). Understanding the complexity of economic, ecological and social systems. *Ecosystems,* (4), 390–405.

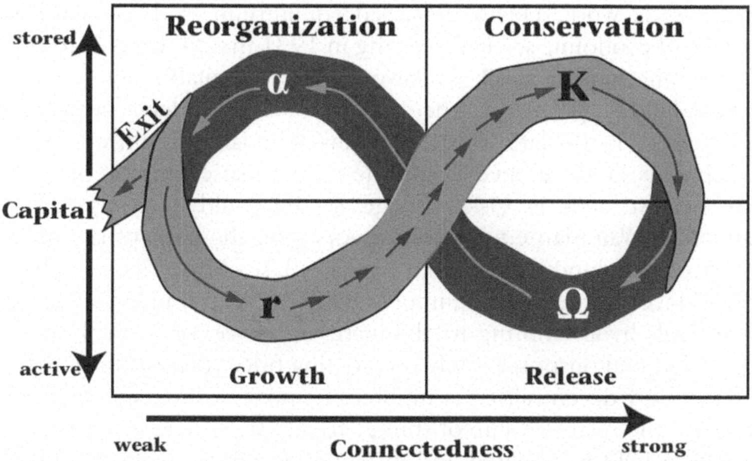

Fig. 14.1 Resilience as 'capital': the inherent potential of a system available for change. (Source: *Panarchy*, by Lance Gunderson & C.S. Holling. Copyright © 2002 Island Press. Reproduced by permission of Island Press, Washington, DC)

forge a broad transdisciplinary philosophy capable of unifying nature and society under a single set of all-encompassing concepts.

Hayek's Legacy: The Market as a Complex Ecological System

It was once common for the academic and activist imagination to conflate and critique neoliberalism as the radical generalisation of neoclassical equilibrium theory to all aspects of social life. Yet as Mirowski and Plehwe have amply documented in their history of the 'neoliberal thought collective', the various scholars associated with the Mont Pèlerin Society (MPS) were at times internally split between the followers of the positivist Chicago School approach of Gary Becker, George Stigler, and Milton Friedman and the advocates of the ultra-subjectivist philosophy of the Austrian school, most famously associated with Friedrich Hayek.[14]

[14] Mirowski, P. & Plehwe, D. (Eds.) (2009). *The road from Mont Pèlerin: the making of the neoliberal thought collective*. Cambridge, MA: Harvard University Press.

In his early work, Hayek defended equilibrium as the constitutional metaphor of economic science, arguing in 1931 that 'if we want to explain economic phenomena at all, we have no means available but to build on the foundations given by the concept of a tendency towards an equilibrium'.[15] By the late 1930s, the signs of his later dissent were becoming visible, as he developed a critique of rational economic planning—which socialists such as Oskar Lange argued could by computationally grounded in Walrasian equilibrium theory—on the basis of the inherent limits to foresight and anticipation.[16] In 'The Use of Knowledge in Society' (1945), Hayek argued for the impossibility of 'central planners' to arrive at their goals by attempting to eliminate, influence, or control prices for rational planning purposes. Only the floating prices constituting 'the market', a radically decentralised computation and signalling system, are able to discover the relative value of things, to adjust, evolve, and incorporate information held by isolated and differentiated individuals. Importantly, 'these adjustments are probably never "perfect" in the sense in which the economist conceives of them in his equilibrium analysis'.[17]

The notion of price formation as distributed computation led Hayek to his mature unified theory of spontaneous order and social evolution, first suggested in his 'Theory of Complex Phenomena'.[18] By the 1980s, he had abandoned the equilibrium analysis of his early career as a 'normal science' economist. Acknowledging that equilibrium analysis permitted the idea that planning was possible, he criticised the Keynesian state for seeing the economy as a hydraulic machine, a 'suction pump' operating on aggregate balances of supply and demand contained in a system of pipes and tanks.[19] Remarking on the sheer complexity of the capital structure, he spoke of multiple 'streams' of value, ebbing and flowing into a river of liquid capital, constantly re-adjusting the production process, coursing down an

[15] Hayek, F. (1931). *Prices and production*. London: Routledge & Kegan Paul, pp. 34–35.

[16] Lange, O. (1938). *On the economic theory of socialism*. Minneapolis: University of Minnesota Press; Lange, O. (1949). The practice of economic planning and the optimum allocation of resources, *Econometrica* (17), 166–171; Hayek, F. (1937). Economics and knowledge. *Economica*, 4(13), 33–54.

[17] Hayek, F. (1945). The use of knowledge in society. *American Economic Review*, 25(4), 519–530.

[18] Hayek, F. (1967). The theory of complex phenomena. In: *Studies in philosophy, politics and economics* (pp. 22–42). London: Routledge & Kegan Paul.

[19] Caldwell, B. (2004). *Hayek's challenge: an intellectual biography of F. A. Hayek*. Chicago: University of Chicago Press, p. 226.

ever-changing river bed.[20] In the 1950s, Hayek had mounted a subjectivist critique of the errors of 'scientism', exemplary of which were 'the various forms of social "energetics" which, from [...] Ernest Solvay, Wilhelm Ostvald and F. Soddy down to our own day have constantly reappeared among scientists and engineers when they turn to the problems of social organization'.[21] In his late career, however, Hayek claimed natural science foundations for his own theory of spontaneous order, aligning his project with 'autopoesis, cybernetics, homeostasis, spontaneous order, synergetics, systems theory' and claiming the far-from-equilibrium thermodynamics of Ilya Prigogine as support for his work.[22] In his final book, Hayek wrote:

> [...] the extended order is perfectly natural, in the sense that it has itself, like similar biological phenomena, evolved naturally in the course of natural selection.[23]

In 1974, Hayek was awarded the 'Nobel' economics prize, an event that signalled the changing fortunes of the Mont Pèlerin Society and terminated his long exile from academic respectability. In the speech he delivered for the occasion, 'The Pretence of Knowledge' (1974), Hayek not only gave voice to his enduring hostility to the state-engineered equilibria of the Keynesian welfare state but also to his dissatisfaction with the equilibrium models favoured by neoclassical economists, including the Chicago school neoliberals.[24]

Hayek's speech focused on the crises of the early 1970s—oil shocks, stagflation, third world and worker militancy—and the various efforts to intervene in them by expanding the regulatory arena of government. These crises, he contended, were symptoms of the intellectual failure of Keynesianism. He was therefore highly sceptical of efforts to respond to

[20] Hayek, F. (1981, Jan 27). The flow of goods and services. Address to the London School of Economics, published in German: (1984) Der strom und der güter und leistungen. Tübingen: Mohr Siebeck.

[21] Hayek, F. (1955). *The counter-revolution of science*. New York: Free Press, p. 51.

[22] Hayek, F. (1988). *The fatal conceit: the errors of socialism—collected works of F.A. Hayek, vol. 1*. London: Routledge, p. 9; Hodgson, G. (1994). Hayek, evolution and spontaneous order. In P. Mirowski (Ed.), *Natural images in economic thought: markets read in tooth and claw* (pp. 409–447). Cambridge: Cambridge University Press.

[23] Hayek (1988, p. 19).

[24] Hayek, F. (1974, Dec 11). The pretence of knowledge. Acceptance speech upon the award of the Sveriges Riksbank Prize in Economics in Memory of Alfred Nobel. Salzburg.

them through techniques of state intervention which, he believed, had engendered them in the first place. Such interventions, Hayek intoned, were at best doomed to failure. The natural complexity of market phenomena was such that no centralised authority could hope to predict, much less control, the precise evolution of individual elements in the system. At worst, such efforts risked inducing long-term crises that would not have occurred without the undue interference of government. Hayek's critique of Keynesian and neoclassical equilibrium theories went well beyond the political sphere. What was at stake for him is no less than a thorough rethinking of epistemology itself. As a counter-argument to the predictive fantasies he saw as integral to the post-WWII Keynesian consensus, Hayek espoused an epistemology of limited knowledge and uncertain futures. 'I confess that I prefer true but imperfect knowledge, even if it leaves much indetermined and unpredictable, to a pretence of exact knowledge that is likely to be false.'[25]

Hayek immediately put this imperfect yet superior epistemic position to good use, repudiating the claims of the nascent environmental movement and the landmark Report to the Club of Rome. The report included Keeling's curve measuring the rapid accumulation of atmospheric carbon dioxide and predicted that exponential growth would catastrophically undermine the regenerative capacities of the biosphere in the absence of a 'controlled, orderly transition from growth to global equilibrium'.[26] Hayek denounced it as exemplary of the hubris of predictive modelling in the face of unknowable complexity:

> The enormous publicity recently given by the media to a report pronouncing in the name of science on The Limits to Growth, and the silence of the same media about the devastating criticism this report has received from the competent experts, must make one feel somewhat apprehensive about the use to which the prestige of science can be put.[27]

[25] Hayek (1974).

[26] Meadows, D. et al. (1972). *The limits to growth: a report for the Club of Rome's project on the predicament of mankind.* New York: Universe Books, p. 71 & p. 84.

[27] Hayek (1974). Hayek cited but two 'competent experts', who were not scientists but economists: Gottfried Haberler (MPS), a fellow student of von Mises then working for the (Atlas-affiliated) American Enterprise Institute, and Atlas fellow-traveller Wilfred Beckerman of Balliol College, Oxford, who assured us as holder of the 'second oldest chair in Political Economy' (first occupied by Nassau Senior) that although 'life on this Earth is very far from perfect there is no reason to think that continued economic growth will make it any worse'. Georgescu-Roegen begins his classic 1975 essay by citing Beckerman thus, with incredulity.

Hayek's drastic critique was informed by a comprehensive ontology, one which would increasingly borrow from complex systems theory. Social systems, Hayek now claimed, are like the biological systems newly defined by scientists as complex, adaptive, and non-linear. They are not subject to quantification or prediction on the basis of the physical laws that govern the well-characterised systems of classical mechanics and thermodynamics (e.g. heat engines).

> While in the physical sciences [...] any important factor which determines the observed events will [...] be directly observable and measurable, in the study of [...] complex phenomena such as the market, [...] the circumstances which will determine the outcome of a process [...] will hardly ever be fully known or measurable.[28]

Hayek went on to observe that '[it is] almost impossible for the layman, to distinguish between legitimate and illegitimate claims advanced in the name of science'—an insight which has since been put to great effect by the Atlas Network in undermining 'the prestige of science' through the mass dissemination of anti-environmental 'business propaganda'.

In rejecting the legacy of classical physics, Hayek's texts of the later 1970s and 1980s deploy an approach to complex adaptive systems that is formally very similar to Holling's but much more radical in its conflation of the financial, social, and biological spheres. Like all ontologies, Hayek's complexity turn generates a number of normative consequences. Reflecting the heritage of Herbert Spencer's ultra-libertarian evolutionism, it assumes that time's arrow moves ever in the direction of greater complexity: evolution occurs spontaneously in far from equilibrium conditions. Perturbations of greater or lesser force are not only inevitable, they are also necessary to the creativity of spontaneously organised complexity. This is a philosophy that does not so much reject natural law as redefine it in immanent, evolutionary terms, as that which is continually created anew and selected by the very exercise of market freedom:

> Like scientific theories, [rules of conduct] are preserved by proving themselves useful, but, in contrast to scientific theories, by a proof which *no one*

Haberler, G. (1974). *Economic growth and stability*. Los Angeles: Nash; Beckerman, W. (1974). *In defence of economic growth*. London: Cape; Georgescu-Roegen, N. (1975). Energy and economic myths. *Southern Economic Journal*, 41(3), 347–381.

[28] Hayek (1974).

> *needs to know*, because the proof manifests itself in the resilience and progressive expansion of the order of society which makes it possible. [my emphasis][29]

Whilst Hayek defines the radical freedom of the market by its indifference to all external limits and laws, he also endows the market itself with immanent law-making powers, to which he then subjects the state. The laws of the market rest on no pre-existing foundation: their very resilience serves as proof of concept, in much the same way as the law of natural selection constantly proves or disproves the viability of chance mutations in nature. On a purely ontological level, Hayek places the immanent laws of market freedom prior to those of the state as an intentional law-making power. In historical terms, however, he recognises that the global order of market freedom has yet to be fully realised. This is a project of constitutional reform which would involve the radical remaking of all law and institutions in accordance with the allegedly 'self-organising' dynamic of the market, a project for which Hayek characteristically enrols the coercive powers of the state, even in its most authoritarian expressions.

Complex Systems, Resilience, and Financial Risk Management

It was once easy to dismiss Hayek's late philosophy as an intellectually interesting but politically inconsequential episode in the convoluted history of neoliberal economic thought. Chicago school neoliberals, including Milton Friedman, routinely derided their Austrian counterparts as too hermetic and subjectivist to deliver any practical dividends in the field of economics. Yet as complex systems theory has itself developed a repertoire of practical methodologies, the force of Hayek's late Austrian philosophy has come into its own, moving beyond the circles of the Santa Fe Institute and the Cato Institute to offer itself up, in the wake of the financial crisis of 2007, as a method of financial risk management. In 2006, the Federal Reserve Bank of New York hosted a conference exploring the usefulness of ecological models for rethinking the dynamics of risk in modern financial markets. Noting that 'systemic risk' in the financial system bears a strong resemblance to the dynamics of complex adaptive systems in the

[29] Hayek, F. (1978). The errors of constructivism. *New studies in philosophy, politics, economics and the history of ideas* (pp. 3–22), Chicago: University of Chicago Press.

biophysical world, the conference report concluded that 'approaches to risk management in natural and physical systems could be pertinent to financial risk management'.[30] Resilience was singled out as the watchword for new models of adaptive risk management sensitive enough to cope with the systemic risks of deregulated finance.

As the global financial crisis deepened, central bankers and financial risk managers increasingly turned to the resources of complex systems theory. In 2008, a group of leading ecologists including Robert May (at the time an advisor to the Bank of England) published a paper in *Nature* which offered the insights of complex ecological systems as a model for bankers during the gathering financial crisis.[31] Andrew Haldane, executive director for financial stability at the Bank of England, was a vocal champion of this complexity turn. In a widely reported paper, he pointed to parallels between the epidemic of the SARS virus and the contagion effects of the collapse of Lehman Brothers:

> Both events were manifestations of the behaviour under stress of a complex, adaptive network. Complex because these networks were a cat's-cradle of interconnections, financial and non-financial. Adaptive because behaviour in these networks was driven by interactions between optimizing, but confused, agents. Seizures in the electricity grid, degradation of ecosystems, the spread of epidemics and the disintegration of the financial system—each is essentially a different branch of the same network family tree.[32]

Pointing to the limitations of stress-testing, with its focus on 'tame' risk and normal distributions, Haldane suggested that financial risk managers should instead look to the strategies of non-predictive futurology (scenario planning) and adaptive anticipatory risk management deployed in the field of ecological management.

To insist on the Austrian influence in the inner sanctums of the world's leading financial institutions and their regulators may seem

[30] Kambhu, J., Weidman, S. & Krishnan, N. (2007). New directions for understanding systemic risk: a report on a conference co-sponsored by the Federal Reserve Bank of New York and the National Academy of Sciences. Washington D.C.: National Academies Press, pp. 5–7.

[31] May, R., Simon Levin, S. & Sugihara, G. (2008). Complex systems: ecology for bankers. *Nature* (451), 893–895.

[32] Haldane, A. (2009). Rethinking the financial network. Speech delivered at the Financial Students Association, Amsterdam, Bank of England.

counter-intuitive. It is more commonly held that the reigning influence on financial risk and price modelling lies not in Hayek's hermetic philosophy but in neoclassical finance: Friedman's 'rational speculators', Arrow-Debreu securities, the Efficient Market Hypothesis, the standardised algorithms of portfolio management software, and the central banker's Computable General Equilibrium Model—all of which presume the formal calculability of relevant states of risk. Again, however, we would contend that a de facto 'division of labour' has established itself between the formalism of equilibrium models, lending the imprimatur of rational calculation to the design of derivative trading instruments, and the Hayekian cosmology of complex systems theory, which both informs a macroeconomic vision of market dynamics in general and, in more recent times, justifies the implementation of new crisis response strategies at the institutional level. What unites both camps is the insistence that the distributed computation of the market always surpasses the state's ability to process information, the neoliberal creed that had licensed the progressive dismantling of the banking reforms legislated in the wake of the crisis of 1929 and the Great Depression.

Knowingly or not, advocates of the complexity turn in financial risk management share Hayek's epistemology of limited foresight. The predictability of future states of the world is, for them, not only an empirical but also a logical impossibility. What is distinctive about the interventions of Haldane and others is the fact that complex systems theory no longer functions for them as an argument against regulation, as it was for Hayek, but as the starting point for a reform of financial risk management, involving the systematic introduction of non-predictive, futurological methods of vulnerability analysis such as scenario planning. In the words of Nout Wellink, President of the Netherlands Bank and Chairman of the Basel Committee on Banking Supervision, 'the goal of regulatory changes should not be to decrease complexity per se, or to return to the financial regulations of the past, but to make complexity more manageable, by constraining systemic risk, and improving the resilience of the financial system as a whole'.[33] It is more than ironic that the influence of the later Hayek should be making itself felt within the walls of the central bank—an institution whose pretensions to centralised knowledge were much maligned by the Austrian neoliberals.

[33] Wellink, N. (2009, Nov 10). Managing complexity. Speech given to the NautaDutilh seminar, Bussum.

The global financial crisis played something like the triggering role that 9/11 represented for security, pushing new methods of futurology, contingency planning, and crisis response onto the policy reform agenda, testament to the growing respectability of the resilience perspective as a framework of crisis management.

Resilience in US National Security: Critical Infrastructure Protection and the Culture of Preparedness

'Resilience' has become ubiquitous as an operational strategy of national security, emergency preparedness, and crisis response. Although by no means absent prior to 2001 or restricted to the US prosecution of the war on terror, the term has proliferated since the establishment of the US Department of Homeland Security (DHS) in 2002 under President George W. Bush. The National Strategy for Homeland Security, published in 2002 and reissued in 2007, brought together the structural resilience of 'critical infrastructures' with the 'operational resilience' of emergency response organisations, government institutions, and private enterprises in the face of crisis. Identifying 'resilience' as the essence of a 'culture of preparedness', it situated its recommendations within a general recognition of the limits to preparation.

> Despite our best efforts, achieving a complete state of [...] protection is not possible in the face of the numerous and varied catastrophic possibilities that could challenge the security of America today. [...] we cannot envision or prepare for every potential threat, we must understand and accept a certain level of risk as a permanent condition.[34]

These 'catastrophic possibilities' span the divide between military and civil threats, encompassing terrorist attacks, natural disasters, climate change, and infectious disease in a non-exhaustive 'full-spectrum' list of contingencies. The strategy is notable for insisting that none of these threats are fully preventable, proposing instead the notion of 'resilience' as a default condition of emergency response.

[34] Department of Homeland Security (2007, Oct). The national strategy for homeland Security. www.dhs.gov/xabout/history/gc_1193938363680.shtm.

In US security policy discourse, the concept of resilience was first deployed in a 1981 proposal to the Federal Emergency Management Agency (FEMA) for a fully decentralised renewable energy grid.[35] During the administration of President Carter, alternative energy development enjoyed federal support, but this was reversed by the Reagan administration. The defence of critical infrastructure as an area of government interest began to crystallise under President Clinton. In 1996, the President's Commission on Critical Infrastructure Protection defined critical infrastructure as national utilities so vital 'that their incapacity or destruction would have a debilitating effect on the defence or economic security of the United States'.[36] It is significant that the emergence of critical infrastructure as a national security concern took place during a period of intense (re)privatisation of formerly public infrastructure services, a move that created an opportunity for secondary financial markets specialising in the income streams (or securitised debts) arising from infrastructure privatisation itself. The categorisation of critical infrastructure protection as a national security concern signalled an at least tacit recognition that the financial and civil risks generated by the widespread privatisation of vital national services could themselves be construed as a significant threat to civil defence. As an optic for assessing and responding to risk, critical infrastructure protection blurs the boundaries between the properly military threat of terrorist attack and civil contingencies such as natural disaster, operational accidents, the failure of financial systems, or civil protest. This strategic conflation of previously separate spheres of action would be institutionalised under the administration of George W. Bush, when both FEMA and the Environmental Protection Agency would be absorbed into the Department of Homeland Security (DHS). At stake in this process of reform was not merely the deregulation of formerly state-controlled services and networks but the transfer of regulatory authority from the civil sectors of public transport, health and safety, environmental protection, and emergency response to a logistics and security sector newly organised around counter-terrorism.

[35] Lovins, A. & Lovins, L. (1981). *Energy policies for resilience and national security*. San Francisco: Friends of the Earth.

[36] Lopez, B. (2006). Critical infrastructure protection in the United States since 1993. In P. Auerswald et al. (Eds.), *Seeds of disaster, roots of response: how private action can reduce public vulnerability* (pp. 37–50). Cambridge: Cambridge University Press.

The National Security Strategy (NSS) of 2007 reasserted the importance of 'resilience' as both a strategic and psychological imperative of national preparedness and more fully incorporated the ecosystemic and financial dimension of crisis into its taxonomy of contingencies. Between the 2002 and 2007 editions of the NSS, Hurricane Katrina had intervened, blurring further the cognitive distinctions between the unpredictable terrorist threat, financial crisis, and environmental disaster. The 2007 NSS combined an almost obsessive focus on the necessity of preparedness with the disarming recognition that anticipation and prevention of future contingencies is a logical impossibility. Within this optic, 'preparedness' demands a generic ability to adapt to unknowable contingencies rather than actual prevention or indeed adaptation to future events of known probability. As in the work of the later Hayek, the catastrophic event here becomes a sign not of the occasional failure to predict, prevent, and manage crisis but of the allegedly ontological limits to public management and the impossibly of state planning for social security. What was called for instead is a 'culture' of resilience that turns crisis response into a strategy of permanent, open-ended responsiveness, integrating emergency preparedness into the infrastructures of everyday life and the psychology of citizens. It is notable that the culture of preparedness envisaged by the Department of Homeland Security sees no end point to emergency. The strategy of resilience replaces the short-term relief effort—with its aim to restore the status quo through post-catastrophe reconstruction—with a call to permanent adaptability in and through crisis. What is resilience, after all, if not the acceptance of disequilibrium itself as a principle of organisation? The permanentisation of crisis response leads to another consequence—the blurring of the boundaries between crisis response, post-catastrophe reconstruction, and urban planning. With Friedman, the DHS lost no time in asserting that climate disasters such as Hurricane Katrina should be seized upon as opportunities for the selective transformation of urban space—a recommendation heeded all too well in the subsequent 'regeneration' of New Orleans, with its selective exclusion of the African-American poor.

Resilient Urbanism, Post-developmental Growth, and Ecological Security

If 'resilience thinking' has effectively pervaded the institutional logic and operational procedures of homeland security in the United States, the conflation of security, environmental disaster response, and critical infrastructure protection under the rubric of resilience should not be accounted as an aberration of US exceptionalism. There exists a real tension between the various factions pushing the policy agenda of resilience in its different aspects (environmentalists vs security interests) and between the precautionary and pre-emptive perspectives on resilience (the European Commission as opposed to the United States); it would be simplifying things to distribute these differences along national fault lines. Such divisions are complicated by the transnational networks of scientific and economic expertise informing institutions such as the Stockholm Resilience Centre, a significant node of contact between the academic world of environmental science and the policy-making world of international development organisations, where multilateral climate agreements and environmental conventions are forged.

Responding to the criticisms of social movements and NGOs, during the 1990s, institutions such as the World Bank, the IMF, and the United Nations rallied around the ethos of 'environmentally sustainable development'. The monolithic industrial modernisation projects of the post-WWII era, designed to replace subsistence agriculture with large export industries, were now discredited by their all too evident environmental costs. The World Bank began hiring environmental NGOs as project consultants and found a new role for itself in using its role as creditor to leverage debtor-state reforms supporting the internationalisation of conservation along neoliberal lines, 'developing' shifting cultivators into park rangers and eco-tour operators while recasting projects such as hydroelectric dams as supporting environmental sustainability. This has occurred in tandem with calls for the 'securitization of the biosphere': the privatisation and trading of the flow of 'ecosystem services' maintained by intact ecosystems, in recognition that rainforests and watersheds are critical 'natural infrastructure assets' that must to be priced in financial markets in order that corporations can 'capture the value' of biodiversity conservation.[37] As institutions have begun to recognise the intensifying socio-economic effects of global heating and ecosystem losses, we have seen a rapid uptake of the adaptive model of resource management offered by resilience science.

The Stockholm Resilience Centre, directly inspired by the work of Holling, serves as a mediator between the theorists of socio-ecological resilience and global development organisations. It performs the work of scaling up and standardising the principles of adaptive management for use in the field. Through this translational work, resilience science as a largely theoretical proposition has become fully operative as a methodology of micro and macro resource management. Yet the Stockholm Resilience Centre aspires to be much more than a platform for the strictly environmental uses of resilience science. Through its publications in journals, symposia, reports, consultancies, and collaborations with international institutions, it also shows ambition to furnish a general systems theory of 'socio-ecological governance' of direct use to policy-makers in the field of development economics.

The international development and environment projects now couched in the language of resilient urbanism are legion. The United Nations advocates the concept of an ecosystem-based approach to the management of urban environments. Operating on the principle that environmental management, urban planning, and infrastructure renewal must be pursued simultaneously, projects in urban regeneration and anticipatory reconstruction outlined by the UN and by the Rockefeller Foundation explicitly invoke the Hollingian principles of adaptive management and resilient infrastructure. A rural development programme co-sponsored by the United Nations Development Programme, United Nations Environment Programme, the World Bank, and the World Resources Institute is even more explicitly indebted to the principles of resilience science, outlining an ambitious model of post-developmental, post-industrial growth financed by payments to rural communities for the maintenance of ecosystem services. In a report entitled 'Roots of Resilience', the ideological project of socio-ecological resilience is summarised in the most succinct of terms:

[37] Chilchilnisky, G. & Heal, G. (2000). Securitizing the biosphere. In G. Chilchilnisky &. G. Heal (Eds.), *Environmental markets* (pp. 169–179). Columbia University Press.

> Resilience is the capacity to adapt and to thrive in the face of challenge. This report contends that when the poor successfully (and sustainably) scale-up ecosystem-based enterprises, their resilience can increase in three dimensions. They can become more economically resilient—better able to face economic risks. They—and their communities—can become more socially resilient—better able to work together for mutual benefit. And the ecosystems they live in can become more biologically resilient—more productive and stable.[38]

In the vision of a post-developmental future offered by these various projects, financial and ecological crises stand in a relationship of mutual determination. The resilient community is better able to weather its chronic exposure to global financial markets through the adoption of a localised, decentralised, post-carbon, ecosystems-based model of growth. Building up resilience to the intensifying climate crisis is not merely analogous to coping with recurrent financial shocks, it is also the means through which economic and social resilience is to be achieved. This is a tacit recognition that 'development' for the post-colonial poor no longer consists in achieving the first-world standards of urban affluence promised by W.W. Rostow's modernisation theory but in merely surviving, preferably on the land instead of the slums, the after-effects of industrial development and the financial conditions imposed by the Washington consensus.

There is a strong selective dimension to the contemporary doctrine of resilient growth, one that both reiterates and modifies the Darwinian law of natural selection. Relying as it does on the non-equilibrium dynamics of complex systems theory, what the resilience perspective demands is not so much progressive adaptation to a continually reinvented norm as permanent adaptability to extremes of turbulence. In this context, the appeal to ecological security is often invoked as a means of distinguishing those who are sufficiently resilient to survive as dignified participants in a globally integrated world from those who are not resilient enough. Thus Holling's later work, relayed by his associates in the Stockholm Resilience Centre, offers a classificatory schema of socio-economic adaptability in which various types of maladaptation can be distinguished. There are those societies that can be compared to depleted ecosystems, whose resilience has been so

[38] United Nations Development Programme, United Nations Environment Programme, World Bank, World Resources Institute (2008). *World resources 2008: roots of resilience—growing the wealth of the poor.* Washington, DC: World Resources Institute.

thoroughly eroded that there is no longer any scope for reorganisation.[39] But there are also those societies that have become so internally integrated that they are now too resistant to perturbation—unable to change in the face of shocks that can be as creative and generative as they are destructive. 'Rigidity traps' occur when 'maladaptive' regimes with 'large bureaucracies' inhibit the chaotic creativity of complex systems evolution. As in Hayek's political philosophy of a fully decentralised liberalism, the morality tale of resilient growth routinely invokes the totalitarian socialism of the Soviet Union as cautionary counter-example. Increasingly articulated within a discourse of 'ecological security', the threat represented by the non-resilient society is routinely linked to the fear of migration. The authors of *Roots of Resilience*, for example, offer the following ominous reflections on the failure to adapt to climate change:

> [...] in the coming decades, the rural poor will be tested as the impacts of climate change manifest. There are no cities in the developing world large enough or wealthy enough to absorb the migration of the poor who have no buffer against these dangers and can find no means to adapt. The political and social instability inherent in such potentially massive movements of people is of increasing concern to the international community. [...] The consequences of not acting may well test the depths of compassion.[40]

The consequences of this logic, of course, go well beyond the arena of strictly environmental politics if, as we have suggested, the dynamics of a stressed biosphere have been rendered indistinguishable from those of world markets in contemporary security policy. Whether we look to the politics of 'sustainable development', the regulation of global finance, or the organised obstruction of climate policy, resilience risks becoming the measure of one's fitness to survive in the turbulent order of things. The criteria of selection may well have shifted. Yet in the last instance, and for all its flexibility, the resilience perspective is no less rigorous in its selective function than Darwinian evolution.

[39] Holling (2001, p. 400).
[40] UNDP, UNEP, WB, WRI (2008: ix).

Resilience as Abandonment: The Neoliberal Response to Planetary Crisis

Complex systems theory, it should be remembered, grew out of libertarian, environmentalist, and often leftist critiques of the 'command and control' hubris of Cold War, first-order cybernetics. In this respect, the conceptual and political career of Holling's concept of 'resilience', developed as a reaction against the 'pathology' of top-down natural resource management, is exemplary.[41] If second-order (or complex) systems theory was advanced by those who opposed the falsely omniscient, commanding vision of the Cold War state, it would appear that the new epistemological realism was achieved by re-absorbing critique into the workings of systems theory itself. The point is underscored in no uncertain terms by Niklas Luhmann, an advocate of complex systems theory as a rigorous sociological method. The complex social system, he remarks, 'feeds upon deviations from normal reproduction'; that is, it thrives upon disruptions to its own state of equilibrium.[42] By metabolising critique into its internal dynamic, the complex adaptive system remains self-referential even when it encounters the most violent of shocks. It is for this reason, Luhmann concludes, that complex adaptive systems defy critique, forcing all would-be critics to inhabit the system they set out to challenge: 'The unity of the system is the self-reference of the system and its change will always require working within, not against the system.'[43]

This logic is exemplified in the clearest of terms by the evolution of Holling's theory of resilience. Originally positioned as an ecological critique of the destructive consequences of orthodox growth economics—in his classic 1973 article Holling defined resilience as essentially 'concerned with the probabilities of extinction'[44]—it has now moved to a position of subordinate collusion with an agenda of resource management which collapses ecological crisis into the 'creative' destruction of a truly Hayekian world order, one too complex and too far beyond equilibrium for any governing body to understand, predict, or regulate in the service of long-term social objectives, such as conserving the thermal and ecological sta-

[41] Holling, C.S. & Meffe, G. (1996). Command and control and the pathology of natural resource management. *Conservation Biology*, 10(2), 328–337.

[42] Luhmann, N. (1990). World society as a social system. In: *Essays on self-reference* (pp. 175–190). New York: Columbia University Press, p. 180.

[43] Luhmann (1990, p. 183).

[44] Holling (1973, p. 20).

bility of the *oikos* upon which all our life depends. In the process, resilience has ceased to operate as a critique emphasising the fragility of complex communities of life. It now functions to naturalise neoliberal strategies of rule and to normalise the events of planetary catastrophe.

This is a post-environmentalist 'economy of nature' which finds its clearest expression in the counter-intuitive claim of Peter Kareiva, senior ecologist at the corporate conservation foundation The Nature Conservancy, that 'Nature is so resilient that it can recover rapidly from even the most powerful human disturbances'.[45] Recall that until the 1990s, the IMF maintained that 'macroeconomics has nothing to do with the environment'.[46] In stark contrast to this early denialist position, and to Kareiva's equally denialist doctrine of the infinite resilience of ecological communities to destruction, an IMF paper released in 2019 reflected on 'a growing agreement between economists and scientists [that] the risk of catastrophic and irreversible disaster is rising, implying potentially infinite costs of unmitigated climate change, including, in the extreme, human extinction'.[47]

For decades, the neoliberal political machine has mounted a counter-science campaign that defines the 'post-truth' era, railing against any effective decarbonisation strategy in the name of the 'free market'. Due to the efforts of engineers, large-scale renewable energy systems combining solar photovoltaics, wind turbines, and batteries now yield clean, dispatchable electricity at a lower cost than maintaining existing coal, nuclear, and gas generation assets in operation, even taking into account the perennial subsidies, indemnities, and general immunity from pollution taxes enjoyed by the thermal power sector. As the renewables revolution gathers momentum with movements for climate justice, financial markets are slowly responding to price signals and market sentiment by withdrawing investment and insurance from these catastrophically risky 'stranded assets'. As the free market begins to abandon the fossil fuel sector, certain voices within the Atlas Network have reciprocated, calling liberalism into question and discovering the virtues of conservative nationalism and

[45] Kareiva, P., Lalasz, R., & Marvier, M. (2011). Conservation in the Anthropocene: beyond solitude and fragility. In: *Love your monsters: post-environmentalism and the Anthropocene*. Oakland: Breakthrough Institute.

[46] Daly, H. (1996). *Beyond growth: the economics of sustainable development*. Boston Beacon Press, p. 144.

[47] Krogstrup, S. & Oman, W. (2019, Sept 4). Macroeconomic and financial policies for climate change mitigation. IMF Working Paper No. 19/185.

nationalist industry policy—as Atlas chair Alejandro Chafuen (MPS) reports in a recent article.[48] This is especially apparent in the fossil fuel sectors of hydrocarbon-rich jurisdictions such as the US, Brazil, Canada, and Australia. States have always underwritten the risky profitability of private fossil fuel extraction and energy distribution through tax holidays, direct subsidies from public revenue, and national security interventions: now neoliberals and industry lobbies are calling for all manner of bailouts, even nationalisations of fossil infrastructure. Increasingly, such states appear as the local legislative arm of fossil capital, as fossil lobbyists occupy administrations, sweep away 'green tape', and designate coal, oil, and gas projects as 'critical infrastructures' vital to the national security of the homeland. Stripped of its cosmopolitan and internationalist garb, neoliberalism appears ever more clearly as what it perhaps has always been at root: an authoritarian project to manage surplus life in terms of private interests invested in the extractive industries.

An exemplary case of the contemporary uses of resilience for the 'management' of ecological risk can be found in the Great Barrier Reef Foundation (GBRF), a publicly funded business 'charity' promising to 'build the resilience' of Australia's coral reef ecosystems in the wake of the unprecedented mass bleaching events of 2016–2018. According to the Intergovernmental Panel on Climate Change (IPCC), coral reefs are likely to decline by 70–90% with global warming of 1.5°C, and more or less consigned to history with warming of 2°C.[49] The physical cause of coral reef destruction is of course well known. Mass bleaching events are the result of increasingly intense marine heatwaves—which kill the photosynthesising microbial symbionts at the basis of the coral reef food chain.[50] The oceans have accumulated more than 90% of the vast quantity of heat gained by the planetary surface as a result of the combustion of fossil fuels.[51] Marine ecosystems are also threatened by the acidification of the oceans through absorption of carbon dioxide from the atmosphere, which

[48] Chafuen, A. (2019, Aug 29). Business, the economy, and the new effort to enshrine nationalism. *Forbes*.

[49] IPCC (2018, Oct 28). Summary for policy makers of IPCC special report on global warming of 1.5°C.

[50] Veron, J., Hoegh-Guldberg, O., Lenton, T., et al (2009). The coral reef crisis: the critical importance of < 350 ppm CO_2. *Marine Pollution Bulletin*, 58(10), 1428–1436.

[51] Zanna, L., Khatiwala, S., Gregory, J., Ison, J. & Heimbach, P. (2019). Global reconstruction of historical ocean heat storage and transport. *Proceedings of the National Academy of Science,* 116(4), 1126–1131.

will increasingly prevent the formation of the calcium carbonates that comprise the skeletons of reef corals and other marine organisms, from shellfish to several classes of the photosynthesising phytoplankton which drive much of the global carbon cycle.

In response to public alarm at the coral reef crisis, in 2018 the Australian government of Liberal Prime Minister Malcolm Turnbull awarded an untendered grant of A$444 million to the GBRF, which was tasked with finding ways to 'boost reef resilience'. How, we might well ask, is this to be achieved? In its publicly funded public relations material—lavishly illustrated with images of healthy and abundant reefs, and not with dead acres of ghost-white corals—we are promised community engagement and education programmes, experimental studies in coral aquaculture and cryopreservation, and the development of wholly speculative geoengineering technologies. These include 'approaches to potentially decrease solar radiation on reefs (e.g. creating shade through clouds, mist, fog, or surface films)' and the prototyping of a 'Swiss army knife-style robot reef protector, the RangerBot Autonomous Underwater Vehicle'. Confusing marine ecosystem collapse with urban planning, the GBRF's Resilient Reefs Program promises to combine 'site-specific coral reef expertise […] with learnings from the proven resilience-building model of the [Rockefeller Foundation's] 100 Resilient Cities initiative'.[52] A huge coal and bauxite bulk transport vessel operated by Rio Tinto—described by the GBRF as a 'ship of opportunity'—has been fitted out with instruments to measure ocean acidification, as it sails past the reef on its ordinary business of contributing to ocean acidification. Whilst duly acknowledging the Paris Climate Treaty and the existential threats to coral reefs posed by global heating, at no point does the GBRF offer any support for the obvious conclusion, that the global collapse of marine ecosystem resilience can only be slowed, if at all, by a worldwide programme of rapid and permanent retrenchment of the fossil fuel sector.

Like chalk in acid, the apparent paradox dissolves when we learn that the board of the GBRF is composed of executives recruited from among the fossil fuel, mining, and other combustion-intensive corporations that, through business and industry associations, the Atlas-affiliated Institute for Public Affairs, a media dominated by Murdoch's News Corporation, and the governing Liberal-National Party, have effectively neutralised all

[52] Great Barrier Reef Foundation (2019). https://www.barrierreef.org/. Accessed 17 Dec 2019.

attempts to establish a national climate policy in Australia. Deservingly or otherwise, Australia in the past has been seen as a beacon of egalitarian democracy and international cooperation. Since the 1996 election of Liberal Prime Minister John Howard (inducted into the MPS in 2011), conservative governments have used Australia's vote in UN Framework Convention on Climate Change (UNFCCC) conferences to frustrate international climate agreements. This is a network of power which seems bent on constituting the coal and gas exporting equivalent of an authoritarian petro-state in a dry, hot continent increasingly vulnerable to the intensifying droughts and summer heatwaves of global heating, manifest in the wholly unprecedented bushfires of 2019–2020. At the time of writing, this unimaginable inferno has reduced a vast portion of the remnant eucalypt and rainforests of south-eastern Australia to ashes, a holocaust of livestock and perhaps a billion native animals, choking cities and towns with smoke-filled air and threatening their dwindling water supplies, burning thousands of houses, farms, and livestock, killing volunteer firefighters, and generating waves of internally displaced climate refugees, with hundreds of wildfires still burning out of control. Some seventeen million hectares are reported to have already burned, an area greater than that of England. The initial response of Liberal Prime Minister Scott Morrison to this unfolding climate emergency was to take a secret holiday in Hawaii, to deny that anything out of the ordinary was going on, and to extol the virtues of the national cricket team. When asked whether the bushfire crisis would prompt a rethink of his government's opposition to climate policy, Morrison replied: 'I think more significantly that resilience and adaptation need an even greater focus. [...] We must build our resilience for the future and that must be done on the science and the practical realities of the things we can do right here to make a difference.'[53] Fittingly, the chair of the GBRF board is a former chief executive of the Australian subsidiary of ExxonMobil—that global private empire synonymous not only with the Atlas Network's decades-long campaign of deception, science denial, and obstruction of climate policy but with the century-old neoliberal project to immunise 'the world economy' from democracy.

* * *

[53] Martin, S. (2020, Jan 14). Scott Morrison to focus on 'resilience and adaptation' to address climate change. *The Guardian*.

Such is one version of the tragic history by which the Earth's inestimably beautiful, diverse and regenerative capacity for intergenerational abundance was exposed to a future of more heat than life, the legacy of a *nomos* that refuses the *oikos*, of an eschatology of infinite combustion beyond all limits of life, law, knowledge, and care. A world deep in ruin: as C.S. Peirce foretold, such are the consequences of a creed given to 'the exaggeration of the beneficial aspects of greed [...]', of a philosophy in which 'greed is the great agent in the elevation of the human race and the evolution of the universe'.[54] The house, as the children born to the millennium see with terrifying moral clarity, is on fire. If there is anything for people of goodwill to strive together for, it is surely for a global Green New Deal capable of bringing the fossil-fuelled neoliberal era to a close.[55] It must of course be admitted that the window of possibility for avoiding the unthinkable fate that the Earth sciences warned us of a lifetime ago seems ever more narrow. Yet there is certainly time enough remaining to us for the truth to be told and—it must resolutely be hoped—for justice to be done.

[54] Peirce, C. S. (1893). Evolutionary love. *The Monist,* 3(2), 176–200.

[55] See e.g.: Klein, N. (2019). *On fire: the (burning) case for a Green New Deal.* Simon & Schuster.

References

Abel, T., & Stepp, J. (2003). A new ecosystems ecology for anthropology. *Ecology and Society, 7*(3). https://doi.org/10.5751/ES-00579-070312.

Adushkin, V., & Reith, W. (2001). *The containment of Soviet underground nuclear explosions.* Open File Report: US Geological survey.

Allchin, J. (1997). James Hutton and coal. *Cadernos IG/UNICAMP, 7*, 167–183.

Allee, W., Emerson, A., Park, O., & Schmidt, K. (1949). *Principles of animal ecology.* Philadelphia: W. B. Saunders.

Altman, J. (2009). Manwurrk (fire drive) at Namilewohwo. In J. Russell-Smith, P. Whitehead, & P. Cooke (Eds.), *Culture, ecology and economy of fire management in North Australian savannas: Rekindling the Wurrk tradition.* Canberra: CSIRO.

Andrews, J., et al. (1990). The vision thing: Conservatives take aim at the '90s. *Policy Review, 52*, 38–43.

Anker, P. (2001). *Imperial ecology: Environmental order in the British Empire, 1895–1945.* Cambridge, MA: Harvard University Press.

Aronova, E., Baker, K., & Oreskes, N. (2010). Big science and big data in biology: From the International Geophysical Year through the International Biological Program to the Long Term Ecological Research (LTER) Network, 1957–present. *Historical Studies in the Natural Sciences, 40*(2), 183–224.

Arrow, K., Bolin, B., Costanza, R., Dasgupta, P., Folke, C., Holling, C. S., Jansson, B., Levin, S., Mäler, K., Perrings, C., & Pimentel, D. (1995). Economic growth, carrying capacity, and the environment. *Ecological Economics, 15*(2), 91–95.

J. Walker, *More Heat than Life: The Tangled Roots of Ecology, Energy, and Economics*,
https://doi.org/10.1007/978-981-15-3936-7

Arthur, W. (1990). *The green machine: Ecology and the balance of nature.* Oxford: Basil Blackwell.
Atlas Network. (2019). Our partners: Global directory. Retrieved from https://www.atlasnetwork.org/partners/global-directory.
Ayache, E. (2010). *The blank swan: The end of probability.* Chichester: John Wiley & Sons.
Ayres, R. (1994). *Information, entropy and progress: A new evolutionary paradigm.* New York: American Institute of Physics.
Bambach, R. (2005). Ecology as an economy: Review of 'Nature: An economic history' by G. J Vermeij, 2004. *American Scientist, 93*(2), 185–187.
Banerjee, N. (2016, October 5). For Exxon, hybrid car technology was another road not taken. *Inside Climate News.*
Bank for International Settlements. (2008). *Basel Committee on banking supervision announces steps to strengthen the resilience of the banking system.* Geneva: BIS.
Barnett, H., & Morse, C. (1963). *Scarcity and growth.* Baltimore: John Hopkins Press.
Barrowman, J. (1897). *Slavery in the coal-mines of Scotland.* Paper presented to the Annual General Meeting of the Federated Institution of Mining Engineers.
Bataille, G. ([1949] 1997). The meaning of general economy; The laws of general economy. In F. Botting & S. Wilson (Eds.), *The Bataille reader.* Oxford: Blackwell.
Bateson, G. (1946). Physical thinking and social problems. *Science, 103,* 717–718.
Bateson, G. (1972). *Steps to an ecology of mind.* New York: Ballantine.
Baynes, C. (2019, May 15). Germany's far-right AfD takes aim at Greta Thunberg as it denies climate emergency. *The Independent.*
Beck, U. (1992). *The risk society: Towards a new modernity.* London: SAGE.
Beck, U. (1995). *Ecological politics in an age of risk.* Cambridge: Polity Press.
Beckerman, W. (1974). *In defence of economic growth.* London: Cape.
Beeman, R. (1995). 'A green and permanent land': Agriculture in the age of ecology, 1935–1985. PhD thesis, Iowa State University.
Begon, M., Harper, J., & Townsend, C. (1995). *Ecology.* Cambridge, MA: Blackwell.
Bell, D. ([1973] 1976). *The coming of post-industrial society.* New York: Basic Books.
Bell, D., Wiener, A., & Kahn, H. (1967). *The year 2000.* London: Macmillan.
Bendell, J. (2018, July 27). Deep adaptation: A map for navigating climate tragedy. IFLAS Occasional Paper 2.
Bigelow, G. (2003). *Fiction, famine, and the rise of economics in Victorian Britain and Ireland.* Cambridge University Press.
Bigelow, G. (2005). Let there be markets. *Harper's Magazine,* 310(1860).
Blake, W. (2008). *The complete poetry and prose of William Blake.* Berkeley: University of California Press.

Blaug, M. (1980). *The methodology of economics.* Cambridge, UK: Cambridge University Press.

Blumenberg, H. (1983). *The legitimacy of the modern age.* Cambridge: MIT Press.

Bobulescu, R. (2015). From Lotka's biophysics to Georgescu-Roegen's bioeconomics. *Ecological Economics, 120,* 194–202.

Bocking, S. (1997). Ecologists, ecosystems and the atom: Environmental research at the Oak Ridge National Laboratory. *Journal of the History of Biology, 28,* 1–47.

Boltzmann, B. ([1886] 1974). The second law of thermodynamics. In B. McGuiness (Ed.), *Theoretical physics and philosophical problems.* Dordrecht: Reidel.

Borges, J. (1964). The fearful sphere of Pascal. In *Labyrinths: Selected stories & other writings.* New Directions.

Bowler, P. (1993). *Norton history of the environmental sciences.* W.W. Norton.

Bowman, D., et al. (2016). Pyrodiversity is the coupling of biodiversity and fire regimes in food webs. *Philosophical Transactions of the Royal Society B, 371,* 20150169.

Boyd, R. (1993). Metaphor and theory change: What is metaphor for? In A. Ortony (Ed.), *Metaphor and thought* (pp. 356–408). Cambridge: Cambridge University Press.

Brannen, P. (2019, August 3). The Anthropocene is a joke. *The Atlantic.*

Braun-Blanquet, J. (1928). *Plant sociology* (G. Fuller & H. Conrad, Trans.). New York: McGraw Hill.

Brown, W. (2015). *Undoing the demos: Neoliberalism's stealth revolution.* MIT Press.

Browne, J. (1995). *Charles Darwin. Voyaging: Vol. 1 of a biography.* London: Pimlico.

Brulle, R. (2014). Institutionalising delay: Foundation funding and the creation of US climate change counter-movements. *Climatic Change, 122,* 681–694.

Bryant, B. (2000). *Nature and culture in the age of cybernetic systems.* Paper delivered to the American Studies Association Annual Conference, Detroit.

Buck-Morss, S. (1995). Envisioning capital: Political economy on display. In L. Cooke & P. Wollen (Eds.), *Visual display: Culture beyond appearances.* Seattle: Bay Press.

Budyko, M. (1986). *The evolution of the biosphere.* Dordrecht: Reidel.

Bull, M. (1999). *Seeing things hidden: Apocalypse, vision and totality.* London: Verso.

Bunker, S., & Ciccantell, P. (2005). *Globalization and the race for resources.* Baltimore: John Hopkins University Press.

Business Roundtable. (1972). The Business Roundtable: Its purpose and program. Retrieved from http://www.documentcloud.org/documents/3903345-Business-Roundtable-1972-Its-Purpose-and-Program.html

Caldwell, B. (2004). *Hayek's challenge: An intellectual biography of F.A. Hayek.* Chicago: University of Chicago Press.

Callahan, G. (2007, October 1). *How a free society could solve global warming*. New York: Foundation for Economic Education.
Callebaut, W. (1992). *Taking the naturalistic turn*. Chicago: University of Chicago Press.
Callendar, G. (1938). The artificial production of carbon dioxide and its influence on climate. *Quarterly Journal of Research in Meteorology, 64*, 223–240.
Callendar, G. (1940). Variations of carbon dioxide in various air currents. *Quarterly Journal of Research in Meteorology, 66*, 395–400.
Carnot, N. L. S. ([1825] 1897). *Reflections on the motive power of heat*. New York: Wiley & Sons.
Centre for Economic Liberty. (2013, August 3). Nassau Senior, praxeology, and John Stuart Mill. Retrieved from https://centerforeconomicliberty.blogspot.com/2013/08/nassau-senior-praxeology-and-john.html.
Chafuen, A. (2019, August 29). Business, the economy, and the new effort to enshrine nationalism. *Forbes*.
Chang, H. J. (2002). *Kicking away the ladder: Development strategy in historical perspective*. Anthem Press.
Chapman, C. (1967). Edward Smith (1818–1874) physiologist, human ecologist, reformer. *Journal of the History of Medicine and Allied Sciences, 22*(1), 1–26.
Chilchilnisky, G., & Heal, G. (2000). Securitizing the biosphere. In G. Chilchilnisky & G. Heal (Eds.), *Environmental markets* (pp. 169–179). Columbia University Press.
Christensen, P. (1994). Fire, motion, and productivity: The proto-energetics of nature and economy in François Quesnay. In P. Mirowski (Ed.), *Natural images in economic thought* (pp. 249–288). Cambridge University Press.
Cirillo, R. (1984). Léon Walras and social justice. *American Journal of Economics and Sociology, 43*(1), 53–60.
Clark, N. (2012). Rock, life, fire: Speculative geophysics and the anthropocene. *Oxford Literary Review, 34*(2), 259–276.
Clark, N. (2015). Fiery arts: Pyrotechnology and the political aesthetics of the Anthropocene. *GeoHumanities, 1*(2), 266–284.
Clark, N. (2017). Pyro-Gaia: Planetary fire as force and signification. *Ctrl-Z: New Media Philosophy*, 7.
Clark, N., & Yusoff, K. (2014). Combustion and society: A fire-centred history of energy use. *Theory, Culture & Society, 31*(5), 203–226.
Clarke, A. (2017). *Principles of thermal ecology: Temperature, energy and life*. Oxford: Oxford University Press.
Clausius, R. (1867). *The mechanical theory of heat*. London: van Voorst.
Clements, F. (1905). *Research methods in ecology*. Lincoln, Nebraska: University Publishing Co..
Clements, F. (1916). *Plant succession: An analysis of the development of vegetation*. Washington: Carnegie Institute.

Clements, F. (1925). Phylogeny and classification of the climax. *Yearbook of the Carnegie Institution, 24*, 334–335.
Clements, F. (1935). Experimental ecology in the public service. *Ecology, 16*, 342–343.
Clements, F. & Chaney, R. (1936). *Environment and life in the Great Plains.* Carnegie Institute: Supplementary publication. no. 24, p. 33.
Cobden, R. (1846, January 15). Free trade with all nations. Speech given to the Anti-Corn Law League at Manchester.
Cohen, I. (1994). An analysis of interactions between the natural sciences and the social sciences. In I. Cohen (Ed.), *The natural sciences and the social sciences* (pp. 1–99). Dordrecht: Kluwer Academic.
Cohen, J. (1971). Mathematics as metaphor: A book review of 'Dynamical system theory in biology' by R. Rosen. *Science, 172*, 674–675.
Coleman, D. (2010). *Big ecology: The emergence of ecosystem science.* Berkeley: University of California Press.
Coll, S. (2012). *Private empire: ExxonMobil and American power.* Penguin.
Connolly, S. (1992). *Religion, law, and power: The making of Protestant Ireland 1660–1760.* Oxford: Clarendon Press.
Contoski, E. (2016, December 29). The sun, not CO_2, determines our climate. *The Heartland Institute.*
Cooper, M. (2008). *Life as surplus: biotechnology and capitalism in the neoliberal era.* Seattle: University of Washington Press.
Cooper, M. (2017). *Family values: Between neoliberalism and the new social conservatism.* New York: Zone Books.
Cooper, R. (1975). Resource needs revisited. *Brookings Papers on Economic Activity, 1*, 238–245.
Coronil, F. (1997). *The magic of the state: Nature, money and modernity in Venezuela.* Chicago: University of Chicago Press.
Coser, L. (1977). *Masters of sociological thought.* New York: Harcourt Brace.
Craige, B.-J. (2002). *Eugene Odum: Ecosystem ecologist & environmentalist.* Athens: University of Georgia Press.
Crosby, A. (1986). *Ecological imperialism: The biological expansion of Europe, 900–1900.* New York: Cambridge University Press.
Crutzen, P. (2002). Geology of mankind. *Nature, 415*(6867), 23–23.
Cuddington, K. (2001). The 'balance of nature' metaphor and equilibrium in population biology. *Biology and Philosophy, 16*, 464.
Daly, H. (1996). *Beyond growth: The economics of sustainable development.* Boston Beacon Press.
Daly, H. (1999). *Ecological economics and the ecology of economics: Essays in criticism.* Cheltenham UK: Edward Elgar.

Darwin, C., & Wallace, A. R. (1858). On the tendency of species to form varieties; and on the perpetuation of varieties and species by natural means of selection. *Journal of the proceedings of the Linnean Society of London, 3*(9), 45–62.

Daschuk, J. (2013). *Clearing the plains: Disease, politics of starvation, and the loss of Aboriginal life*. Regina, Saskatchewan: University of Regina Press.

Dasgupta, P., & Heal, G. (1974). The optimal depletion of exhaustible resources. *Review of Economic Studies, 41*(128), 3–26.

Davies, R. (2008). *The Darwin conspiracy: Origins of a scientific crime*. London: Golden Square.

Davis, M. (2001). *Late Victorian holocausts: El Nino famines and the making of the third world*. London: Verso.

de Laplante, K. (2005). Is ecosystem management a postmodern science? In K. Cuddington & B. Beisner (Eds.), *Ecological paradigms lost: Routes of theory change* (pp. 398–416). Amsterdam: Elsevier.

Department of Homeland Security. (2007, October). The national strategy for homeland Security. Retrieved from www.dhs.gov/xabout/history/gc_1193938363680.shtm.

Diesel, R. (1893). *Theory and construction of a rational heat-engine to replace the steam engine and combustion engines known today*. Berlin: Springer-Verlag.

Dobbs, B. (1983). *The foundations of Newton's alchemy* (pp. 6–9). Cambridge University Press.

Douglas, M. (1970). *Purity and danger: An analysis of concepts of pollution and taboo*. New York: Routledge.

Douglas, M. (1973). *Natural symbols* (p. 93). London: Penguin.

Douglas, M., & Wildavsky, A. (1983). *Risk and culture: An essay on the selection of technological and environmental dangers*. University of California Press.

Durkheim, E. (2001). *The elementary forms of religious life*. London: Oxford World Classics.

Durkheim, E., & Mauss, M. ([1903] 1963). *Primitive classification*. Chicago: University of Chicago Press.

Eco, U., Gould, S. J., Carriere, J.-C., & Delumeau, J.-C. (1999). *Conversations about the end of time*. Penguin.

Eddington, A. (1920). The internal constitution of the stars. *Nature, 106*, 14–20.

Eddington, A. (1928). *The nature of the physical world*. Cambridge: Cambridge University Press.

Edwards, G. (1978, June). A metaphorical framework: The ecosystem versus the machine. Part 1: CCNR Final Submission to the Ontario Royal Commission on Electric Power Planning.

Egerton, F. (1973). Changing concepts of the balance of nature. *Quarterly Review of Biology, 48*, 323–350.

Egerton, F. (2012). *Roots of ecology: Antiquity to Haeckel*. Berkeley: University of California Press.

Egerton, F. (2019). History of ecological sciences. Part 63: Biosphere ecology. *Bulletin of the Ecological Society of America, 00*(0), e01568.
Elton, C. (1927). *Animal ecology*. New York: Macmillan.
Fama, E. (1970). Efficient capital markets: A review of theory and empirical work. *Journal of Finance, 25*(2), 383–417.
Fang, L. (2017, August 9). Sphere of influence: How American libertarians are remaking Latin American politics. *The Intercept*.
Farrant, A. (2019). What should Knightian economists do? James M. Buchanan's 1980 visit to Chile. *Southern Economic Journal, 85*(3), 691–714.
Ferro, M. (1997). *Colonization: A global history*. London: Routledge.
Feynman, R., Leighton, R., & Sands, M. (1989). *The Feynman lectures on physics*. Addison Wesley.
Fisher, D. (1980). American philanthropy and the social sciences in Britain, 1919–1939. *Sociological Review, 28*(2), 277–315.
Fleck, L. ([1935] 2012). *Genesis and development of a scientific fact*. Chicago: University of Chicago Press.
Fonseca, G. (2006). Neoclassical optimal growth: Golden rules and the Ramsey exercise. History of Economic Thought Website. Retrieved from http://cepa.newschool.edu/het/essays/growth/optimal.htm.
Foucault, M. (1970). *The order of things: An archaeology of the human sciences*. London: Pantheon.
Frank, A. (2009). The petroleum war of 1910: Standard Oil, Austria, and the limits of the multinational corporation. *American Historical Review, 114*(1), 16–41.
Frank, L. (1948). Foreword. *Annual of the New York Academy of Sciences, 50*, 192–195.
Franta, B. (2018, January 1). On its 100th birthday in 1959, Edward Teller warned the oil industry about global warming. *The Guardian*.
Franta, B. (2018, September 19). Shell and Exxon's secret 1980s climate change warnings. *The Guardian*.
Franta, B. (2018). Early oil industry knowledge of CO_2 and global warming. *Nature Climate Change, 8*(12), 1024–1025.
Freese, B. (2003). *Coal: A human history*. Cambridge, MA: Perseus Press.
Friedman, M. (1951). Neoliberalism and its prospects. *Formand, 17*, 89–93.
Friedman, M. (1953). *Essays in positive economics*. Chicago: University of Chicago Press.
Friedman, M. (1970, September 13). The social responsibility of business is to increase profits. *New York Times Magazine*.
Fukuyama, F. (1992). *The end of history and the last man*. Avon Books.
Gaddis, J. L. (2005). *The Cold War*. London: Allen Lane.
Graeber, D. (2012). *Debt: The first 5000 years*. London: Penguin.

Gaffney. M. (1994). Neo-classical economics as a stratagem against Henry George. In M. Gaffney, & F. Harrison (Eds.), *The corruption of economics* (pp. 29–164). London: Shepheard-Walwyn, p. 89.

Galbraith, J. K. (1972). *Power and the useful economist.* Presidential address to the American Economic Association.

Galvez, M., & Gaillardet, J. (2012). Historical constraints on the origins of the carbon cycle concept. *Comptes Rendu Geoscience, 344*, 549–567.

Gammage, B. (2011). *The biggest estate on Earth: How Aborigines made Australia.* Sydney: Allen & Unwin.

Garrett, K. (2006, January15). Interview with J. K. Galbraith. Background briefing. Sydney: Australian Broadcasting Corporation.

Gentner, D., & Jeziorski, M. (1993). The shift from metaphor to analogy in Western science. In A. Ortony (Ed.), *Metaphor and thought* (pp. 447–480). Cambridge: Cambridge University Press.

Georgescu-Roegen, N. (1971). *The entropy law and the economic process.* Cambridge, MA: Harvard University Press.

Georgescu-Roegen, N. (1975). Energy and economic myths. *Southern Economic Journal, 41*(3), 374–381.

Georgescu-Roegen, N. ([1970] 1993). The entropy law and the economic problem. In H. Daly & K. Townsend (Eds.), *Valuing the earth: Economics, ecology, ethics* (pp. 75–88). Cambridge, MA: MIT Press.

Gill, S. (2002). Constitutionalizing inequality and the clash of globalizations. *International Studies Review, 4*(2), 47–65.

Gleick, J. (1987). *Chaos.* London: Abacus.

Goeller, H., & Weinberg, A. (1978). The age of substitutability. *American Economic Review, 68*(6), 1–11.

Golley, F. (1993). *A history of the ecosystem concept in ecology.* New Haven: Yale University Press.

Gonce, R. (1973). Natural law and Ludwig von Mises' praxeology and economic science. *Southern Economic Journal, 39*(4), 490–507.

Goodchild, P. (2005). *Edward Teller: The real Dr. Strangelove.* Harvard University Press.

Gould, S. J. (1988). Kropotkin was no crackpot. *Natural History, 7*(97), 12–21.

Gowdy, J., Dollimore, D., Wilson, D., & Witt, U. (2013). Economic cosmology and the evolutionary challenge. *Journal of Economic Behavior & Organization, 90*, S11–S20.

Gowdy, J., & Krall, L. (2013). The ultrasocial origin of the Anthropocene. *Ecological Economics, 95*, 137–147.

Gowdy, J., & Messner, S. (1998). The evolution of Georgescu-Roegen's bioeconomics. *Review of Social Economy, 56*(2), 136–156.

Granjou, C., & Walker, J. (2016). Promises that matter: Reconfiguring ecology in the ecotrons. *Science & Technology Studies, 29*(3), 49–67.

Granjou, C., Walker, J., & Salazar, J. (2017). The politics of anticipation: On knowing and governing environmental futures. *Futures, 92*, 5–11.

Grattan-Guiness, I. (1984). Work for the workers. *Annals of Science, 41*, 1–33.

Gray, C. (1995). *The cyborg handbook*. New York: Routledge.

Gray, P. (1999). 'Shovelling out your paupers': The British state and Irish famine migration 1846–50. *Patterns of Prejudice, 33*(4), 47–65.

Great Barrier Reef Foundation. (2019). Retrieved from https://www.barrierreef.org/.

Greenspan, A. (2005, March 10). Globalization. Speech to the Council on Foreign Relations, New York. Landsfeld, S. (2000). GDP: One of the great inventions of the 20th Century. *2000 Survey of Current Business*, U.S. Bureau of Economic Analysis.

Grinevald, J. (1992). De Carnot à Gaïa: Histoire de l'effet de serre. *La Recherche, 23*(243), 532–538.

Grinevald, J. (1998). Introduction: The invisibility of the Vernadskyan revolution. In V. Vernadsky (Ed.), *The biosphere* (pp. 20–32). New York: Copernicus.

Gunderson, L., & Holling, C. S. (Eds.). (2002). *Panarchy: Understanding transformations in human and natural systems*. Washington, DC: Island Press.

Haberler, G. (1974). *Economic growth and stability*. Los Angeles: Nash.

Haeckel, E. (1866). *Generelle Morphologie der Organismen. Allgemeine Grundzüge der organischen Formen-Wissenschaft, mechanisch begründet durch die von C. Darwin reformirte Descendenz-Theorie, etc.* (vol. 2). Berlin: G. Reimer.

Haeckel, E. (1899). *Die Welträtsel*. Stuttgart.

Haldane, A. (2009). Rethinking the financial network. Speech delivered at the Financial Students Association, Amsterdam, Bank of England.

Hardin, G. (1990, November 10). Perpetual growth: The next dragon facing biology teachers. Address to the National Association of Biology Teachers, Houston, Texas.

Hayek, F. (1931). *Prices and production*. London: Routledge & Kegan Paul.

Hayek, F. (1937). Economics and knowledge. *Economica, 4*(13), 33–54.

Hayek, F. (1945). The use of knowledge in society. *American Economic Review, 4*(35), 519–530.

Hayek, F. (1955). *The counter-revolution of science: Studies on the abuse of reason*. New York: Free Press.

Hayek, F. (1967). The theory of complex phenomena. In *Studies in philosophy, politics and economics* (pp. 22–42). London: Routledge & Kegan Paul.

Hayek, F. (1974, December 11). The pretence of knowledge. Acceptance speech upon the award of the Sveriges Riksbank Prize in Economics in Memory of Alfred Nobel. Salzburg.

Hayek, F. (1978). The errors of constructivism. In *New studies in philosophy, politics, economics and the history of ideas* (pp. 3–22). Chicago: University of Chicago Press.

Hayek, F. (1981, January 27). The flow of goods and services. Address to the London School of Economics, published in German: (1984) Der strom und der güter und leistungen. Tübingen: Mohr Siebeck.

Hayek, F. (1988). *The fatal conceit: The errors of socialism—collected works of F.A. Hayek, vol. 1.* London: Routledge.

Hayles, K. (1999). *How we became posthuman.* Chicago: University of Chicago Press.

Hegel, G. W. ([1837] 1980). *Lectures on the philosophy of world history.* Cambridge University Press.

Heilemann, J. (1997, January 7). The integrationists vs. the seperatists. *Wired.*

Hesse, M. (1955). *Science and the human imagination.* New York: Philosophical Library.

Hesse, M. (1972). The explanatory function of metaphor. In Y. Bar-Hillel (Ed.), *Logic, methodology, and philosophy of science* (pp. 249–259). Amsterdam: Elsevier.

Hetherington, N. (1983). Isaac Newton's influence on Adam Smith's natural laws in economics. *Journal of the History of Ideas, 44*(3), 497–505.

Ho, M. W. (1999). Are economic systems like organisms? In P. Koslowski P. (Ed.), *Sociobiology and bioeconomics* (pp. 237–258). Berlin: Springer.

Hodgson, G. (1993). *Economics and evolution: Bringing life into economics.* Cambridge: Polity Press.

Hodgson, G. (1994). Hayek, evolution and spontaneous order. In P. Mirowski (Ed.), *Natural images in economic thought: Markets read in tooth and claw* (pp. 409–447). Cambridge: Cambridge University Press.

Hoffman, D. (2006). Structural logos in Heraclitus and the Sophists. *Advances in the History of Rhetoric, 9*(1), 1–32.

Holleman, H. (2017). De-naturalizing ecological disaster: Colonialism, racism and the global Dust Bowl of the 1930s. *Journal of Peasant Studies, 44*(1), 234–260.

Holling, C. S. (1973). Resilience and stability of ecological systems. *Annual Review of Ecology and Systematics, 4*, 1–23.

Holling, C. S. (2001). Understanding the complexity of economic, ecological and social systems. *Ecosystems, 4*, 390–405.

Holling, C. S., & Meffe, G. (1996). Command and control and the pathology of natural resource management. *Conservation Biology, 10*(2), 328–337.

Holt, C., & Roth, A. (2004). The Nash equilibrium: A perspective. *Proceedings of the National Academy of Science, 101*(12), 3999–4002.

Hornborg, A. (2001). *The power of the machine: Global inequalities of economy, technology and environment.* California: Altamira Press.

Hotelling, H. (1931). The economics of exhaustible resources. *Journal of Political Economy, 39*(2), 137–175.

Hutchinson, G. E. (1940). Review: Bio-Ecology. *Ecology, 21*, 267–268.

Hutchinson, G. E. (1946). Social theory and social engineering. *Science, 104*(2694), 166–167.

Hutchinson, G. E. (1948). Circular causal systems in ecology. *Annual of the New York Academy of Sciences, 50*, 221–246.

Hutchinson, G. E. (1970). The biosphere. *Scientific American, 223*(3), 45–53.

Hutchinson, G. E. (1978). *Introduction to population ecology*. New Haven: Yale University Press.

Hutton, J. (1792). A chemical dissertation concerning phlogiston, of the principle of fire: Part 2. In *Dissertations on different subjects in natural philosophy*. Edinburgh.

Hutton, J. (1794). *Dissertation on the philosophy of light, heat, and fire*. Edinburgh.

Hutton, J. (1795). *Theory of the earth* (vols. 1 & 2). London: Cadell, Junior and Davies, & Edinburgh: William Creech.

Ingerson, A. (2002). *A critical users' guide to 'ecosystem' and related concepts in ecology*. Institute for Cultural Landscape Studies, Arnold Arboretum, Harvard University.

Ingrao, B., & Israel, G. (1990). *The invisible hand: Economic equilibrium in the history of science*. Cambridge, MA: MIT Press.

IPCC. (2018, October 28). Summary for policy makers of IPCC special report on global warming of 1.5°C.

Jackson, T. (2003). Sustainability and the 'struggle for existence': The critical role of metaphor in society's metabolism. *Environmental Values, 12*(3), 289–316.

James, L. (1998). *Raj: The making and unmaking of British India*. London: Little, Brown & Co..

Jevons, W. (1866). Brief account of a general mathematical theory of political economy. *Journal of the Royal Statistical Society, 29*, 282–287.

Jevons, W. S. (1871). *Theory of political economy*. London: Macmillan.

Jevons, W. S. (1876). The periodicity of commercial crises, and its physical explanation. *Journal of the Statistical and Social Inquiry Society of Ireland, 7*, 334.

Jevons, W. S. (1905). *The principles of economics: A fragment of a treatise on the industrial mechanism of society and other papers*. London: Macmillan.

Jones, R. (1969). Firestick farming. *Australian Natural History, 16*, 224–231.

Kaiser, D., & Wasserman, L. (2016, December 8). The Rockefeller Family Fund vs. Exxon. *New York Review of Books*.

Kambhu, J., Weidman, S., & Krishnan, N. (2007). New directions for understanding systemic risk: A report on a conference co-sponsored by the Federal Reserve Bank of New York and the National Academy of Sciences. National Academies Press, Washington, DC.

Kareiva, P., Lalasz, R., & Marvier, M. (2011). Conservation in the Anthropocene: Beyond solitude and fragility. In *Love your monsters: Post-environmentalism and the Anthropocene*. Oakland: Breakthrough Institute.
Karier, T. (2010). *Intellectual capital: Forty years of the Nobel Prize in economics.* Cambridge University Press.
Keynes, J. M. (1947). Newton, the man. In *Newton, Tercentenary Celebrations.* London: Royal Society.
Kingsland, S. (1994). Economics and evolution: Alfred Lotka's economy of nature. In P. Mirowski (Ed.), *Natural images in economic thought: Markets read in tooth and claw* (pp. 231–246). Cambridge, UK: Cambridge University Press.
Kingsland, S. (2005). *The evolution of American ecology 1890–2000.* Baltimore: John Hopkins University Press.
Klamer, A., & Leonard, T. (1994). So what's an economic metaphor? In P. Mirowski (Ed.), *Natural images in economic thought* (pp. 20–51). Cambridge: Cambridge University Press.
Klein, N. (2019). *On fire: The (burning) case for a Green New Deal.* Simon & Schuster.
Knapp, A., et al. (1999). The keystone role of bison in North American tallgrass prairie. *BioScience, 49*(1), 39–50.
Kosolopav, V. (1976). *Mankind and the year 2000.* Moscow: Progress.
Krogstrup, S., & Oman, W. (2019, September 4). *Macroeconomic and financial policies for climate change mitigation.* IMF Working Paper No. 19/185.
Kuhn, T. (1962). *The structure of scientific revolutions.* Chicago: University of Chicago Press.
Kümmel, R. (2011). *The second law of economics: Energy, entropy, and the origins of wealth.* Springer Science & Business Media.
Kuznets, S. (1948). National income: A new version. *Review of Economics and Statistics, 30*, 151–179.
Kwa, C. (1993). Modelling technologies of control. *Science as Culture, 4*(3), 363–391.
Kwa, C. (1993). Modelling the grassland. *Historical Studies in the Physical and Biological Sciences, 24*(1), 125–155.
Kwa, C. (2001). The rise and fall of weather modification. In C. Miller & P. Edwards (Eds.), *Changing the atmosphere: Expert knowledge and environmental governance* (pp. 135–165). Cambridge: MIT Press.
Lakoff, G., & Johnson, M. (1987). *Metaphors we live by.* Chicago: University of Chicago Press.
Lane, R. (2014). Resources For the Future, resources for growth: The making of the 1975 growth ban. In B. Stephan & R. Lane (Eds.), *The politics of carbon markets.* London: Routledge.
Lane, R. (2019). The American Anthropocene: Economic scarcity and growth during the great acceleration. *Geoforum, 99*, 11–21.

Lange, O. (1938). *On the economic theory of socialism.* Minneapolis: University of Minnesota Press.

Lange, O. (1949). The practice of economic planning and the optimum allocation of resources. *Econometrica, 17,* 166–171.

Lavoisier, A. ([1789] 1965). *Elements of chemistry.* Courier Corporation.

Lazard. (2019, November). Lazard's levelized cost of energy analysis, version 13.0. Retrieved from https://www.lazard.com/media/451086/lazards-levelized-cost-of-energy-version-130-vf.pdf.

Lecuyer, B.-P. (1994). Probabilistic thinking. In I. Cohen (Ed.), *The natural sciences and the social sciences: Some critical and historical perspectives* (pp. 135–152). Dordrecht: Kluwer Academic.

Lehninger, A., Nelson, D., & Cox, M. (2005). *Lehninger principles of biochemistry.* Macmillan.

Lenin, V. (1913, March 13). A 'scientific' system of sweating. *Pravda,* 60.

Lenin, V. (1914, March 13). The Taylor system—man's enslavement by the machine. *Put Pravdy,* 35.

Lévêque, C. (2003). *Ecology: From ecosystem to biosphere.* New Hampshire: Science Publishers.

Lewontin, R. (1969). The meaning of stability. In *Diversity and stability of ecological systems* (vol. 22, pp. 13–24). Brookhaven, NY: Brookhaven symposia in biology.

Liebig, J. (1852). *Agricultural chemistry.* Philadelphia: TB Peterson.

Lindeman, R. (1942). The trophic-dynamic aspect of ecology. *Ecology, 23*(4), 399–417.

Lloyd, D., Wimpenny, J., & Venables, A. (2010). Alfred Russel Wallace deserves better. *Journal of Bioscience, 35*(3), 339–349.

Lopez, B. (2006). Critical infrastructure protection in the United States since 1993. In P. Auerswald et al. (Eds.), *Seeds of disaster, roots of response: How private action can reduce public vulnerability* (pp. 37–50). Cambridge: Cambridge University Press.

Lorenz, E. (1963). Deterministic non-periodic flow. *Journal of the Atmospheric Sciences, 20,* 130–141.

Lotka, A. (1914). An objective standard of value derived from the principle of evolution. *Journal of the Washington Academy of Science, 4,* 409–418.

Lotka, A. (1925). *Elements of physical biology.* Baltimore: William & Wilkins.

Lotka, A. J. (1922). Contribution to the Energetics of Evolution. *Proceedings of the National Academy of Sciences 8*(6), 147–151.

Lotka, A. J. (1922). Natural Selection as a Physical Principle. *Proceedings of the National Academy of Sciences 8*(6), 151–154.

Lovelock, J. (1979). *Gaia: A new look at life on Earth.* London: Oxford University Press.

Lovelock, J., & Margulis, L. (1974). Atmospheric homeostasis by and for the biosphere: The Gaia hypothesis. *Tellus, 26*(1–2), 2–10.

Lovins, A., & Lovins, L. (1981). *Energy policies for resilience and national security*. San Francisco: Friends of the Earth.
Luhmann, N. (1990). World society as a social system. In *Essays on self-reference* (pp. 175–190). New York: Columbia University Press.
MacKenzie, D. (1996). *Knowing machines: Essays on technical change*. Cambridge, MA: MIT Press.
MacLean, N. (2017). *Democracy in chains: The deep history of the radical right's stealth plan for America*. Penguin.
Magalhãesa, N., Fressoz, J.-B., Jarrige, F., Le Roux, T., Levillain, G., Lyautey, M., & Bonneuil, C. (2019). The physical economy of France (1830–2015): The history of a parasite? *Ecological Economics, 157*, 291–300.
Malm, A. (2016). *Fossil capital: The rise of steam power and the roots of global warming*. London: Verso.
Malthus, T. ([1798] 1970). *An essay on the principle of population*. Penguin: Hammondsworth.
Manssen, B., & McGlade, J. (1993). Ecology, thermodynamics and H.T. Odum's conjectures. *Oecologia, 93*(4), 582–596.
Margulis, L., & Lovelock, J. E. (1974). Biological modulation of the Earth's atmosphere. *Icarus, 21*(4), 471–489.
Margulis, L., & Sagan, D. (1997). *Microcosmos: Four billion years of microbial evolution*. Berkeley: University of California Press.
Marsh, G. P. ([1874] 1965). *Man and nature: The Earth as modified by human action*. Cambridge, MA: Harvard University Press.
Marshall, A. (1890). *Principles of economics*. Book 1, Ch. 4. Retrieved from https://www.econlib.org/library/Marshall/marP.html.
Marx, K. (1861–1863). The Physiocrats. *Economic manuscripts: Theories of surplus value*. Retrieved from www.marxists.org.
Marx, K. ([1908] 1972). Notes on machines. *Economy & Society*, 1(3), 244–254.
Marx, K., & Engels, F. (1957–1968). *Werke* (Vol. 31). Berlin: Dietz.
Marx, K. & Engels, F. (1996). *Collected works: Karl Marx: Capital* (vol. 2). International Publishers.
Masutti, C. (2006). Frederic Clements, climatology, and conservation in the 1930s. *Historical Studies in the Physical and Biological Sciences, 37*(1), 27–48.
May, R. (1973). *Complexity and stability in model ecosystems*. Princeton, NJ: Princeton University Press.
May, R., Simon Levin, S., & Sugihara, G. (2008). Complex systems: Ecology for bankers. *Nature, 451*, 893–895.
Mayor, A. (2008). Suppression of Indigenous fossil knowledge. In R. Proctor & L. Schiebinger (Eds.), *Agnotology: The making and unmaking of ignorance* (pp. 163–182). Stanford CA: Stanford University Press.
Mayr, O. (1971). Maxwell and the origins of cybernetics. *Isis, 62*(4), 425–444.

Mayr, O. (1986). *Authority, liberty and automatic machinery in early modern Europe*. Baltimore: John Hopkins University Press.

McClare, C. (1971). Chemical machines, Maxwell's demon and living organisms. *Journal of Theoretical Biology, 30*, 1–34.

McCloskey, D. (1983). The rhetoric of economics. *Journal of Economic Literature, 31*, 434–461.

McCulloch, W. (1948). A recapitulation of theory with a forecast of several extensions. *Annual of the New York Academy of Science, 50*(4), 259–277.

McGee, J. (1956). Price discrimination and predatory effects: The Standard Oil of Indiana case. *University of Chicago Law Review, 23*, 398–473.

McKenzie, J. (1997). Empire and the ecological apocalypse: The historiography of the imperial environment. In T. Griffiths & L. Robin (Eds.), *Ecology and empire: Environmental history of settler societies*. Melbourne University Press.

McKibben, B. (1990). *The end of nature*. London: Viking.

Meadows, D., Meadows, D., Randers, J., & Behrens, W. (1972). *The limits to growth: A report to the Club of Rome*. New York: Universe.

Menger, C. ([1870] 1976). *Principles of economics* (p. 117). Alabama: Ludwig von Mises Institute.

Meyerson, E. ([1908] 1930). *Identity and reality*. London: Allen & Unwin.

Midgely, M. (2002). *Science as salvation: A modern myth*. London: Routledge.

Mirowski, P. (1984). Macroeconomic instability and the 'natural' economic process in early neoclassical economics. *Journal of Economic History, 44*(2), 345–354.

Mirowski, P. (1988). *Against mechanism, protecting economics from science*. New Jersey: Rowman & Littlefield.

Mirowski, P. (1988). Energy and energetics in economic theory: A review essay. *Journal of Economic Issues, 22*(3), 811–830.

Mirowski, P. (1989). *More heat than light: Economics as social physics, physics as nature's economics*. Cambridge: Cambridge University Press.

Mirowski, P. (1992). *Against mechanism: Protecting economics from science*. Totowa, NJ: Rowman & Littlefield.

Mirowski, P. (1994). The realms of the natural. In P. Mirowski (Ed.), *Natural images in economic thought: Markets read in tooth and claw* (pp. 451–483). New York: Cambridge University Press, p. 453.

Mirowski, P. (1996, October 6). Ecology in the mirror of economics. *History of Economics Society*. Retrieved from http://www.eh.net/lists/archives/hes/oct-1996/0042.php.

Mirowski, P. (2009). Postface: Defining neoliberalism. In P. Mirowski & D. Plehwe (Eds.), *The road from Mont Pèlerin: The making of the neoliberal thought collective* (pp. 418–451). Cambridge, MA: Harvard University Press.

Mirowski, P. (2009). Postface: Defining neoliberalism. In P. Mirowski & D. Plehwe (Eds.), *The road from Mont Pelerin: The making of the neoliberal thought collective* (pp. 418–451). Cambridge, MA: Harvard University Press.

Mirowski, P. (2013). *Never let a serious crisis go to waste: How neoliberalism survived the financial meltdown.* London: Verso.

Mirowski, P., & Plehwe, D. (Eds.). (2009). *The road from Mont Pèlerin: The making of the neoliberal thought collective.* Cambridge MA: Harvard University Press.

Mirowski, P., Walker, J., & Abboud, A. (2013). Beyond denial: Neoliberalism, climate change, and the left. *Overland, 210*, 80–86.

Mitchell, T. (2011). *Carbon democracy: Political power in the age of oil.* London: Verso.

Modhrizinska, Y., & Stephanyan, C. (1973). *The future of society.* Moscow: Progress.

Monbiot, G. (2004). *The age of consent.* London: Flamingo.

Müller, J. (2014). *On conceptual history: Rethinking modern European intellectual history.* Oxford: Oxford University Press.

Mumford, L. (1967). *The myth of the machine: Technics and human development.* Michigan: Harcourt Brace.

National Public Radio. (2019, September 23). Transcript: Greta Thunberg's speech at the UN Climate Action Summit.

Neufeld, H. (2001). The history of the ecosystem concept: Part 3, post-WWII ecology. Retrieved from http://www.appstate.edu/~neufeldhs/Ecosystems/OriginsPart3.htm.

Niele, F. (2005). *Energy: Engine of evolution.* Amsterdam: Elsevier.

Nietzsche, F. (1979). *Philosophy and truth: Selections from Nietzsche's notebooks of the early 1870s.* New Jersey: Humanities Press.

Nsouli, S., et al. (1995, December 1). Resilience and growth through sustained adjustment: The Moroccan experience. IMF Occasional Paper No. 117.

Nuri, V. (2001). Fractional reserve banking as economic parasitism: A scientific, mathematical, and historical exposé, critique and manifesto (No. 0203005): University Library of Munich.

Nye, D. (1994). *American technological sublime.* Cambridge, MA: MIT Press.

O'Boyle, E. (2006). Classical economics and the Great Irish Famine: A study in limits. *Forum for Social Economics, 35*(2), 21–53.

Odum, E. P. (1969). The strategy of ecosystem development. *Science, 164*(3877), 262–270.

Odum, E. P. (1970). Introduction: The attitude revolution. In F. Scott et al. (Eds.), *The crisis of survival.* Glenview, Ill: Scott, Foresman & Co..

Odum, E. P. (1971). *Fundamentals of ecology* (3rd ed.). Philadelphia: W.B. Saunders.

Odum, H. T. (1971). Radiation ecology. In E. P. Odum (Ed.), *Fundamentals of ecology* (pp. 451–467). W.B. Saunders.

Odum, H. T. (1971). *Environment, power and society.* New York: Wiley Interscience.

Odum, H. T. (1973). Energy, ecology, and economics. *Ambio, 2*(6), 220–227.

Odum, H. T. (1995). Self-organization and maximum empower. In C. Hall (Ed.), *Maximum power: The ideas and applications of H.T. Odum* (pp. 311–330). Niwot, Colorado: University of Colorado Press, p. 311

Oreskes, N., & Conway, E. (2011). *Merchants of doubt: How a handful of scientists obscured the truth on issues from tobacco smoke to global warming*. New York: Bloomsbury.

Ormerod, P. (1994). *The death of economics*. Boston: Faber & Faber.

Owen, T & Kiernan, B. (2007, May 12). Bombs over Cambodia: New light on US air war. *Third World Traveller*.

Pack, S. (1996). Slavery, Adam Smith's economic vision, and the invisible hand. *History of Economic Ideas, 4*(1/2), 253–269.

Pancaldi, G. (1994). The technology of nature: Marx's thoughts on Darwin. In I. Cohen (Ed.), *The natural and the social sciences: Some critical and historical perspectives* (pp. 257–273). Dordrecht: Kluwer.

Pascoe, B. (2014). *Dark emu. Black seeds: Agriculture or accident?* Broome: Magabala.

Peirce, C. S. (1893). Evolutionary love. *The Monist, 3*(2), 176–200.

Ponting, C. (1991). *A green history of the world*. London: Penguin.

Popper, K. ([1957] 2002). *The poverty of historicism*. London: Routledge.

Powell, L. (1971, August 23). Attack on American free enterprise system—memorandum to Eugene B. *Sydnor*.

President's Materials Policy Commission. (1952). *Resources for freedom* (Vol. 1–5). Washington, DC: US Government Printing Office.

Prigogine, I. (1986). The chaotic universe. In L. Wijers (Ed.), *Art meets science and spirituality in a changing economy*. London: Academy Editions.

Prigogine, I., Nicolas, G., & Bablyoantz, A. (1972). Thermodynamics of evolution. *Physics Today, 23* (11 & 12), 22–28 & 38–44.

Prigogine, I., & Stengers, I. (1984). *Order out of chaos: Man's new dialogue with nature*. London: Heinemann.

Proctor, R., & Schiebinger, L. (Eds.). (2008). *Agnotology: The making and unmaking of ignorance*. Stanford University Press.

Purdy, J. (2014). Neoliberal constitutionalism: Lochnerism for a new economy. *Law and Contemporary Problems, 77*, 195–213.

Pyne, S. (1990). Firestick history. *Journal of American History, 76*(4), 1132–1141.

Pyne, S. (1997). *Vestal fire: An environmental history, told through fire, of Europe and Europe's encounter with the world*. University of Washington Press.

Pyne, S. (2013). A new epoch of fire: The Anthropocene. In A. Scott et al. (Eds.), *Fire on Earth (Ch. 12)*. Wiley and Sons.

Quesnay, F. (1747). *Essai physique sur l'oeconomie animale* (Vol. 3). Paris: Guillaume Cavelier.

Rabinbach, A. (1992). *The human motor: Energy, fatigue, and the origins of modernity*. University of California Press.

Raitt, S. (2003). Psychic waste: Freud, Fechner and the principle of constancy. In G. Hawkins & S. Muecke (Eds.), *Culture and waste: The creation and destruction of value*. Roman and Littlefield: Lanham, MA.

Rankine, W. (1853). On the general law of the transformation of energy. *Proceedings of the Philosophical Society of Glasgow, 3*(5), 276–280.

Rappaport, P. (1959). The photovoltaic effect and its utilization. *Solar Energy, 3*(4), 8–18.

Rehder, J. (2000). *The mastery and uses of fire in antiquity.* Montreal: McGill-Queens University Press.

Reuleaux, F. (1894). *The constructor: A handbook of machine design.* Philadelphia.

Revelle, R., Broeker, W., Keeling, C. D., Craig, H., & Smagorinsky, J. (1965). Atmospheric carbon dioxide. In President's Scientific Advisory Committee, Restoring the quality of our environment: Report of the environmental pollution panel (pp. 111–133). Washington DC: The White House.

Revelle, R., & Suess, H. (1957). Carbon dioxide exchange between atmosphere and ocean and the question of an increase of atmospheric CO2 during the past decades. *Tellus, 9*(1), 18–27.

Ricardo, D. ([1817] 2001). *On the principles of political economy and taxation.* Kitchener, Ontario: Batoche.

Richter, D., & Billings, S. (2015). 'One physical system': Tansley's ecosystem as Earth's critical zone. *New Phytologist, 206*, 900–912.

Ricklefs, R. (1997). *The economy of nature: A textbook in basic ecology.* New York: W.W. Freeman & Co.

Rifkin, J. (1998). *The biotech century.* New York: Tarcher/Putnam.

Robbins, L. (1932). *An essay on the nature and significance of economic science.* New York: New York University Press.

Robins, N. (2012). The corporation that changed the world: How the East India Company shaped the modern multinational. *Asian Affairs, 43*(1), 12–26.

Ropke, I. (2004). The early modern history of ecological economics. *Ecological Economics, 50*, 293–314.

Rose, D. B. (1996). *Nourishing terrains: Australian Aboriginal views of landscape and wilderness.* Australian Heritage Commission.

Rosenblueth, A., Wiener, N., & Bigelow, J. (1943). Behaviour, purpose and teleology. *Philosophy of Science, 10*(1), 18–24.

Rostow, W. (1960). *The stages of economic growth: A non-communist manifesto.* Cambridge: Cambridge University Press.

Rowe, F. (1984). The decline of antitrust and the delusions of models. *Georgetown Law Journal, 72*, 1511–1570.

Rusbridger, A. (1987, December 22). The Adam Smith Institute's sense and nonsense. *The Guardian.*

Samuelson, P. (1970). *Economics.* New York: McGraw-Hill.

Schabas, M. (1992). The greyhound and the mastiff: Darwinian themes in Mill and Marshall. In P. Mirowski (Ed.), *Natural images in economic thought: Markets read in tooth and claw* (pp. 322–335). Cambridge: Cambridge University Press.

Scheuerman, W. (1997). The unholy alliance of Carl Schmitt and Friedrich A. Hayek. *Constellations, 4*(2), 172–188.
Schmitt, C. ([1950] 2003). *The nomos of the Earth in the international law of the Jus Publicum Europeaum* (G. Ulmen, Trans.). New York: Telos Press.
Schneidermann, D. (2013). *Constitutionalizing economic globalisation: Investment rules and democracy's promise.* Cambridge, UK: Cambridge University Press.
Schrödinger, E. ([1944] 1967). *What is life?* Cambridge: Cambridge University Press.
Schumpeter, J. ([1954] 1994). *History of economic analysis.* Oxford University Press.
Schurr, S. (1972). *Energy, economic growth, and the environment.* Baltimore: Johns Hopkins University Press.
Scott, A., Bowman, D., Bond, W., Pyne, S., & Alexander, M. (2013). *Fire on earth: An introduction.* Wiley & Sons.
Sears, P. (1964). Ecology—a subversive subject. *BioScience, 14*, 11–13.
Selgin, G. (2012). Those dishonest goldsmiths. *Financial History Review, 19*(3), 269–288.
Seligman, B. (1962). *Main currents in modern economics.* Glencoe: Free Press.
Shiva, V. (1997, June). How free is free India? *Resurgence.*
Sieferle, R. ([1982] 2000). *The subterranean forest: Energy systems and the industrial revolution.* White Horse Press.
Simberloff, D. (1980). A succession of paradigms in ecology: Essentialism to materialism and probabilism. *Synthese, 43*(1), 3–39.
Singer, K. (1958). Oikonomia: An inquiry into beginnings of economic thought and language. *Kyklos, 11*(1), 29–57.
Skousen, M. (2009). *The making of modern economics.* Armonk, NY: M.E. Sharpe.
Slobodian, Q. (2018). *Globalists: The end of empire and the birth of neoliberalism.* Cambridge, MA: Harvard University Press.
Sloterdijk, P. (2011). *Bubbles: Spheres vol I: Microspherology.* Los Angeles: Semiotext(e).
Smith, A. ([1776] 1965). *The wealth of nations.* New York: Modern Library.
Smith, A. ([1759] 1976). *The theory of moral sentiments.* Indianapolis: Liberty Fund.
Smith, C. (2004). Wallace's unfinished business: The 'other man' in evolutionary theory. *Complexity, 10*(2), 25–32.
Smith, F. (1968). The International Biological Program and the science of ecology. *Proceedings of the National Academy of Sciences, 60*(1), 5–11.
Soddy, F. (1920). *Science and life.* New York: Dutton.
Soddy, F. (1922). *Cartesian economics.* London: Hendersons.
Solow, R. (1956). A contribution to the theory of economic growth. *Quarterly Journal of Economics, 70*(1), 65–94.
Solow, R. (1974). The economics of resources or the resources of economics. *American Economic Review, 64*(2), 1–14.

Spash, N. (2010). The brave new world of carbon trading. *New Political Economy, 12*(2), 169–195.

Spencer, W., & Gillen, F. (1899). *The native tribes of Central Australia*. London: Macmillan.

Steffenson, V. (2016, February 15). Indigenous fire manager: Our country needs to burn more. *SBS Insight: Special Broadcasting Service*. Retrieved from http://www.sbs.com.au/news/article/2016/02/15/our-country-needs-burn-more-indigenousfire-manager.

Steffensen, V. (2020). *Fire country: how Indigenous fire management could help save Australia*. Wurundjeri (Melbourne): Hardie Grant Travel.

Stern, N. (1981). *From ENIAC to UNIVAC: An appraisal of the Eckert-Mauchly computers*. Digital Press.

Stern, N. (2006). *Stern review on the economics of climate change*. HM Treasury.

Stewart, J., & Kemp, J. (Eds.) (1943). Conversation between d'Alembert and Diderot, 1769. In *Diderot, interpreter of nature*. International Publishers.

Stewart, O. (1951). Burning and natural vegetation in the United States. *Geographical Review, 41*(2), 317–320.

Stewart, O. (2002). *Forgotten fires: Native Americans and the transient wilderness*. University of Oklahoma Press.

Stewart, O. ([1963] 2010). Barriers to understanding the influence of use of fire by aborigines on vegetation. *Fire Ecology*, 10(2). https://doi.org/10.4996/fireecology.1002001.

Stigler, G. (1984). Economics: The imperial science? *Scandinavian Journal of Economics, 86*(3), 301–313.

Stiglitz, J. (1974). Growth with exhaustible natural resources. *Review of Economic Studies, 41*(128), 139–145.

Stone, J. (2016, December 6). Britain could slash environmental and safety standards 'a very long way' after Brexit, Tory MP Jacob Rees-Mogg says. *The Independent*.

Supran, G., & Oreskes, N. (2017). Assessing ExxonMobil's climate change communications (1977–2014). *Environmental Research Letters, 12*, 1–18.

Szostak, R. (2009). *The causes of economic growth*. Berlin: Springer.

Tansley, A. (1935). The use and abuse of vegetational concepts and terms. *Ecology, 16*(3), 284–307.

Tarbell, I. ([1904] 2009). *The history of the Standard Oil company*. New York: Cosimo.

Taylor, P. (1988). Technocratic optimism: H.T. Odum and the partial transformation of ecological metaphor after World War II. *Journal of the History of Biology, 21*(2), 221–246.

Teller, E. (1960). Energy patterns of the future. In *Energy and man: A symposium* (pp. 54–72). New York: Appleton-Century Crofts.

Tienhaara, K. (2017). Regulatory chill in a warming world: The threat to climate policy posed by Investor-State Dispute Settlement. *Transnational Environmental Law, 7*(2), 229–250.

Tomashot, S. (2015). Selling peace: The history of the International Chamber of Commerce, 1919–1925. PhD thesis: Georgia State University.

UK Cabinet Office. (2007). Operations in the UK: The defence contribution to resilience. Joint Doctrine Publication, 2e.

United Nations. (1972). Stockholm Declaration of the UN Conference on the Human Environment.

United Nations Development Program, United Nations Environment Program, World Bank, World Resources Institute. (2008). *World resources 2008: Roots of resilience—growing the wealth of the poor.* Washington, DC: World Resources Institute.

US Congress. (2017). Bill H.R.861—To terminate the Environmental Protection Agency. 115th Congress.

Vallette, J. (1999, June 15). Larry Summers' war against the Earth. *Counterpunch.*

Veblen, T. (1909). The limitations of marginal utility. *Journal of Political Economy, 17*(9), 620–636.

Vernadsky, V. (1924). *La géochimie.* Paris: Alcan.

Vernadsky, V. ([1926] 1998). *The biosphere* (D. Langmuir, Trans.). New York: Copernicus.

Veron, J., Hoegh-Guldberg, O., Lenton, T., et al. (2009). The coral reef crisis: The critical importance of <350 ppm CO_2. *Marine Pollution Bulletin, 58*(10), 1428–1436.

Voigt, A. (2011). The rise of systems theory in ecology. In A. Schwarz & K. Jax (Eds.), *Ecology revisited: Reflecting on concepts, advancing science* (pp. 183–194). New York: Springer.

von Böhm-Bawerk, E. ([1890] 1959). *Capital and interest* (vol. 2, book 3). South Holland: Libertarian Press.

von Foerster, H. (Ed.) (1954). *Cybernetics: Circular causal and feedback mechanisms in biological and social systems.* Proceedings of the Macy Conferences, Josiah Macy Jr. Foundation.

von Helmholtz, H. (1863). On the conservation of force. Introduction to a series of lectures delivered at Carlsruhe in the winter of 1862–1863.

von Humboldt, A. (1845–1862). *Kosmos.* Stuttgart: Cotta.

von Mises, L. ([1949] 1996). *Human action.* Irvington, NY: Foundation for Economic Education.

von Neumann, J., & Morgenstern, O. (1944). *Theory of games and economic behaviour.* Princeton University Press.

Waddington, C. (1975). The origin. In E. Worthington (Ed.), *The evolution of IBP* (pp. 4–11). Cambridge: Cambridge University Press.

Walker, J. (2013). Worlds to endure: Weathering disorder from Arnhem Land to Chicago. *Global Networks, 13*(3), 391–409.
Walker, J. (2016). Bringing liquidity to life: Markets for ecosystem services and the new political economy of extinction. In K. Kohli & M. Menon (Eds.), *Business interests and the environmental crisis* (pp. 5–37). New Delhi: SAGE.
Walker, J., & Cooper, M. (2011). Genealogies of resilience: From systems ecology to the political economy of crisis adaptation. *Security Dialogue, 42*(2), 143–160.
Walker, J., & Granjou, C. (2017). MELiSSA the minimal biosphere: Human life, waste and refuge in deep space. *Futures, 92*, 59–69.
Walker, J., & Johnson, M. (2018). On mineral sovereignty: Towards a political theory of geological power. *Energy Research & Social Science, 45*, 46–66.
Wall, J., & Slezak, M. (2016, June 7). Great Barrier Reef: Diving in the stench of millions of rotting animals. *The Guardian.*
Wallace, A. R. (1855). On the law which has regulated the introduction of new species. *Annals & Magazine of Natural History, 16*(93), 184–196.
Walras, L. ([1874] 1954). *Elements of pure economics.* Homewood, IL: Richard Irwin.
Weart, S. (2010). The idea of anthropogenic global climate change in the 20th century. *Wiley Interdisciplinary Reviews: Climate Change, 1*(1), 67–81.
Weingart, P., & Massen, S. (1997). The order of meaning: The career of chaos as a metaphor. *Configurations, 5*(3), 463–520.
Wellink, N. (2009, November 10). Managing complexity. Speech given to the NautaDutilh seminar, Bussum.
Wertime, T. (1973). Pyrotechnology: Man's first industrial uses of fire. *American Scientist, 61*(6), 670–682.
White, M. (2004). In the lobby of the energy hotel: Jevon's formulation of the postclassical 'economic problem'. *History of political economy, 36*(2), 227–271.
Whyte, J. (2019). *The morals of the market: Human rights and the rise of neoliberalism.* London: Verso.
Wiener, N. (1948). *Cybernetics: Or control and communication in the animal and the machine.* Cambridge: MIT Press.
Wiener, N. ([1946] 1983). A scientist rebels: Letter to Mr. George Forsythe. *Science Technology, and Human Values*, 8(3), 36–38.
Wildavsky, A. (1982). Pollution as moral coercion: Culture, risk perception and libertarian values. *Cato Journal, 2*(1), 304–325.
Wilson, E. O. (1999). *Consilience: The unity of knowledge.* Cleveland: Vintage.
Wilson, K. (2014, July 25). Justus von Liebig and the birth of modern biochar. *Biochar Journal.* Retrieved from www.biochar-journal.org/en/ct/5.
Wilson, W. (1908). *Constitutional government in the United States.* Columbia University Press.
Wohlstetter A. (1959, January). The delicate balance of terror. *Foreign Affairs.*
Woodham-Smith, C. (1968). *The Great Hunger: Ireland 1845–1849.* Hamilton: London.

Wootton, D. (2006). Liberty, metaphor, and mechanism: Checks and balances and the origins of modern constitutionalism. In D. Womersley (Ed.), *Liberty and American experience in the eighteenth century* (pp. 209–275). Amagi.

World Bank. (2006, April 27). *Social resilience and state fragility in Haiti: A country social analysis.* Caribbean Country Management Unit, Latin America and the Caribbean Region Report No. 36069–HT.

World Bank. (2008). *Climate resilient cities: A primer on reducing vulnerabilities to climate change impacts and strengthening disaster risk management in East Asian cities.* Washington, DC: World Bank.

Worster, D. (1977). *Nature's economy: The roots of ecology.* San Francisco: Sierra Club.

Worster, D. (1993). *The wealth of nature: Environmental history and the ecological imagination.* New York: Oxford University Press.

Zanna, L., Khatiwala, S., Gregory, J., Ison, J., & Heimbach, P. (2019). Global reconstruction of historical ocean heat storage and transport. *Proceedings of the National Academy of Science, 116*(4), 1126–1131.

Index

J. Walker, *More Heat than Life: The Tangled Roots of Ecology, Energy, and Economics*,
https://doi.org/10.1007/978-981-15-3936-7

The manufacturer's authorised representative in the EU is Springer
Nature Customer Service Centre GmbH, Europaplatz 3, 69115 Heidelberg,
Germany. If you have any concerns regarding our products, please
contact ProductSafety@springernature.com

Printed and bound by CPI Group (UK) Ltd, Croydon, CR0 4YY
25/09/2026
02222839-0001